Emerging Electronics Technologies and Solutions for Eco-Friendly Cities

Emerging Electronics Technologies and Solutions for Eco-Friendly Cities

Editors

Darius Andriukaitis

Yongjun Pan

Peter Brida

MDPI • Basel • Beijing • Wuhan • Barcelona • Belgrade • Manchester • Tokyo • Cluj • Tianjin

Editors

Darius Andriukaitis
Department of Electronics
Engineering
Kaunas University of
Technology
Kaunas
Lithuania

Yongjun Pan
College of Mechanical and
Vehicle Engineering
Chongqing University
Chongqing
China

Peter Brida
Faculty of Electrical and
Information Technology
University of Zilina
Zilina
Slovakia

Editorial Office
MDPI
St. Alban-Anlage 66
4052 Basel, Switzerland

This is a reprint of articles from the Special Issue published online in the open access journal *Electronics* (ISSN 2079-9292) (available at: www.mdpi.com/journal/electronics/special_issues/EFC).

For citation purposes, cite each article independently as indicated on the article page online and as indicated below:

LastName, A.A.; LastName, B.B.; LastName, C.C. Article Title. *Journal Name* **Year**, *Volume Number*, Page Range.

ISBN 978-3-0365-6522-4 (Hbk)
ISBN 978-3-0365-6521-7 (PDF)

Contents

Preface to "Emerging Electronics Technologies and Solutions for Eco-Friendly Cities"

The rising demand for power, water supply and waste management, pedestrian safety, the efficient use of electric vehicles, traffic jams, commuting, the identification of all road users and prediction of their travel paths, and increasing air and noise levels are the most pressing issues in today's cities. This Special Issue aims to present high-quality research and recent advances of technologies towards eco-friendly cities.

Darius Andriukaitis, Yongjun Pan, and Peter Brida
Editors

Editorial

Emerging Electronics Technologies and Solutions for Eco-Friendly Cities

Darius Andriukaitis [1,*], Yongjun Pan [2] and Peter Brida [3]

1 Department of Electronics Engineering, Kaunas University of Technology, Studentu St. 50–438, LT-51368 Kaunas, Lithuania
2 College of Mechanical and Vehicle Engineering, Chongqing University, Chongqing 400030, China
3 Faculty of Electrical Engineering and Information Technologies, University of Zilina, 010 26 Zilina, Slovakia
* Correspondence: darius.andriukaitis@ktu.lt; Tel.: +370-37-300-519

Citation: Andriukaitis, D.; Pan, Y.; Brida, P. Emerging Electronics Technologies and Solutions for Eco-Friendly Cities. *Electronics* **2023**, *12*, 476. https://doi.org/10.3390/electronics12030476

Received: 6 January 2023
Accepted: 12 January 2023
Published: 17 January 2023

Introduction

The development of electronic solutions and their application to smart cities are an inevitability. On the other hand, the growing population is forcing us find new, more effective solutions. According to the preliminary model forecasts, population growth is expected to stop at the end of this century (source: Pew Research Centre (https://www.pewresearch.org/fact-tank/2019/06/17/worlds-population-is-projected-to-nearly-stop-growing-by-the-end-of-the-century/ (accessed on 9 October 2020))). The increasing demand for power, water supply and waste management, pedestrian safety, the efficient use of electric vehicles, traffic jams and long commutes, the identification of road users and travel path predictions, and rising air and noise pollution levels are the most pressing issues in today's cities. The place we live in has a huge impact on our lives. Advanced planning is a feature of green development which reduces our dependence on vehicles that produce greenhouse gas emissions. The development of eco-friendly cities includes the development and application of new efficient solutions and technologies for transport management, emission control and pollution control, energy efficiency and the usage of renewable energy, and resource efficiency, etc. These solutions would ensure a better quality of life for the growing population.

This Special Issue in *Electronics* reports on some of the recent research efforts on this important topic. The eleven papers in this issue cover various aspects of emerging electronics technologies and solutions for eco-friendly cities.

The authors of [1] propose a method for predicting the bucket fill factor of a loader based on the three-dimensional information of the material surface. Firstly, a co-simulation model of loader shovelling material was established by using the multi-body dynamics software RecurDyn and the discrete element method software (DEMS) EDEM, and the co-simulation was conducted under different excavation trajectories. Then, before the shovel excavation, three-dimensional material surface information was obtained from DEMS, and the surface function of the material contour was fitted based on the corresponding shovel excavation trajectory information. Meanwhile, they obtained the volume of the material excavated by the loader by employing the numerical integration method, and it was divided by the rated bucket volume to obtain the estimated bucket fill factor. Finally, the actual volume of the material after the shovel excavation was divided by the rated bucket volume to obtain the accurate bucket fill factor. Thus, the prediction model of the bucket fill factor was built. The experimental results show that the proposed method is feasible, with a maximum error of 4.3%, a root mean square error of 0.025, and an average absolute error of 0.021.

In [2], simple mathematical model of a seismocardiogram (SCG) was developed, and three algorithms were created to explain the processes and behaviours of the model in detail. Using this algorithm, the processing program can be written in several programming languages, not only in MATLAB. This seismocardiogram model can be used to obtain the optimal parameters (fifth-order delay before adaptive filter, 892 filter order) for adaptive filters that can perform real cardio-mechanical vibration signal processing in order to estimate the AO peaks. As a result, the heart rate was calculated. However, filter orders 200 and 50 were evaluated again, and so the process was faster for both of the filters; the duration periods decreased from 1.20 to 1.09 and from 1.19 to 1.11 s, respectively. In each case, the signal processing and the necessary calculations were performed using MATLAB programs.

In [3], the authors analysed the excavator system's energy flow under a typical working condition load. In operation conditions, the output energy of the engine only accounts for 50.21% of the engine's fuel energy, and the actuation and the swing system account for 9.33% and 4% of it, respectively. In transportation conditions, the output energy of the engine only accounts for 49.80% of the engine's fuel energy, and the torque converter efficiency loss and excavator driving energy account for 15.09% and 17.98% of it, respectively. The research results show that the energy flow analysis method based on a typical working condition load can accurately obtain each excavator component's energy margin, which provides a basis for designing energy-saving schemes and control strategies.

The authors of [4] present a performance analysis of the overall ergodic outage probability of the optimal RS scheme for a low-power energy harvesting (LPEH) wireless sensor network (WSN). Since the simulations correlate with the theory, each relay was equipped with a battery that consisted of an on/off (1/0) decision scheme, according to the Markov property. In this context, an optimal loop interference relay selection was proposed and investigated. Moreover, the log-normal distribution method is crucial for characterizing the LPEH WSN's constraints. The system's performance was evaluated analytically and numerically in terms of the overall ergodic outage probability (OP) with the Monte Carlo simulation. The system had the lowest overall ergodic OP, thus, it performed the best, with an energy harvesting switch time of 0.175. Following the increase in the signal-to-noise ratio (SNR), the system without a direct link performed the worst. Furthermore, the system performed better as more relays were deployed. Finally, more than 80% of the data can be obtained under the household condition, without the need for additional bandwidth and power supply.

To improve the sound source identification performance in low-SNR cabin environments, the authors of [5] introduce cross-spectral matrix (CSM) reconstruction methods such as diagonal reconstruction (DRec), robust principal component analysis (RPCA), and probabilistic factor analysis (PFA), which have been widely studied in planar arrays, into spherical arrays-based functional delay and sum (FDAS). Three enhancement methods, namely EFDAS-DRec, EFDAS-RPCA, and EFDAS-PFA, were established. The main conclusions obtained through the simulations and experiments are as follows: (1) EFDAS can significantly improve the sound source identification performance of FDAS under low SNRs, which effectively suppress the sidelobe contamination, shrink the mainlobe contamination, and maintain the strong localization capability for weak sources. (2) Compared with FDAS at SNR = 0 dB and when the number of snapshots = 1000, the average MSLs of EFDAS-DRec, EFDAS-RPCA, and EFDAS-PFA were reduced by 6.4 dB, 21.6 dB, and 53.1 dB, respectively, and the mainlobes of the sound sources shrunk by 43.5%, 69.0%, and 80.0%, respectively. (3) The three EFDAS methods improved the quantification accuracy of FDAS when there was a large number of snapshots.

The authors of [6] applied cardiac biosignals, an excited accelerometer, and a gyroscope for the prevention of accidents on the road. This paper adopts the seismocardiogram hypothesis which involves using measurements from a seismocardiogram to identify the drivers' heart problems and safely stop the vehicle before they enter into a critical condition by informing the relevant departments in a nonclinical manner. The proposed system works without an electrocardiogram, and it detects heart rhythms more easily. The estimation of the heart rate (HR) is calculated through automatically detected aortic valve opening (AO) peaks. The system is composed of two micro-electromechanical systems (MEMSs) which evaluate the physiological parameters and eliminate the effects of external interference on the entire system. A few digital filtering methods are discussed and benchmarked to increase the seismocardiogram efficiency. As a result, the fourth adaptive filter obtains the estimated HR = 65 beats per min (bmp) in a still noisy signal (SNR = -11.32 dB). In contrast with the low processing benefit (3.39 dB), 27 AO peaks were detected with a 917.56 ms peak interval mean over 1.11 s, and the calculated root mean square error (RMSE) was 0.1942 m/s^2 when the adaptive filter order was 50 and the adaptation step was equal to 0.933.

In [7], the paper investigates the rescheduling strategy and algorithm for DDBFSP, in which machine breakdown events are categorised as disruptions in manufacturing sites. Firstly, the mathematical model of DDBFSP, including the event simulation mechanism, was constructed. The makespan and stability as the objectives were considered. The goal of the paper was to achieve two objectives. An "event-driven" policy in response to the disruption was applied. A two-stage "predictive-reactive" rescheduling strategy was proposed. In the first stage, a static environment, a distributed blocking flowshop scheduling problem (DBFSP) without machine breakdown, was considered, and the global initial schedules were generated, while in the second stage, after the machine breakdown occurs, the initial schedule is locally optimized by a hybrid repair policy based on "right-shift repair + local reorder", and the discrete memetic algorithm (DMA) reordering algorithm based on differential evolution was proposed for the local reorder operation. To test the effectiveness of the DMA, comparisons with mainstream algorithms were conducted on different scales. The statistical results show that the ARPDs obtained from DMA were improved by 88%.

In [8], a method of coal thickness prediction using VMD and LSTM is presented. Firstly, empirical mode decomposition (EMD) and VMD methods were used to denoise simple signals, and the denoising effect of the VMD method was verified. Then, the wedge-shaped coal thickness model was constructed, and the seismic forward modelling and analysis were carried out. The results show that coal thickness predictions based on seismic attributes are feasible. On the basis of the VMD denoising of the original 3D seismic data, VMD-LSTM was used to predict the coal thickness. The data were compared with the prediction results of a traditional BP neural network. The coal thickness prediction method proposed in this paper has high accuracy and coincides with the coal seam information about existing boreholes. The minimum absolute error of the predicted coal thickness is 0.08 m, and the maximum absolute error is 0.48 m. This indicates that VMD-LSTM has high accuracy in predicting coal thickness.

In [9], the authors present a mobile sensor node for monitoring air and noise pollution, and the developed system was installed on an remote control drone, which could quickly monitor large areas. It relies on a Raspberry Pi Zero W board and a wide set of sensors (i.e., NO_2, CO, NH_3, CO_2, VOCs, PM2.5, and PM10) to sample the environmental parameter at regular time intervals. A proper classification algorithm was developed to quantify the traffic level from the noise level acquired by the onboard microphone. Additionally, the drone is equipped with a camera, and it implements a visual recognition algorithm (Fast R-CNN) to detect waste fires and mark them using a GPS receiver. Furthermore, firmware for managing the sensing unit operation was developed, as well as the power supply section. In particular, the node's consumption was analysed in two use cases, and the battery capacity needed to power the designed device was determined. The on-field tests demonstrated the proper operation of the developed monitoring system. Finally, a cloud application was developed to remotely monitor the information acquired by the sensor-based drone and upload them to a remote database.

In [10], a beacon-based hybrid routing protocol was designed to adapt to the new forms of intelligent warfare, accelerate the application of unmanned vehicles in the military field, and solve problems such as high maintenance costs, path failures, and repeated routing pathfinding in large-scale unmanned vehicle network communications in new battlefields. This protocol used the periodic broadcast pulses initiated by the beacon nodes to provide synchronization and routing to the network and established a spanning tree, which the nodes used to communicate with each other. An NS3 platform was used to build a dynamic simulation environment with service data to evaluate the network performance. The results showed that when it was used in a range of 5~35 communication links, the beacon-based routing protocol's PDR was approximately 10% higher than that of AODV routing protocol. At 5~50 communication links, the result was approximately 20% higher than the DSDV routing protocol.

The authors of [11] aimed to explore a more ecological and sustainable solution to the problems of cities around the world. Particularly, this paper presents a conceptual design of the main sterilization chamber for infectious waste. The Design Thinking (DT) method was used, since it has a user-centred approach which allows for the co-design and inclusion of the target population. This study demonstrates the possibility of obtaining feasible results based on the user's needs through the application of DT as a framework for engineering designs.

Author Contributions: Writing—Original draft preparation, P.B., D.A. and Y.P.; writing—Review and editing, D.A., Y.P. and P.B. All authors have read and agreed to the published version of the manuscript.

Acknowledgments: Finally, the guest editors of this Special Issue would like to thank all of the authors who have submitted their manuscripts to *Sensors* journal and the reviewers for their hard work during the review process. Furthermore, we sincerely thank the editors of *Sensors* for their kind help and support. We hope that the readers enjoy reading the articles in this Special Issue. Finally, the guest editors wish to acknowledge partial support from the Slovak VEGA grant agency, Project No. 1/0626/19, "Research of mobile objects localization in IoT environment".

Conflicts of Interest: The authors declare no conflict of interest.

References

1. Wang, S.; Yu, S.; Hou, L.; Wu, B.; Wu, Y. Prediction of Bucket Fill Factor of Loader Based on Three-Dimensional Information of Material Surface. *Electronics* **2022**, *11*, 2841. [CrossRef]
2. Uskovas, G.; Valinevicius, A.; Zilys, M.; Navikas, D.; Frivaldsky, M.; Prauzek, M.; Konecny, J.; Andriukaitis, D. A Novel Seismocardiogram Mathematical Model for Simplified Adjustment of Adaptive Filter. *Electronics* **2022**, *11*, 2444. [CrossRef]
3. Su, D.; Hou, L.; Wang, S.; Bu, X.; Xia, X. Energy Flow Analysis of Excavator System Based on Typical Working Condition Load. *Electronics* **2022**, *11*, 1987. [CrossRef]
4. Nguyen, H.-S.; Sevcik, L.; Van, H.-P. Performance Analysis on Low-Power Energy Harvesting Wireless Sensors Eco-Friendly Networks with a Novel Relay Selection Scheme. *Electronics* **2022**, *11*, 1978. [CrossRef]
5. Zhao, Y.; Chu, Z.; Li, L. Performance Enhancement of Functional Delay and Sum Beamforming for Spherical Microphone Arrays. *Electronics* **2022**, *11*, 1132. [CrossRef]
6. Uskovas, G.; Valinevicius, A.; Zilys, M.; Navikas, D.; Frivaldsky, M.; Prauzek, M.; Konecny, J.; Andriukaitis, D. Driver Cardiovascular Disease Detection Using Seismocardiogram. *Electronics* **2022**, *11*, 484. [CrossRef]
7. Zhang, X.; Han, Y.; Królczyk, G.; Rydel, M.; Stanislawski, R.; Li, Z. Rescheduling of Distributed Manufacturing System with Machine Breakdowns. *Electronics* **2022**, *11*, 249. [CrossRef]
8. Huang, Y.; Yan, L.; Cheng, Y.; Qi, X.; Li, Z. Coal Thickness Prediction Method Based on VMD and LSTM. *Electronics* **2022**, *11*, 232. [CrossRef]
9. De Fazio, R.; Dinoi, L.M.; De Vittorio, M.; Visconti, P. A Sensor-Based Drone for Pollutants Detection in Eco-Friendly Cities: Hardware Design and Data Analysis Application. *Electronics* **2022**, *11*, 52. [CrossRef]
10. Mu, W.; Li, G.; Ma, Y.; Wang, R.; Li, Y.; Li, Z. Beacon-Based Hybrid Routing Protocol for Large-Scale Unmanned Vehicle Ad Hoc Network. *Electronics* **2021**, *10*, 3129. [CrossRef]
11. Castiblanco Jimenez, I.A.; Mauro, S.; Napoli, D.; Marcolin, F.; Vezzetti, E.; Rojas Torres, M.C.; Specchia, S.; Moos, S. Design Thinking as a Framework for the Design of a Sustainable Waste Sterilization System: The Case of Piedmont Region, Italy. *Electronics* **2021**, *10*, 2665. [CrossRef]

Article

Prediction of Bucket Fill Factor of Loader Based on Three-Dimensional Information of Material Surface

Shaojie Wang [1,2,*], Shengfeng Yu [1], Liang Hou [1], Binyun Wu [1] and Yanfeng Wu [1]

[1] Department of Mechanical and Electrical Engineering, Xiamen University, Xiamen 361000, China
[2] Shenzhen Research Institute, Xiamen University, Shenzhen 518057, China
* Correspondence: wsj@xmu.edu.cn

Abstract: The bucket fill factor is a core evaluation indicator for the optimization of the loader's autonomous shoveling operation. Accurately predicting the bucket fill factor of the loader after different excavation trajectories is fundamental for optimizing the loader's efficiency and energy cost. Therefore, this paper proposes a method for predicting the bucket fill factor of the loader based on the three-dimensional information of the material surface. Firstly, the co-simulation model of loader shoveling material is established based on the multi-body dynamics software RecurDyn and the discrete element method software (DEMS) EDEM, and the co-simulation is conducted under different excavation trajectories. Then, the three-dimensional material surface information before shovel excavation is obtained from DEMS, and the surface function of the material contour is fitted based on the corresponding shovel excavation trajectory information. Meanwhile, the volume of the material excavated by the loader is obtained by the numerical integration method, and it is divided by the rated bucket volume to obtain the estimated bucket fill factor. Finally, the actual volume of the material after the shovel excavation is divided by the rated bucket volume to obtain the accurate bucket fill factor. Based on this, the prediction model of the bucket fill factor is built. The experimental results show that the proposed method is feasible, with a maximum error of 4.3%, a root mean square error of 0.025 and an average absolute error of 0.021. The research work lays the foundation for predicting the bucket fill factor of construction machinery such as loaders and excavators under real working conditions, which is conducive to promoting the development of autonomous, unmanned, and intelligent construction machinery.

Keywords: loader; bucket fill factor; material surface three-dimensional information; regression prediction; RecurDyn; EDEM

Citation: Wang, S.; Yu, S.; Hou, L.; Wu, B.; Wu, Y. Prediction of Bucket Fill Factor of Loader Based on Three-Dimensional Information of Material Surface. *Electronics* **2022**, *11*, 2841. https://doi.org/10.3390/electronics11182841

Academic Editor: Cheng-Chi Lee

Received: 9 August 2022
Accepted: 7 September 2022
Published: 8 September 2022

Publisher's Note: MDPI stays neutral with regard to jurisdictional claims in published maps and institutional affiliations.

1. Introduction

A loader is a piece of earth and stone construction machinery and equipment widely used in mining, construction, water conservancy, urban construction, among other contexts. Its main function is to shovel and transport bulk materials such as soil, sand, and gravel and complete engineering tasks such as bulldozing and lifting. It can not only reduce the labor intensity of construction personnel but also improve the construction speed and the project quality.

The related research on loaders can be divided into two major directions. The first is system control optimization, performance optimization of parts and components, electrification, etc., which are carried out to improve the economy of the loader, save energy, and protect the environment. Jun G et al. [1] proposed a new energy alternate recovery and utilization system to recover the potential energy generated in the lifting and lowering of excavators, Yang Y et al. [2] used an objective optimization algorithm to optimize the parameters of a new continuously variable transmission system for a loader to obtain better power transmission performance, and He X et al. [3] mentioned the importance of hybrid (HES) construction machinery to protect the environment and introduced the control strategies and challenges of hybrid construction machinery. The second is online fault diagnosis, intelligent shifting, vibration and noise reduction, etc., which are carried out to improve the intelligence of the loader and improve the operational efficiency, safety, and driver comfort of the loader. Chen Z et al. [4] extracted the signs of loader gear noise and made fault diagnosis of the loader gearbox based on ICA and SVM algorithms, Wu G [5] used a neural network approach to determine the complex mapping relationship between loader gears and the current operating conditions, which improved the shift response speed and shift quality, Zhao H et al. [6] optimized the parameters of the loader suspension to improve the driving comfort of the loader driver. Therefore, compared with the early loaders, the current loaders have been greatly improved in economy, safety, and comfort. In recent years, the electrification, intelligence, and unmanned operation of loaders have become a new research hotspot. Dadhich et al. [7] gave a detailed introduction to the prospects and challenges faced by the intelligent and unmanned construction machinery. when the loader is unmanned, the bucket fill factor becomes an important optimization objective of the autonomous excavation strategy of the loader, and its importance is self-evident. Therefore, this paper focuses on how to effectively predict the bucket fill factor of a loader.

Bucket fill factor prediction is a prerequisite for unmanned construction machinery and has attracted the attention of scholars. In recent years, Chen Yu et al. [8] proposed the SPC (Statistical Process Control) method to judge whether the test process of the loader bucket fill factor is stable and whether the test data are available. This method helps to ensure the validity of the loader bucket fill rate test data and correctly evaluate the bucket performance. Pengpeng Huang et al. [9] tried to mathematize the shoveling process to establish a mathematical model of the bucket fill factor of the loader during the shovel loading process. However, many idealistic assumptions were made during the study, so the applicability of the model is not good. Dadhich Set al. [7] mentioned in their previous work that the pressure in the loader cylinder can be used to represent the mass of the material in the bucket, but if the density information of the material to be excavated is unknown in advance, it is not possible to obtain the volume of the material in the bucket and the corresponding value of the bucket fill factor. To address this issue, Anwar H et al. [10] proposed to use a camera to photograph the bucket and estimate the volume of the material inside the bucket based on an image processing algorithm. However, the accuracy of the algorithm heavily depends on the bucket at the exact constraint position. Therefore, Guevara Jet al. [11] proposed to estimate the volume of the material in the bucket through segmentation, matching, and volume calculation based on the three-dimensional point cloud data of the material in the bucket. However, when the bucket full rate is high, this method cannot accurately extract the edge contour of the bucket, and the prediction error of the bucket fill factor increases up to 20%. Similarly, Lu J et al. [12] proposed to identify the bucket fill factor of a loader based on machine vision and bucket position information, but the method is not robust to environmental changes. Subsequently, Lu J et al. [13] proposed a neural-network-based method for predicting bucket fill factor, which relies on the classification of environmental factors to improve the robustness of the model to the environment. It is worth noting that all the previously mentioned methods for identifying bucket fill factor can only work after the loader has completed the shoveling action. R.J. Sandzimier R J et al. [14] used statistical methods to find the relationship between the excavator shovel trajectory and the bucket fill factor and tried to obtain the optimal excavator trajectory planning based on this. However, this method requires a lot of experimental data, and the experiment needs to be repeated when the size of the attachment changes. Filla R [15] et al. simulated the material shoveling process with the EDEM software and summarized the shovel trajectory offset map for a specific bucket to obtain a certain bucket fill factor. However, the offset method proposed by the authors not only requires a large number of simulations but also cannot summarize well the relationship between the bucket fill factor and the shovel trajectory. From the above analysis, it can be seen that the predicted completion time of the bucket fill factor of the loader can be divided into after the completion of shoveling and before the completion of shoveling. Additionally, the prediction of bucket fill factor before shovel excavation completion is more suitable for the automation of the loader and is the focus of this paper. Based on the analysis of the existing studies, this paper proposes a method for predicting the bucket fill factor of loader based on the three-dimensional information of material surface. Our method can achieve accurate bucket fill factor prediction before excavation and provide technical support for optimizing the operation trajectory of the loader under autonomous excavation.

The rest of this paper is organized as follows. Section 2 describes the bucket fill factor prediction method proposed in this paper. Section 3 introduces the construction of the co-simulation model of shovel. Section 4 presents the implementation of the proposed bucket fill factor prediction method. Finally, Section 5 concludes the research work of this paper.

2. Prediction Method of Bucket Fill Factor

The bucket fill factor is an important optimization target for the loader under autonomous excavation. The research on the prediction method of the bucket fill factor of the loader helps to promote the development of unmanned loaders. Considering the complexity of the experimental process of the loader, the difficulty of obtaining the trajectory, and the high cost of the experiment, the multi-body dynamics software RecurDyn and the discrete element software EDEM can reproduce the complete material shoveling process. Therefore, to verify the feasibility of the proposed bucket fill factor prediction method, the method will be tested in a virtual simulation environment, and the specific flow chart is shown in Figure 1. Firstly, the three-dimensional virtual model of the loader is created in RecurDyn and the material pile model is created in DEMS. Then, the RecurDyn and DEMS software is used to perform co-simulation of multiple shovel trajectories. Next, the material volume in the bucket after the loader digging is completed is divided by the volume of the bucket to obtain the accurate bucket fill factor; meanwhile, the shovel trajectory and the three-dimensional surface of the material pile are obtained for each shovel excavation process. Based on this, the volume between the shoveled path and the material surface profile is integrated to obtain the estimated bucket fill factor. Finally, the regression analysis method is used to find and verify the mapping relationship between the estimated bucket fill factor and the accurate bucket fill factor obtained in the simulation.

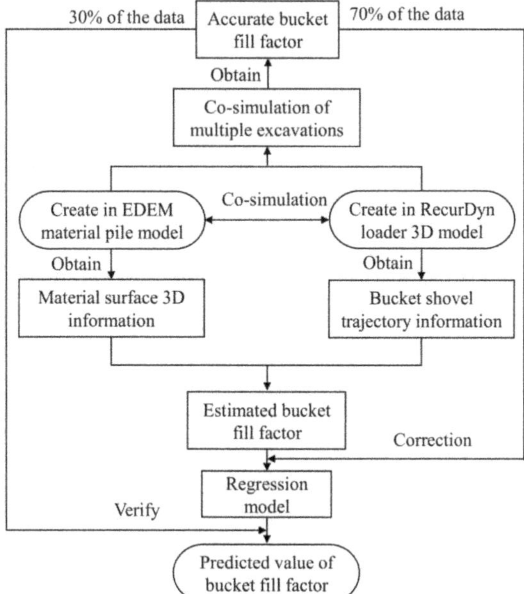

Figure 1. Flow chart of prediction method of bucket fill factor.

3. Co-Simulation of Loader Excavation Process Based on EDEM and RecurDyn

In order to carry out the co-simulation of the loader excavation process, firstly, the material pile model is created in the DEMS, secondly, the loader model is constructed in the multi-body dynamics software, and, finally, the co-simulation of the loader excavation process is carried out.

3.1. The Stockpile Model

The working conditions of the loader are complex, and the operation objects are mainly loose granular materials, such as sand, cinder, loose soil, and gravel [16]. Therefore, in this paper, a typical material such as sand is selected as the operation object to build a sandpile model based on DEMS, and the model is finally used as the material pile model to simulate the shoveling process of the loader. It is worth mentioning that DEMS divides the whole simulation time into several small time steps and then calculates and updates the motion state of all the material particles in each time step. Therefore, a too-large material pile will directly affect the subsequent simulation time and data storage, and a too-small material pile cannot reflect the real shoveling process [17]. In summary, the sandpile model constructed in this paper has a base radius of about 6 m and a height of about 2.3 m, as shown in Figure 2. The material pile was created on a workstation equipped with Intel Xeon Silver 4210 CPU (20 cores, 2.19 GHz), 128 GB main memory, and NVIDIA Quadro P620 and running Windows 10 operating system. The creation process takes about two weeks. To build a larger pile of material, it is recommended to build the material on an inclined slope to reduce the number of particles and the simulation calculation overhead. This is because the innermost layer of material does not affect the simulation of the entire shovel excavation, but it increases the calculation time.

Figure 2. Flow chart of prediction method of bucket fill factor.

The intrinsic and contact parameters [18] of the sand are shown in Table 1. Meanwhile, to reflect the phenomenon that the particle size of sand varies in practice, the particle size distribution of sand is set in EDEM as follows: 1 mm (30%), 1.5 mm (40%), and 2 mm (30%).

Table 1. Material properties and parameters.

Properties/Parameters	Value
Density/(kg·m^{-3})	1900
Poisson's ratio	0.25
Modulus of shear/MPa	1.6×10^4
Sand-Sand coefficient of restitution	0.62
Sand-Steel coefficient of restitution	0.42
Sand-Sand coefficient of static friction	0.74
Sand-Sand coefficient of rolling friction	0.12
Sand-Steel coefficient of static friction	0.42
Sand-Steel coefficient of rolling friction	0.01

3.2. The Loader Model

In this paper, a 958 N loader is selected as the research object. It is also feasible to select the other models of loader, and the three-dimensional virtual model of the loader is established with the multi-body dynamics software RecurDyn, as shown in Figure 3. The main performance parameters of the loader are listed in Table 2.

Figure 3. The three-dimensional model of the 958 N loader.

Table 2. The performance parameters of the 958 N loader.

Performance Parameters	Value
Rated loading capacity/kg	5000
Rated power/kw	162
Digging power/kn	175
Peak traction/kn	160
Minimum turning radius/mm	5910
Unloading height/mm	3167
Rated bucket capacity/m^3	2.4

During the shoveling process of the loader, the movement of the bucket is subject to the joint action of three factors: the movement direction of the loader, the expansion and contraction of the turning bucket cylinder, and the expansion and contraction of the lifting cylinder. Therefore, after constructing the three-dimensional virtual model of the loader with RecurDyn, it is necessary to add motion driving functions for the forward movement of the loader, the telescoping of the bucket cylinder, and the telescoping of the lifting cylinder to simulate the shoveling action of the loader. The above-mentioned three types of motion can be performed using the step-driven function [19] with the following expressions.

$$\text{step}(t, t_0, x_0, t_1, x_1) = \begin{cases} x_0 & t \leq t_0 \\ x_0 + (x_1 - x_0)\left(\frac{t-t_0}{t_1-t_0}\right)^2 [3 - \frac{2(t-t_0)}{(t_1-t_0)}] & t_0 \leq t \leq t_1 \\ x_1 & t_0 \leq t \leq t_1 \end{cases} \tag{1}$$

In Equation (1), t is the time; t_0 is the start time of the motion; t_1 is the end time of the motion; x_0 is the initial position of the motion, and x_1 is the end position of the motion.

3.3. Co-Simulation of Loader Shovel Excavation Process

After the sandpile model and the three-dimensional virtual model of the loader are created, the co-simulation of the excavation process of the loader can be conducted. First, the WALL file containing the three-dimensional model information that can be recognized by DEMS is exported from RecurDyn. Then, the WALL file is loaded into the DEMS, and the two pieces of software exchange data through the data interface file. During the co-simulation process, RecurDyn controls the shovel movement of the loader and calculates and updates the position, velocity, and other motion states of the loader at each time point; meanwhile, DEMS calculates and updates the motion states of all material particles at each time point, and it feedbacks the forces generated by the interaction between the material and the bucket to RecurDyn [20]. The two together simulate the process of the loader shoveling the material. The schematic diagram of the co-simulation is shown in Figure 4.

Figure 4. Schematic diagram of co-simulation.

In the actual operation of the loader, the loader can use different shoveling methods according to the material density, particle size, and the variability of the driver's operating proficiency. The three common shovel excavation methods [19] are shown in Figure 5, i.e., one excavation, segmental excavation, and slicing excavation. Considering the huge calculation overhead of the simulation, only the first half of the loader model can be imported into DEMS, and the shoveling time of each trajectory is within 5 to 6.5 s depending on the shoveling depth and shoveling method. Even in this case, the simulation of each shoveling trajectory still took about 2 days, and the volume of the data generated by the simulation was about 1.5 TB. Therefore, to save time and storage space, a total of 30 simulations of the shoveling process was carried out for the three shoveling methods in this paper. Meanwhile, to ensure the validity of the method, the accurate bucket fill factor values obtained for all three shoveling methods were in the range of 60% to 110%, thus ensuring a reasonable distribution of samples.

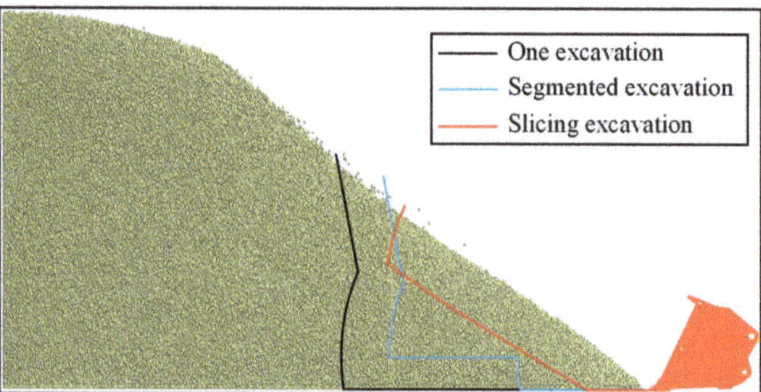

Figure 5. Schematic diagram of the shoveling method.

4. Prediction of Bucket Fill Factor

4.1. Calculation of the Accurate Bucket Fill Factor

As shown in Equation (2), the bucket fill factor is defined as the ratio η of the volume V_m of the material shoveled into the bucket to the rated bucket volume V_b.

$$\eta = \frac{V_m}{V_b} \tag{2}$$

where η is the bucket fill factor, and V_m is the volume of the material in the bucket. According to the performance parameters listed in Table 2, the rated bucket volume V_b is 2.4 m^3.

The forces on a bucket after a shovel excavation are shown in Figure 6. It can be seen that the resultant force on the bucket at the last moment is equal in magnitude to that on the bucket in the vertical direction.

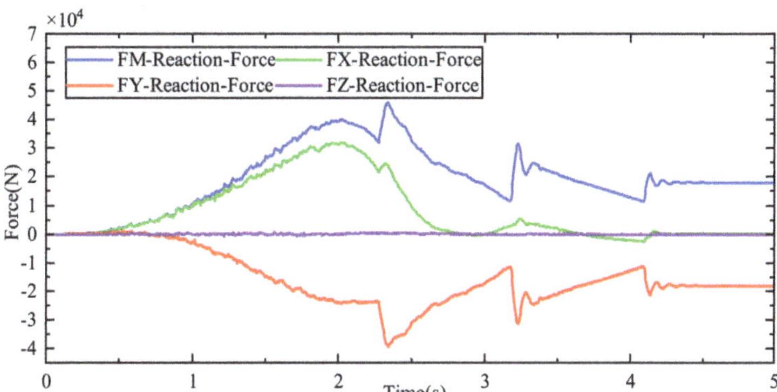

Figure 6. Graph of the force on the bucket.

This means that the bucket at the last moment is only subject to the gravity of the material in the bucket. Based on this, it is possible to calculate the volume of the material in the bucket at the end of each excavation according to Equation (3), and then the bucket fill factor can be calculated by substituting into Equation (2).

$$V_m = \frac{F}{g\rho_m} \tag{3}$$

where F is the magnitude of the resultant force on the bucket at the last moment, g is the acceleration of gravity, and ρ_m is the bulk density of the material.

4.2. Calculation of the Estimated Bucket Fill Factor

4.2.1. Acquisition of Three-Dimensional Information on the Material Surface

The three-dimensional coordinates (X, Y, Z) of the particles on the surface of the pile can be derived from the DEMS simulation results. Based on the three-dimensional coordinates of the particles on the surface of the pile, the three-dimensional surface information of the pile can be expressed by surface fitting or neural network fitting.

First, the polynomial function shown in Equation (4) is used to perform surface fitting, and the fitting result is shown in Figure 7. In practical applications, lidar, binocular camera, or depth camera [11] can be used to obtain the three-dimensional point cloud data of the material pile surface, and surface fitting can be performed on this point cloud data to obtain the three-dimensional surface information of the material pile.

$$f(x,y) = a_{00} + a_{10}x + a_{20}x^2 \cdots + a_{n0}x^n + a_{11}xy + a_{21}x^2y \cdots a_{n1}x^ny + \cdots + a_{n1}x^ny + \cdots + a_{nn}x^ny^n \tag{4}$$

Then, considering the good data fitting ability of the neural network, the relationship between X, Y, and Z is established by building the neural network shown in Figure 8. Therefore, the three-dimensional information of the surface of the material pile can be expressed by the constructed neural network.

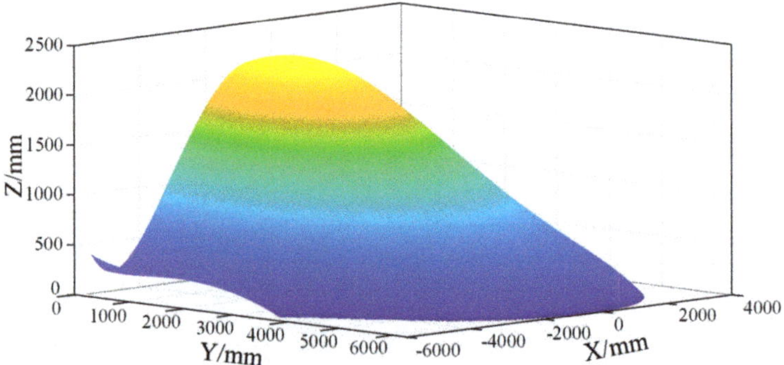

Figure 7. The polynomial surface fitting results of the sandpile surface.

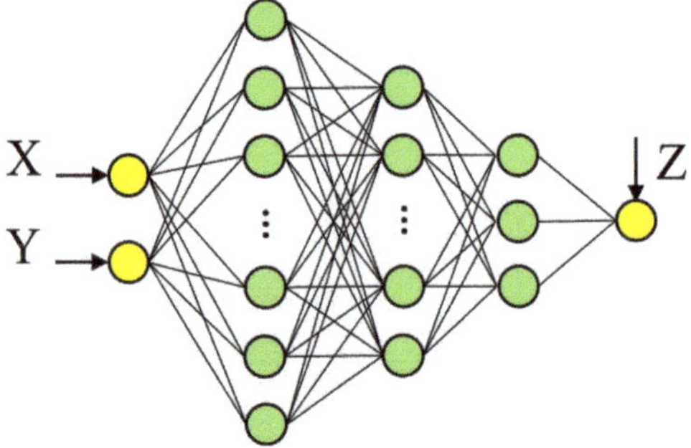

Figure 8. The structure of the neural network.

Finally, according to Equations (5) and (6), the coefficient of determination R^2 and root mean square error $RMSE$ of the fitting results obtained by the above two methods are calculated, and the results are shown in Table 3. The results show that both methods can well express the three-dimensional surface information of the material pile.

$$R^2 = 1 - \frac{\sum_{i=1}^{n}(z_i - \hat{z}_i)^2}{\sum_{i=1}^{n}(z_i - \hat{z})^2} \tag{5}$$

$$RMSE = \sqrt{\frac{\sum_{i=1}^{n}(z_i - \hat{z}_i)^2}{n}} \tag{6}$$

where z_i represents the true value, $\hat{z}$ represents the mean value of the true value, and $\hat{z}_i$ represents the predicted value.

Table 3. The fitting error of sandpile surface.

Surface Fitting Method	Coefficient of Determination (R^2)	Root Mean Squared Error (RMSE)
Polynomial function	0.9927	51.26
Neural networks	0.9969	33.86

4.2.2. Volume Integration Based on Shovel Trajectory and Material Surface Information

After the three-dimensional information of the material surface is obtained, the information of bucket excavation trajectory can be obtained directly from RecurDyn. Then, based on the three-dimensional surface information of the material and the excavation trajectory information, the material volume between the excavation trajectory and the material surface contour can be integrated, and the integration area is shown in Figure 9.

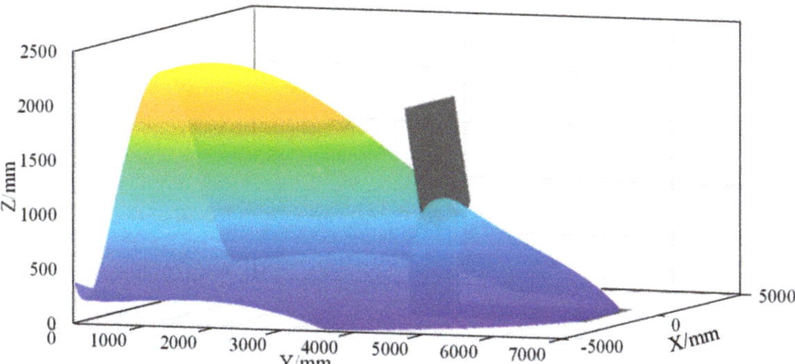

Figure 9. The integration area between the material surface profile and the shovel trajectory.

An integral unit in the shoveling process is shown in Figure 10. The integral unit at different times is slightly different. The process of integrating the material volume is as follows. Firstly, all the integration units between the shovel tooth position section at T_n and the shovel tooth position section at T_{n+1} of the loader are summed up. Then, the material volume under the corresponding shovel trajectory is obtained, as shown in Equation (7).

$$V = \iint f(x,y)dxdy \tag{7}$$

where $f(x,y)$ is the acquired three-dimensional information of the material surface. The integration area is the area between the digging trajectory and the material surface contour.

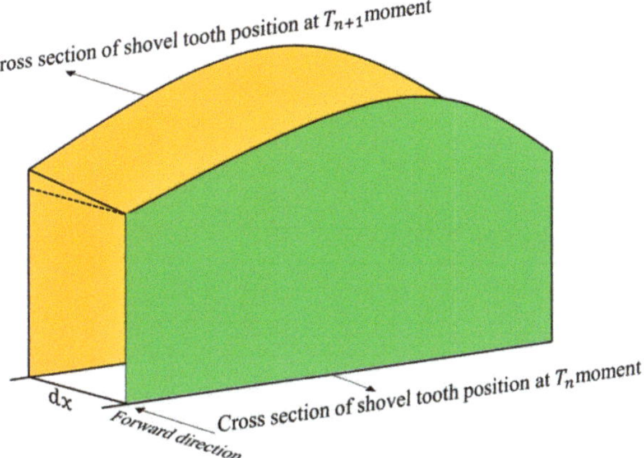

Figure 10. A volume integration unit.

Finally, following the above method, each shovel trajectory is calculated to obtain the volume of the material, and then the volume is divided by the rated bucket volume of the bucket to obtain the estimated bucket fill factor.

Table 4 shows the calculated bucket fill factor and estimated bucket fill factor obtained through the above-mentioned methods. Using the estimated bucket fill factor as the horizontal coordinate and the calculated bucket fill factor as the vertical coordinate, the scatter plot is shown in Figure 11. It can be seen from Table 4 that the difference between the calculated accurate bucket fill factor and the estimated bucket fill factor is large. This is because the estimated bucket fill factor is obtained by adding up all the material on the excavation path, which is not consistent with the actual shoveling process. Figure 11 shows that there is a clear linear regression relationship between the calculated bucket fill factor and the estimated bucket fill factor. Therefore, a regression prediction model can be constructed so that the predicted bucket fill factor can be obtained by inputting the estimated fill bucket fill factor to this model.

Table 4. Calculation results of estimated bucket fill factor.

Trajectory Type	Sample Number	Accurate Bucket Fill Factor/%	Estimated Bucket Fill Factor/%	
			Polynomial	**Neural Networks**
One excavation	1	67.99	92.98	86.88
	2	64.02	85.17	83.07
	3	97.65	143.06	135.87
	4	94.15	133.33	130.70
	5	89.62	125.70	126.09
	6	87.32	120.79	122.40
	7	84.28	119.17	120.09
	8	116.87	205.05	195.48
	9	115.29	193.23	189.17
	10	112.68	184.39	183.98
	11	111.15	179.17	180.38
	12	108.83	178.20	178.82
	13	120.25	234.15	223.25
	14	39.08	52.83	47.67
Segmented excavation	1	75.80	99.09	91.93
	2	64.67	81.73	82.18
	3	59.09	75.18	76.21
	4	104.27	152.29	142.65
	5	93.96	131.61	131.19
	6	88.92	125.43	126.03
	7	114.47	177.43	166.62
Slicing excavation	1	30.75	27.45	23.81
	2	40.79	42.83	35.70
	3	51.27	59.72	49.87
	4	59.03	77.12	65.55
	5	69.01	95.01	82.28
	6	78.16	113.41	99.83
	7	99.08	161.51	146.78
	8	109.86	189.27	175.64
	9	123.09	236.45	221.85

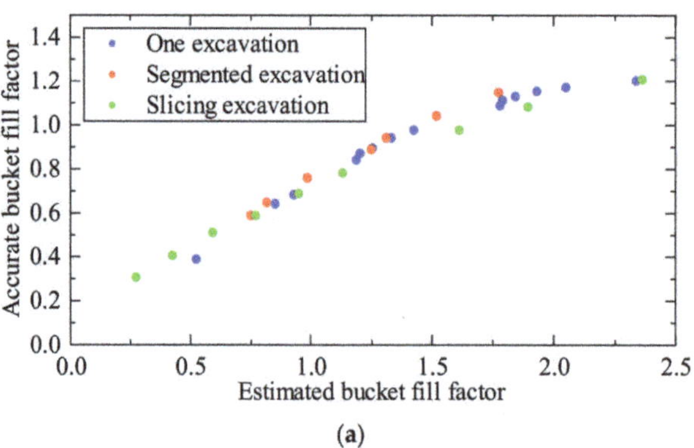

(a)

Figure 11. *Cont.*

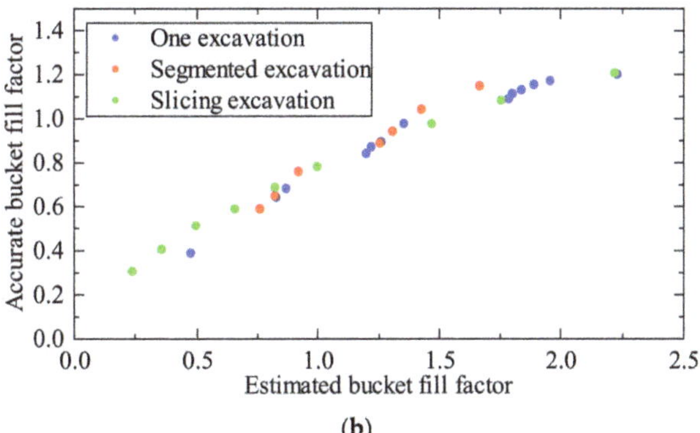

(**b**)

Figure 11. Scatter plot of predicted bucket fill factor under different fitting surface methods. (**a**) Polynomial surface fitting; (**b**) Neural network surface fitting.

4.3. Prediction of Bucket Fill Factor Based on Regression Model

After the calculated bucket fill factor and the estimated bucket fill factor are obtained, the latter is taken as the input of the prediction model, and the former is taken as the output of the prediction model. Therefore, the mapping relationship between the estimated bucket fill factor and the accurate bucket fill factor can be constructed using regression algorithms such as support vector machine regression and polynomial regression. The obtained regression curves are shown in Figure 12, and the specific bucket fill factor prediction results are shown in Table 5. Based on the constructed regression prediction model, it is possible to predict the bucket fill factor after shoveling on a predefined shovel trajectory. To ensure the validity of the method, the values of the accurate bucket fill factor for all the three shoveling methods were in the range of 60% to 110%, thus ensuring a reasonable distribution of the samples.

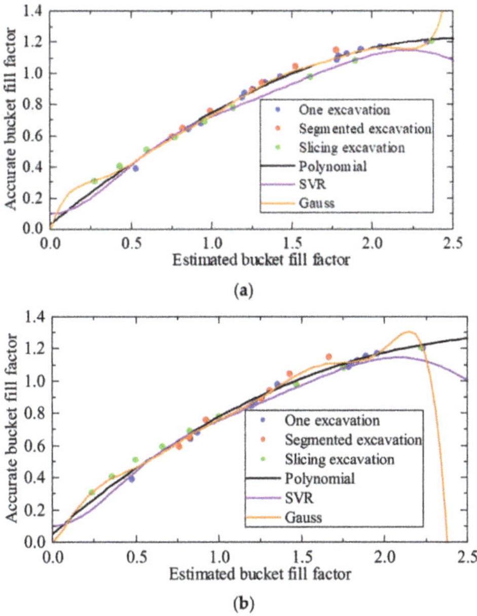

Figure 12. Prediction regression curve of shovel full rate under different fitting surface methods. (**a**) Polynomial Surface Fitting; (**b**) Neural Network Surface Fitting.

Table 5. Predicted results of bucket fill factor.

Trajectory Type	Sample Number	Accurate Bucket Fill Factor/%	Predicted Bucket Fill Factor of Different Regression Method/%						Error/%					
			Polynomial		SVR		Gauss		Polynomial		SVR		Gauss	
			Poly	NN	Poly	NN	Poly	NN	Poly	NN	Poly	NN	Poly	NN
One excavation	1	67.99	70.53	70.25	69.40	69.50	69.10	69.27	2.54	2.26	1.41	1.51	1.10	1.28
	2	64.02	65.74	67.94	65.68	67.54	64.60	66.98	1.72	3.92	1.66	3.52	0.58	2.96
	3	97.65	96.79	95.58	90.33	91.27	98.15	96.57	0.86	2.07	7.32	6.38	0.50	1.08
	4	94.15	92.32	93.27	86.18	88.98	93.46	93.09	1.83	0.88	7.97	5.17	0.69	1.06
	5	89.62	88.60	91.14	82.98	86.95	89.16	90.08	1.02	1.52	6.64	2.67	0.46	0.46
	6	87.32	86.10	89.39	80.96	85.33	86.18	87.80	1.22	2.07	6.36	1.99	1.14	0.48
	7	84.28	85.26	88.27	80.30	84.32	85.18	86.44	0.98	3.99	3.98	0.04	0.90	2.16
	8	116.87	117.24	116.30	112.93	113.45	116.62	117.02	0.37	0.57	3.94	3.42	0.25	0.15
	9	115.29	114.46	114.60	109.85	112.06	114.59	113.65	0.83	0.69	5.44	3.23	0.70	1.64
	10	112.68	112.03	113.12	106.96	110.61	111.93	112.16	0.65	0.44	5.72	2.07	0.75	0.52
	11	111.15	110.45	112.05	105.09	109.48	110.23	111.63	0.70	0.90	6.06	1.67	0.92	0.48
	12	108.83	110.15	111.58	104.73	108.96	109.92	111.49	1.32	2.75	4.10	0.13	1.09	2.66
	13	120.25	121.63	122.40	113.63	112.71	120.20	119.89	1.38	2.15	6.62	7.54	0.05	0.36
	14	39.08	44.09	44.16	43.73	42.09	44.15	45.42	5.01	5.08	4.65	3.01	5.07	6.34
Segmented excavation	1	75.80	74.15	73.22	72.08	71.94	72.69	72.09	1.65	2.58	3.72	3.86	3.11	3.71
	2	64.67	63.58	67.39	63.88	67.07	62.61	66.43	1.09	2.72	0.79	2.40	2.06	1.76
	3	59.09	59.36	63.67	60.14	63.71	58.73	62.60	0.27	4.58	1.05	4.62	0.36	3.51
	4	104.27	100.73	98.48	94.27	94.27	101.72	101.07	3.54	5.79	10.00	10.00	2.55	3.20
	5	93.96	91.50	93.50	85.45	89.20	92.54	93.41	2.46	0.46	8.51	4.76	1.42	0.55
	6	88.92	88.47	91.11	82.87	86.92	89.00	90.05	0.45	2.19	6.05	2.00	0.08	1.13
	7	114.47	109.91	107.60	104.44	104.44	109.67	110.65	4.56	6.87	10.03	10.03	4.80	3.82
Slicing excavation	1	30.75	25.21	25.64	21.84	20.78	31.23	31.01	5.54	5.11	8.91	9.97	0.48	0.26
	2	40.79	36.84	35.12	34.97	31.03	37.97	39.54	3.95	5.67	5.82	9.76	2.82	1.25
	3	51.27	48.93	45.76	49.37	44.06	48.77	46.50	2.34	5.51	1.90	7.21	2.50	4.77
	4	59.03	60.61	56.71	61.29	56.75	59.89	55.55	1.58	2.32	2.26	2.28	0.86	3.48
	5	69.01	71.75	67.46	70.31	67.12	70.28	66.50	2.74	1.55	1.30	1.89	1.27	2.51
	6	78.16	82.20	77.70	77.96	75.53	81.55	76.12	4.04	0.46	0.20	2.63	3.39	2.04
	7	99.08	104.36	100.18	98.14	96.08	104.71	103.58	5.28	1.10	0.94	3.00	5.63	4.50
	8	109.86	113.41	110.58	108.61	107.85	113.47	111.29	3.55	0.72	1.25	2.01	3.61	1.43
	9	123.09	121.83	122.15	113.12	113.06	123.14	123.54	1.26	0.94	9.97	10.03	0.05	0.45

Description: 1, Poly represents the polynomial surface fitting method; 2, NN represents the neural networks fitting method.

After the regression model on the dataset in Table 4 is established, to verify the reliability of the model, this paper conducts the excavation simulation of nine trajectories. Table 6 shows the predicted bucket fill factor and the error under the polynomial regression method. The analysis of the prediction results under each regression method is shown in Table 7. It can be seen from Table 7 that the polynomial regression method performs the best with a root mean square error of 0.025, a lowest mean absolute error of 0.021, and a lowest maximum error of 4.36% on the validation set. Therefore, our proposed prediction method of the bucket fill factor based on three-dimensional information of material surface achieves a good performance in the simulation.

Table 6. Statistics of the results under the polynomial regression method on the validation set.

Trajectory Type	Sample Number	Accurate Bucket Fill Factor/%	Estimated Bucket Fill Factor/%		Predicted Bucket Fill Factor/%		Error/%	
			Polynomial	Neural Networks	Polynomial	Neural Networks	Polynomial	Neural Networks
One excavation	1	53.47	71.67	66.02	57.05	57.03	3.58	3.56
	2	83.35	116.77	110.19	84.00	83.28	0.65	0.07
	3	114.12	191.45	182.47	114.00	112.68	0.12	1.44
Segmented excavation	1	46.23	57.79	51.72	47.59	47.11	1.36	0.88
	2	90.86	124.70	116.50	88.10	86.49	2.76	4.37
	3	98.78	14044	136.37	95.62	95.80	3.16	2.98
Slicing excavation	1	53.92	68.36	57.54	54.83	51.23	0.91	2.68
	2	73.98	104.16	90.92	77.07	72.63	3.09	1.35
	3	101.96	171.46	156.61	107.93	104.00	5.97	2.04

Table 7. Analysis of regression results.

Surface Fitting Method	Regression Method	RMSE	MAE	Maximum Error
Polynomial function	Polynomial	0.029	0.024	5.97%
	SVR	0.050	0.039	9.57%
	gauss	0.025	0.018	5.80%
Neural Networks	Polynomial	0.025	0.021	4.36%
	SVR	0.044	0.038	8.10%
	gauss	0.036	0.031	6.44%

5. Discussion

In practical applications, the use of lidar, binocular camera, or depth camera is essential to realize the unmanned construction machinery. At present, the use of these sensors for the 3D reconstruction of the global or local operating environment of the loader has been one of the hot spots of research and has made some progress. Combined with the method proposed in this paper, the bucket fill factor of the loader after shoveling on this preset trajectory can be calculated based on the surface 3D information of the material pile and the preset planning trajectory, and then the trajectory planning algorithm is prompted to find a suitable operating trajectory with bucket fill factor, which has potential in this aspect.

Due to the complexity of the actual shovel loading process and the limitations of the shovel loading simulation, the research work of this paper still needs to be deepened and expanded. Specifically, the experimental data are obtained from the simulation results of the loader shoveling a material, and the simulation verification of a variety of materials needs to be performed. Moreover, the proposed method is only verified in the simulation environment, and a large number of verification experiments are required. Therefore, in future research, simulations of the shoveling process for many different materials and verification experiments will be conducted to optimize this research work.

6. Conclusions

To accurately predict the bucket fill factor of a loader under different shovel trajectories, this paper proposes a prediction method based on three-dimensional information of the material surface. Firstly, the multi-body dynamics software RecurDyn and the discrete element software EDEM are used to create a co-simulation model of the loader excavating materials, and a total of 30 excavation processes are co-simulated. Then, the material volume in the bucket after the loader excavation is divided by the rated volume of the bucket to obtain the accurate bucket fill factor, the shovel trajectory information, and the three-dimensional surface information of the material pile in each shovel excavation process. Based on this, the volume between the shoveled path and the material surface profile is integrated to obtain an estimated bucket fill factor. Finally, a variety of regression algorithms are used to find the mapping relationship between the estimated bucket fill factor and the accurate bucket fill factor. The verification of the mapping relationship shows that the maximum error between the predicted bucket fill factor and the accurate bucket fill factor under the polynomial regression algorithm is 4.3%, the root mean square error is 0.025, and the average absolute value error is 0.021. The results indicate that the method proposed in this paper can well predict the bucket fill factor of the loader after excavation for the preset trajectory. This plays a key role in optimizing the loader unmanned algorithm and in evaluating the merits of the unmanned strategy.

Author Contributions: Conceptualization, S.W. and L.H.; Methodology, S.W. and S.Y.; Investigation, S.Y. and Y.W.; Validation, S.Y.; Writing—original draft preparation, S.Y. and B.W.; Writing—review and editing, S.W. and L.H.; funding acquisition, S.W. and L.H. All authors have read and agreed to the published version of the manuscript.

Funding: This work was supported by the National Key Research and Development Program of China (Grant No. 2020YFB1709904, 2020YFB1709901), National Natural Science Foundation of China

(Grant No. 51905460,51975495), Natural Science Foundation of Fujian (Grant No.2022J01060), and Guangdong Basic and Applied Basic Research Foundation (Grant No. 2021A1515012286).

Institutional Review Board Statement: The study did not require ethical approval.

Informed Consent Statement: The study did not involve humans.

Data Availability Statement: The study did not report any data.

Conflicts of Interest: The authors declare no conflict of interest.

References

1. Jun, G.; Daqing, Z.; Yong, G.; Zhongyong, T.; Changsheng, L.; Peng, H.; Yuming, Z.; Weicai, Q.; Yongping, J. Potential energy recovery method based on alternate recovery and utilization of multiple hydraulic cylinders. *Autom. Construct.* **2020**, *112*, 103105. [CrossRef]
2. You, Y.; Sun, D.; Qin, D.; Wu, B.; Feng, J. A new continuously variable transmission system parameters matching and optimization based on wheel loader. *Mech. Mach. Theory* **2020**, *150*, 103876. [CrossRef]
3. He, X.; Jiang, Y. Review of hybrid electric systems for construction machinery. *Autom. Construct.* **2018**, *92*, 286–296. [CrossRef]
4. Chen, Z.; Zhao, F.; Zhou, J.; Huang, P.; Zhang, X. Fault diagnosis of loader gearbox based on an Ica and SVM algorithm. *Int. J. Environ. Res. Public Health* **2019**, *16*, 4868. [CrossRef] [PubMed]
5. Wu, G.; Ma, W.; Liu, C.; Wang, S. IOT and cloud computing based parallel implementation of optimized RBF neural network for loader automatic shift control. *Comput. Commun.* **2020**, *158*, 95–103. [CrossRef]
6. Zhao, H.; Wang, G.; Lv, W.; Cao, Y.; Li, X. Optimization of hydropneumatic suspension for articulated wheel loader based on kriging model and particle swarm algorithm. *Adv. Mech. Eng.* **2018**, *10*, 1687814018810648. [CrossRef]
7. Dadhich, S.; Bodin, U.; Andersson, U. Key challenges in automation of earth-moving machines. *Autom. Construct.* **2016**, *68*, 212–222. [CrossRef]
8. Yu, C.; Jiang, S.; Chu, C. Application of SPC Method to Loader Bucket's Fill Factor Test. *Construct. Mach. Equip.* **2014**, *45*, 13–19+7. [CrossRef]
9. Huang, P.; Liao, X. Study on the Mathematical Model of Bucket Filling Rate of Bucket Shoveling and Loading Process. *Met. Min.* **2017**, *3*, 126–130. [CrossRef]
10. Anwar, H.; Abbas, S.M.; Muhammad, A.; Berns, K. Volumetric estimation of contained soil using 3d sensors. *Int. Commer. Veh. Technol. Symp.* **2014**, 11–13. Available online: https://cyphynets.lums.edu.pk/images/SoilEstimCVT2014.pdf (accessed on 8 August 2022).
11. Guevara, J.; Arevalo-Ramirez, T.; Yandun, F.; Torres-Torriti, M.; Cheein, F.A. Point cloud-based estimation of effective payload volume for earthmoving loaders. *Autom. Construct.* **2020**, *117*, 103207. [CrossRef]
12. Lu, J.; Bi, Q.; Li, Y.; Li, X. Estimation of fill factor for earth-moving machines based on 3D point clouds. *Measurement* **2020**, *165*, 108114. [CrossRef]
13. Lu, J.; Yao, Z.; Bi, Q.; Li, X. A neural network–based approach for fill factor estimation and bucket detection on construction vehicles. *Comput. Aided Civ. Infrastruct. Eng.* **2021**, *36*, 1600–1618. [CrossRef]
14. Sandzimier, R.J.; Asada, H.H. A data-driven approach to prediction and optimal bucket-filling control for auonomous excavators. *IEEE Robot. Autom. Lett.* **2020**, *5*, 2682–2689. [CrossRef]
15. Filla, R.; Obermayr, M.; Frank, B. A study to compare trajectory generation algorithms for automatic bucket filling in wheel loaders. In Proceedings of the 3rd Commercial Vehicle Technology Symposium, Kaiserslautern, Germany, 11–13 March 2014; pp. 588–605. [CrossRef]
16. Frank, B.; Kleinert, J.; Filla, R. Optimal control of wheel loader actuators in gravel applications. *Autom. Construct.* **2018**, *91*, 1–14. [CrossRef]
17. Coetzee, C.J. Calibration of the discrete element method. *Powder Technol.* **2017**, *310*, 104–142. [CrossRef]
18. Shao-Jie, W.; Yue, Y.; Sheng-Feng, Y.; Liang, H. Dynamic analysis on loader coupling based on RecurDyn-EDEM. *J. Mach. Des.* **2021**, *38*, 1–6. [CrossRef]
19. Yu, X.J.; Huai, Y.H.; Li, X.-F.; Wang, D.-W.; Yu, A. Shoveling trajectory planning method for wheel loader based on kriging and particle swarm optimization. *J. Jilin Univ.* **2022**, *50*, 437–444. [CrossRef]
20. Richter, C.; Roessler, T.; Otto, H.; Katterfeld, A. Coupled discrete element and multibody simulation, part I: Implementation, verification and validation. *Powder Technol.* **2021**, *379*, 494–504. [CrossRef]

Article

A Novel Seismocardiogram Mathematical Model for Simplified Adjustment of Adaptive Filter

Gediminas Uskovas [1], Algimantas Valinevicius [1], Mindaugas Zilys [1], Dangirutis Navikas [1], Michal Frivaldsky [2], Michal Prauzek [3], Jaromir Konecny [3] and Darius Andriukaitis [1,*]

[1] Department of Electronics Engineering, Kaunas University of Technology, Studentu St. 50-438, LT-51368 Kaunas, Lithuania

[2] Department of Mechatronics and Electronics, Faculty of Electrical Engineering and Information Technologies, University of Zilina, 010 26 Zilina, Slovakia

[3] Department of Cybernetics and Biomedical Engineering, VSB—Technical University of Ostrava, 708 00 Ostrava, Czech Republic

* Correspondence: darius.andriukaitis@ktu.lt; Tel.: +370-37-300-519

Abstract: Nonclinical measurements of a seismocardiogram (SCG) can diagnose cardiovascular disease (CVD) at an early stage, when a critical condition has not been reached, and prevents unplanned hospitalization. However, researchers are restricted when it comes to investigating the benefits of SCG signals for moving patients, because the public database does not contain such SCG signals. The analysis of a mathematical model of the seismocardiogram allows the simulation of the heart with cardiovascular disease. Additionally, the developed mathematical model of SCG does not totally replace the real cardio mechanical vibration of the heart. As a result, a seismocardiogram signal of 60 beats per min (bpm) was generated based on the main values of the main artefacts, their duration and acceleration. The resulting signal was processed by finite impulse response (FIR), infinitive impulse response (IRR), and four adaptive filters to obtain optimal signal processing settings. Meanwhile, the optimal filter settings were used to manage the real SCG signals of slowly moving or resting. Therefore, it is possible to validate measured SCG signals and perform advanced scientific research of seismocardiogram. Furthermore, the proposed mathematical model could enable electronic systems to measure the seismocardiogram with more accurate and reliable signal processing, allowing the extraction of more useful artefacts from the SCG signal during any activity.

Keywords: mathematic model; cardiovascular system; seismography; modeling; adaptive digital filter; noninvasive method; heart rate

Citation: Uskovas, G.; Valinevicius, A.; Zilys, M.; Navikas, D.; Frivaldsky, M.; Prauzek, M.; Konecny, J.; Andriukaitis, D. A Novel Seismocardiogram Mathematical Model for Simplified Adjustment of Adaptive Filter. *Electronics* **2022**, *11*, 2444. https://doi.org/10.3390/electronics11152444

Academic Editor: João Soares

Received: 16 June 2022
Accepted: 3 August 2022
Published: 5 August 2022

Publisher's Note: MDPI stays neutral with regard to jurisdictional claims in published maps and institutional affiliations.

1. Introduction

The heart is one of the most complex and hardest-working organs in the physiological system of a living organism and determines the quality of human life. Proper cardiac diagnosis allows one to notice heart problems in a timely manner. Various cardiac tests are among the most important, but the number of patients with cardiovascular disease remains high [1,2]. One of the most popular heart work measurements is electrocardiogram measurement, but the measurement of electrocardiogram is a clinical method and is not suitable for monitoring the patient's daily or active activities due to the need to carry uncomfortable electrodes [3,4]. The need for nonclinical monitoring is growing to detect changes in cardiac work at an early stage, to facilitate treatment in areas where cardiac disease experts are missing [5,6]. Additionally, hospitals want to minimize unplanned visits to clinics or hospitals, when majority visits are minute-time and, as a result, not cost beneficial. For this group, the challenge is how to help them to interpret symptoms correctly and alert an ambulance or use medicaments instead of waiting for symptoms to disappear [7,8]. Therefore, the main objective is to reduce mortality by replacing stationary diagnostic devices with portable devices, which can alert to possible disease.

These portable or wearable electronic systems do not disturb the daily activity of patients, have evaluated data noninvasively and extract various information on heart functionality [9,10]. The seismocardiogram signal represents low-frequency (0–50 Hz) cardio cycle mechanic vibrations corresponding to cardiac events [11,12], and contains mitral valve opening (MO) and closing (MC), isovolumetric contraction, ejection, opening of the aortic valve (AO) and closing (AC), and cardiac filling [13]. The heart monitoring system is complicated due to its nonstationary nature [14] and the presence of noises such as muscle movement noise, respiration vibration noise [8], and environment acoustic noise [6,15] in the seismocardiogram (SCG) signal [16,17]. These signals correspond to the surfaces generated by cardiac activities, which we can measure with the accelerometer micromechanical system (electromechanical control system-MEMS) [18]. In addition, we aim to search for other measurement methods and methods, such as forcecardiography (FCG), which describes a combination of force and accelerometer sensors that can monitor weak mechanical oscillations of the heart's work [19,20].

This article is one of the articles which deals with a moving patient without simultaneously measuring the ECG. Therefore, the development of seismocardiogram measurement devices requires a good understanding of this signal morphology. Using mathematical models of heart and seismocardiogram signals reduces investigation time and resources. The article [21] describes the mechanical morphology of the seismocardiogram and suggests an analytic seismocardiogram model. This model can be complicated, because it requires good heart physiological mechanics and liquid mechanics. Such heartbeat signals can be simulated using an electric circuit heart model, which researchers call a cardiovascular functional avatar [22]. Furthermore, the numeric seismocardiogram model suggests using the artefacts that share the fiducial points of the seismocardiogram from early published studies [11,23]. As a result, the mathematical signal results in further progression of the investigation. The mathematical representation of the human heart allows more accurate and better-quality analysis of the cause of cardiovascular diseases.

The rest of the article is organized as follows. Section 2 discusses related works and methods, describes the mathematical model of seismocardiogram signal and algorithms, and introduces the use of the SCG model and the processing of a real signal with adaptive filters. Section 3 presents the experimental data and analysis. Section 4 concludes the work of this paper and briefly introduces future works.

2. Materials and Methods

The cardiovascular system is complex and, in order to create a mathematical model describing it, it is also complicated. Thus, systematic analysis methods are useful when attention first focuses on the structure of the system, then on the interrelationship of the individual components, and finally on the internal structure of individual parts. For this reason, regarding the cardiovascular system, it is appropriate to investigate the heart separately from blood vessels. In addition, the seismocardiogram investigation is related to the transfer of vibrations from the heart muscles to the chest. Unlike the most popular clinical heart rate monitoring methods, such as electrocardiogram, ultrasound cardiogram, phonocardiogram, or photoplethysmogram, mechanocardiogram measurement is focused on recording and analyzing mechanical vibrations in the heart muscle with micro-electromechanical system (MEMS) sensors [24]. This list can be supplemented with a force sensor consisting of an accelerometer and a force-sensitive resistor [25,26].

2.1. Related Works

The complexity of the human cardiovascular system involves many control mechanisms, which can mathematically explain the proposed nature of the investigation. The heart modeling process has a few directions, mathematical and physical, which have to be followed before real experiments occur [22,27]. Apart from that, the cardiovascular system is generally hydraulic, but also can be analyzed in the electrical circuit [28]. The concept that one heart work cycle or a heartbeat duration combines systolic and diastolic durations unites the hemodynamic and electrical nature of development of an artificial heart or a mathematical model. Due to both models, the general scopes demonstrate parts of impulses rising and falling [27].

Examining the cardiovascular system, researchers in papers point out the heart as a pump forcing blood flow through the blood vessels. Hence, the heart is emphasized as a hydraulic system, which is defined using fluid mechanics equations, quantities, and coefficients such as: vessel volume, blood flow rate, blood pressure, wall elasticity and strength, cross-section area, vessels length, blood viscosity, and blood density [29,30].

The main advantage of this model is that it is directly related to the patient, whose several required quantities (blood viscosity, blood density, blood pressure, etc.) can be measured by laboratory tests inside a medical institution.

Another part of the heart mathematical models is related to the electrical nature of the heart, when an electrical impulse is generated as the synodic node, and as a result, it forces the heart muscle to move periodically [22,31]. These pulses can be measured using electrodes and an electrocardiogram. For this reason, the heart mathematical model is based on an electrical circuit. The aim of this model is to reproduce and simulate an electrocardiogram.

The third part of the mathematical models is the result of changing the hemodynamical representation of the heart to electrical [32,33]. So, the hydraulic system interconnects mechanical and electrical quantities, coefficients and constants [28]. The element of each mechanical system is described by parameters corresponding to the analogy of their mechanical nature. Then, the blood flow in the blood vessels is changed to the electrical current, the volume is changed to voltage, the resistance to blood flow (R) Equation (1) [27] is changed to electrical resistance, blood flow inertia (L) Equation (2) [27] is changed to electrical induction, and the storage (C) Equation (3) [27] is changed to electrical capacity. Hence, nonelectrical quantities are replaced by electrical ones.

$$R = \frac{8l\mu}{\pi r^4} \tag{1}$$

$$L = \frac{8l\rho}{\pi r^2} \tag{2}$$

$$C = \frac{3\pi r^3 l}{2Eh} \tag{3}$$

where: l—length of the artery; μ—blood viscosity; r—radius; ρ—blood viscosity; h—thickness; and E—Jung module.

The proposed seismocardiogram model in [21] represents blood flow through large vessels in the upper part of the body. The written mathematical model explains the systolic and diastolic phases and forces [34]. This mathematical model is helpful and suited to explaining heart work from a mechanical perspective.

This article is closely related to the previous article of this research group, and will continue the work of improving the quality of the study of nonclinical SCG signals [35]. Also, the work relates to investigations of using wireless sensors and them effective energy consumption [36–38]. Direct comparison of a few different models is difficult, it requires a deep understanding of each model. All models discussed require a good understanding of the cardiovascular system and individual parameters of the patients that the mathematical model performs correctly. Thus, a simpler mathematical model is required, which can support researchers in investigating the seismocardiogram or developing new devices for ensuring quality of life and safety in roads [39–42].

2.2. Mathematical Modeling

A single heartbeat signal mathematical model is created according to the data provided in Table 1. For that purpose, a heartbeat cycle is split into artefacts, symbolizing the duration between the heart valves' state moments. These durations can be recorded in the aria of fiducial points values (4).

$$FV_{SCG} = \left\{ T^{MC,\,AO}, T^{A0,\,AC}, T^{A0,\,AC}, T^{AC,MO}, T^{RBE}, T^{RBF} \right\} \tag{4}$$

where:

FV_{SCG}—seismocardiogram signal fiducial points values aria;
$T^{MC,AO}$—time between the aorta opening and closing;
$T^{A0,AC}$—time between the aorta opening and closing;
$T^{MC,MO}$—time between the closing and opening of the mitral valves;
$T^{AC,MO}$—time between the closure of the aorta and the opening of the mitral valve;
T^{RBE}—duration of systole;
T^{RBF}—duration of diastole.

Table 1. Seismocardiogram artefacts fiducial points and duration rangers.

Mark	Description	Points from	Points to	Duration, ms
$T^{MC,AO}$	Duration from MC to AO	16 ($\pm$14)	59 ($\pm$11)	43 ($\pm$14)
$T^{AO,AC}$	Duration from AO to AC	59 ($\pm$11)	348 ($\pm$37)	332 ($\pm$37)
$T^{MC,MO}$	Duration from MC to MO	16 ($\pm$14)	437 ($\pm$37)	378 ($\pm$37)
$T^{AC,MO}$	Duration from AC to MO	348 ($\pm$37)	437 ($\pm$37)	89 ($\pm$37)
T^{RBE}	Duration of Systole		-111 ($\pm$24)	
T^{RBF}	Duration of Diastole	505 ($\pm$41)		

This passion in the MATLAB program is converted into a single line matrix with additional variables to mark the relative acceleration of the accelerometer $\pm\,ag$, where a is the measured multiplier and g is the free acceleration equal to 9.8 [m/s^2]. The mathematical model of the theoretical SCG signal has been modeled by taking conditional acceleration values from graphics and fiducial points data from other research publications, such as [11] and [15].

The graphics shown in the second picture reflect the main artefacts of the seismocardiogram signal corresponding to the heart working moments specified in Table 1.

The widely visible artefacts of the time values and duration ranges of the seismocardiogram between two fiducial points are shown in the first table, which was created based on [11,23].

The acceleration amplitude values are tabled in Table 2 and explain all the meanings of the abbreviations in more detail.

Table 2. Modeled SCG signal acceleration amplitude values and description.

Mark	Acceleration Amplitude Value, m/s^2	Description
$ascg_1$	0.5	Arteries systole
$ascg_2$	0.3	Mitral valve closing
$ascg_3$	-0.9	Isovolumetric moment
$ascg_4$	1.8	Aorta valve opening
$ascg_5$	-1.65	Isovolumic contraction
$ascg_6$	1.2	Rapid systole ejection
$ascg_7$	-0.2	Rapid filling
$ascg_8$	0.25	Aorta valve closing
$ascg_9$	-0.55	Mitral valve opening

The proposed time values for each fiducial points are tabled and more details are described in Table 3.

Table 3. Modeled SCG signal fiducial points time values and description.

Mark	Fiducial Points Time Values, ms	Description
tas	1	Arteries systole
tmc	16	Mitral valve closing
tim	21.5	Isovolumetric moment
tao	59	Aorta valve opening
tic	26	Isovolumic contraction
trbe	111	Rapid systole ejection
trbee	141	Rapid filling
tac	348	Aorta valve closing
tmo	437	Mitral valve opening

The mathematical model of the SCG signal was formed based on the Heaviside function, which marks the concrete fiducial points impulse activation sequence in the time domain.

$$sg_1(t) = ascg_1 e^{-10t} \cos(w_0 60t - 6.45)(x_1 - x_2) \tag{5}$$

$$sg_2(t) = ascg_3 \cos(w_0 23t - 1.2)(x_3 - x_2) \tag{6}$$

$$sg_3(t) = ascg_4 e^{-0.98t} \sin(w_0 3t - 2.7) \cos(w_0 16.5t - 5.5)(x_6 - x_3) \tag{7}$$

$$sg_4(t) = ascg_6 e^{-6.5t} 1.5 \cos(w_0 9.23t)(x_6 - x_7) \tag{8}$$

$$sg_5(t) = \left(t + ascg_7 + ascg_8 e^{-2.5t} \cos(w_0 12.8t - 2.7)\right)(x_7 - x_8) \tag{9}$$

$$sg_6(t) = ascg_9 3.3 e^{-2.85t} \cos(w_0 4.1t + 3.14) \cos(w_0 14.2t - 0.1)(x_8 - x_{12})^2 \tag{10}$$

where:

$w_0 = 2\pi f$ angular frequency rad/s when $f = 1$ Hz; $sg_1(t)$–$sg_6(t)$—intermediate signal parts, which represent each acceleration artifact of the modeled SCG signal.

The presented equations are intermediate, and form the final expression of the modeled signal $scg(t)$ Equation (11). Each of these formulas is directly related to the seismocardiogram artifacts described in Tables 2 and 3. On the basis of these artifact values, the artificial signal is developed. Additionally, the synthesis of this signal is made possible, according to investigation requirements, by changing these values.

Adding all the expressions of the signal pulse sequences, a mathematical model of a single heartbeat seismocardiogram signal is obtained, as is the algorithm described in Figure 1, for which the signal is shown in Figure 2.

Following the algorithm described in Figure 1a, the seismocardiogram of a single heartbeat is visualized graphically, so Figure 2 shows the generated theoretical seismocardiogram signal with the main artefacts of a seismocardiogram. So, using this signal the sequence can be produces (Figure 3).

As a result, the mathematical model has been developed on the basis of these artefacts points. In this article, the general Equation (11) is presented and used for the described investigation.

$$scg(t) = sg_1(t) + sg_2(t) + sg_3(t) + sg_4(t) + sg_5(t) + sg_6(t) \tag{11}$$

More detail and a clear explanation of the numerically modeled seismocardiogram signal, sequence creation, and management possibilities with additional noises is described using three algorithms (Figures 1 and 4).

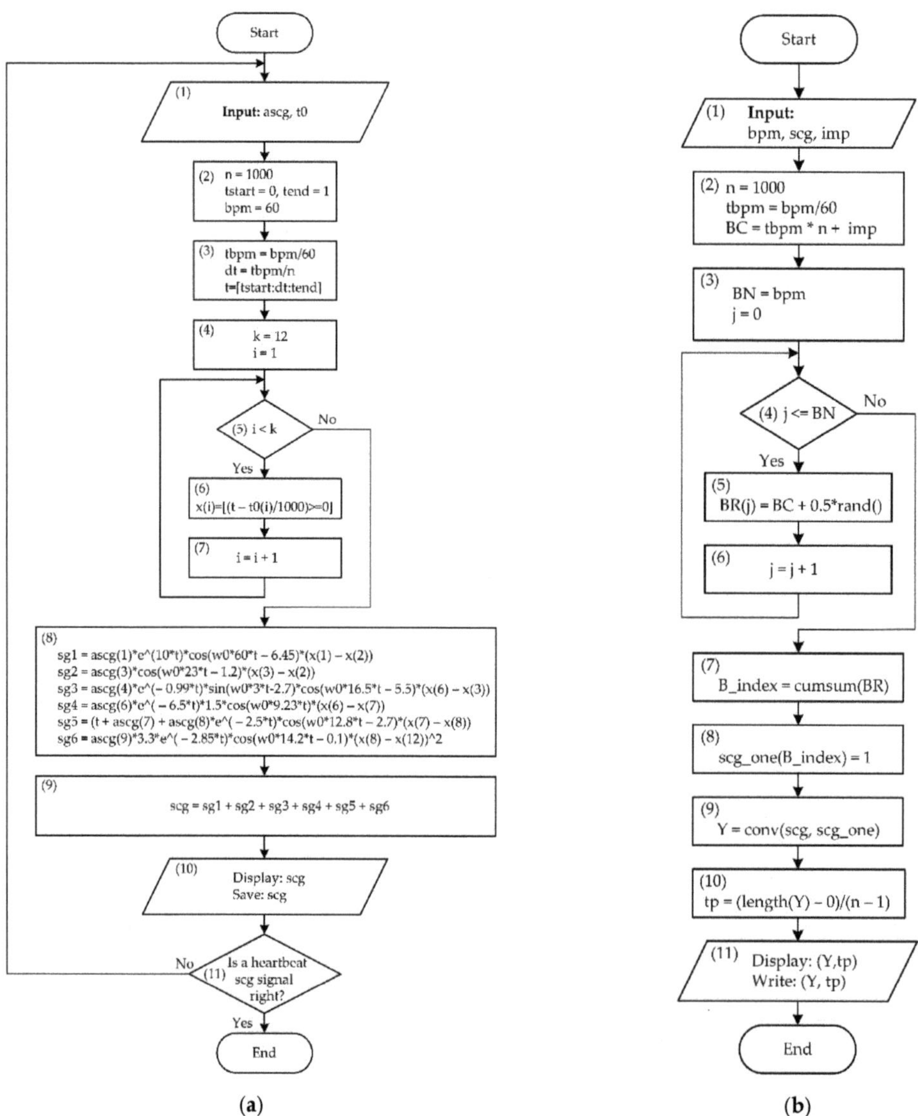

Figure 1. Modeling the seismocardiogram signal sequence without noise algorithms: (**a**) Creation of a heartbeat seismocardiogram signal algorithm; (**b**) modeling the SCG signal without noise algorithm. Where: **ascg** = {**ascg1, ascg2, ascg3, ascg4, ascg5, ascg6, ascg7, ascg8, ascg9**}—SCG data artifacts acceleration values set; **t0** = {**tas, tmc, tim, tao, tic, trbe, trbee, tac, tmo, trbf, trbfe, te**}—SCG data fiducial points time values set; **tstart**—start time in seconds; **tend**—end of one heartbeat in seconds; **n**—number of samples equal to 1000 ms; **dt**—step of time difference; **bpm** = heart rate beats per minute; **imp**—additional impulses for anti-alliancing heartbeats; **BC**—beat countdown; **BN**—beats number for generation one minute sequence; **BR**—set of random beats; **Y**—SCG signal sequence; **tp**—duration of SCG sequence.

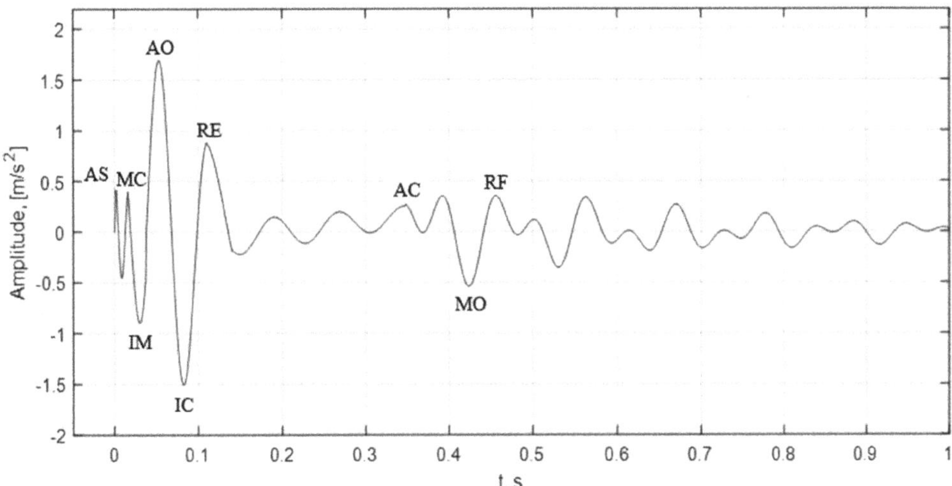

Figure 2. The mathematical model of the SCG signal. Where: AS—atrial systole; MC—closing of the mitral valve; AO—aorta opening; IM—isovolumic movement; IC—isovolumic contraction; RE—rapid systolic ejection; AC—aorta closing; MO—opening of the mitral valve; RF—rapid filing.

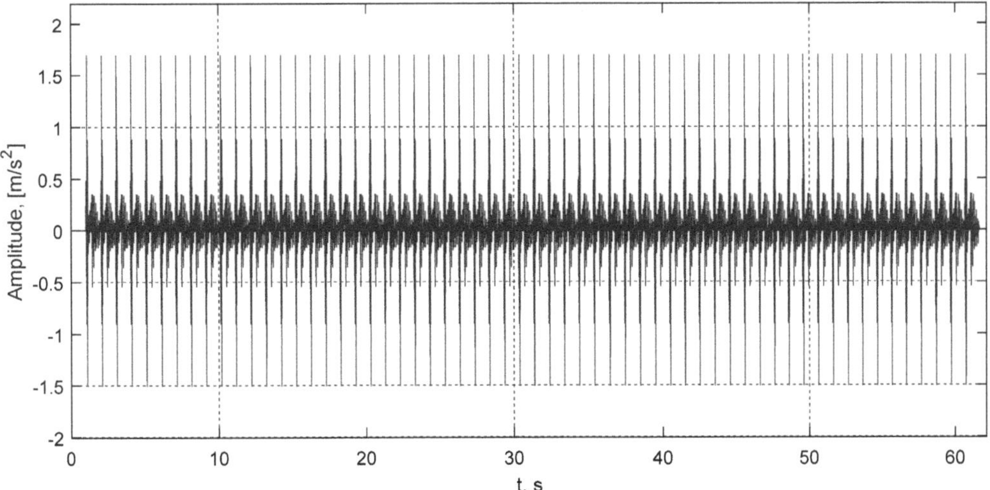

Figure 3. Theoretical sequence of SCG signals.

The algorithm shown in Figure 1a explains the mathematical model of the seismocardiogram. The main aim of this algorithm is to generate a single-heartbeat SCG signal with one-second duration, then the heart rate is 60 bpm. In the first step of the algorithm, two sets of acceleration values **ascg** and the corresponding fiducial points time values **t0** are entered. These values are the main artifacts of the simulated SCG signal, and sample values for these sets, respectively, are given in Tables 2 and 3. The researcher can modify these values according to the investigation objectives.

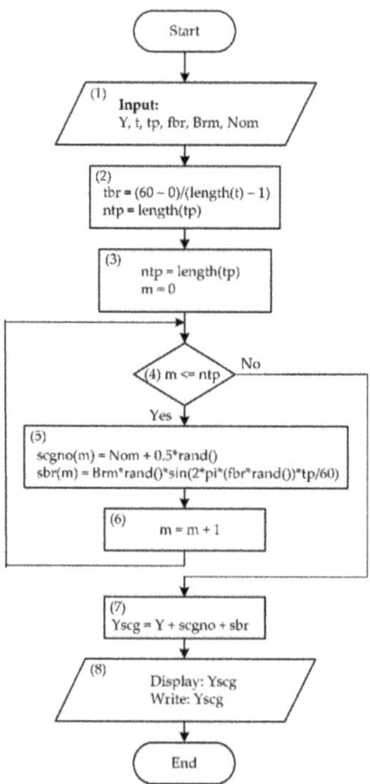

Figure 4. Adding breath noise and white noise signals to the modeled sequence of the seismocardiogram signal algorithm.

The second step specifies the time constant of the start and end of the heartbeat signal (tstart), the heart rate of 60 bpm and the number of samples number **n** = 1000. In the third step, the calculations of the required pulse durations are performed and the duration of the sampling step **dt** creates a 1000 ms time domain.

The 4th, 5th, 6th, and 7th steps of the algorithm show how to find the unit impulse values (**x1: x12**) corresponding to the time references of the **t0** set for creation of a desired shape in the 8th and 9th step of the seismocardiogram signal. The SCG signal is displayed graphically, and all values are stored in the memory during the 10th step of the algorithm. The researcher qualifies a visually received SCG signal in the 11th step and, if required, changes the values of **ascg** and **t0** sets. The researcher can modify the t0 set time values of each fiducial point and parallelly acg set the amplitude values that allow synthesis of the desired SCG signal and allow random changes to be set.

The second algorithm, shown in Figure 1b, describes the simulation of a sequence of SCG signals. This sequence is an analogy of a repetitive heart rhythm. In the first step of the algorithm, the **scg** signal of one heartbeat obtained in the first algorithm is used. In addition, the heart rate **bpm** and the number of additional pulses are examined, so that the signals of individual beats do not overlap in the sequence. The next step of the algorithm is to calculate the beat count **BC** of the sequence. In the following 4th, 5th, and 6th steps, a different number of pulses are counted in the sequence of the SCG signal for each heartbeat. In the seventh step, each heartbeat is given an index and is equated to a unit in Step 8. In Step 9, the sequence **Y** of the SCG signals is found after the convulsion between the single heartbeat signal **scg** and the new unit pulse sequences **scg_one (B_index)** formed in the 9th step, the time counts of which are calculated in the 10th step. In the 10th step of the algorithm, the sequence of the noise-free SCG signal shown in Figure 3 is represented graphically.

From one generated theoretical seismocardiogram cycle, we can create a sequence of desired duration SCG signal without noise, which is shown in Figure 3.

The sequence of SCG signals without noise is not informative and needs to be added with some real noise, which makes the artificial SCG signal more realistic. So, the desired sequence has been constructed with different types of noises, which changes randomly and reflects the real situation. Figure 4 shows the next step, which describes the SCG signal required to perform the sequence with added noise. Thus, the real situation is reflected as a seismocardiogram is measured from a patient and has noncardiac signals that are considered noise. These may be random vibrations in the surrounding environment or chest vibrations caused by breathing.

The algorithm shown in Figure 4 describes the acquisition of an SCG signal with total noise and will be further used to find the optimal parameters of the adaptive filter. First, the parameter values require the **Y** and **t** of SCG signal sequences to be entered without the noise, in addition to the respiration rate **fbr** and its amplitude **Brm**, and the white noise amplitude **Nom**. The following four, five, and six steps of the algorithm describe the finding of random values for white noise and respiration. Additionally, the respiration amplitude and frequency vary randomly. In the seventh step, an SCG signal with noise Yscg is obtained, displayed graphically (Figure 5), and saved. Figure 5 shows an example of such a signal. The mathematical model created first allows us to model the desired seismocardiogram signals, monitoring the reaction of the designed system to them, adding various environmental noise signals, and taking into account the research objectives. On the basis of the obtained results, it is possible to develop and research real seismocardiogram measuring devices, which are capable of processing nonclinical data much more efficiently. The legal regulations related to testing medical devices before they are approved and approved are avoided. Additionally, research becomes safer and more economical in this way because optimal data obtained from many simulations are used to realize a real electronic device.

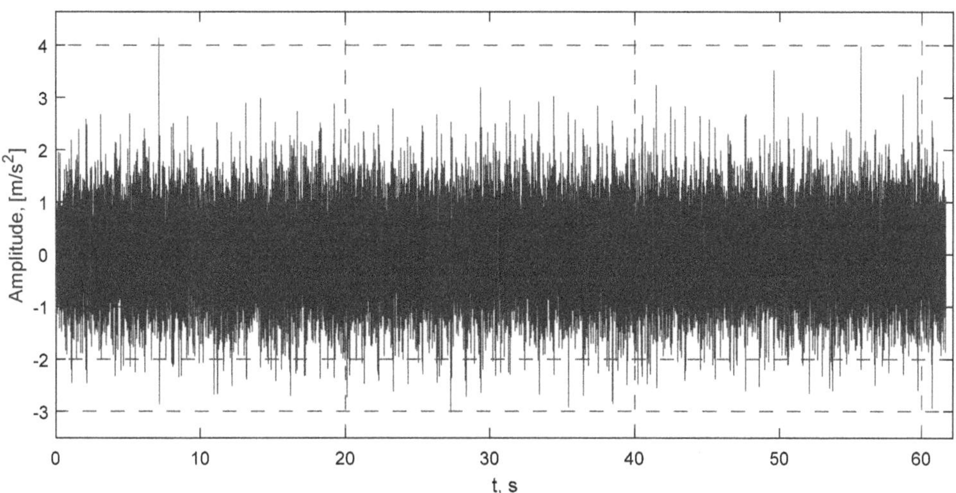

Figure 5. The theoretical seismocardiogram signal with different types of noises.

Therefore, an example is presented that examines how the settings of an adaptive filter operating with the least mean squares (LMS) algorithm are obtained after performing signal processing simulations. On the basis of these results, a real electronic measuring device is established, which measures the heart work of one of the authors. Figure 6 shows the method how accelerometer is placed.

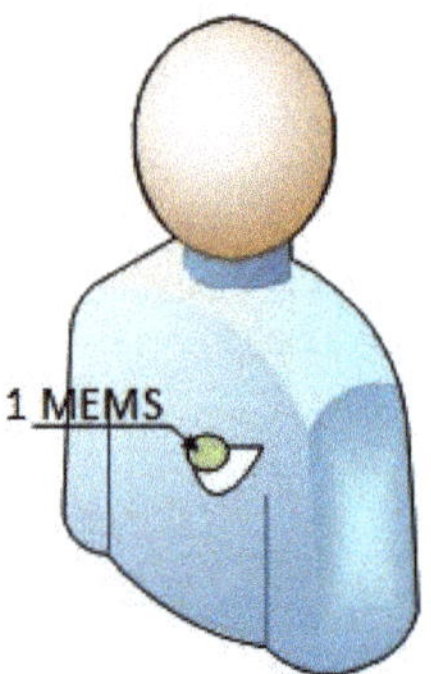

Figure 6. Position of accelerometer.

Adaptive filtering has many advantages and is useful for biomedical applications when the filter coefficients self-adjust to a rapidly and unpredictably changing signal. The principle of operation of the adaptive filter is defined by Equation (12) [43,44].

$$e(n) = d(n) - y(n) \tag{12}$$

where: $e(n)$—adaptation error, $y(n)$—denoised signal in output, $d(n)$—desired signal.

The operation of this filter is based on the tendency of the filter output signal $y(n)$ defined by Equation (13) to correspond as closely as possible to the affected signal $d(n)$ through the feedback response of the error signal $e(n)$ to the coefficients H(z) of the filter transfer function, so that $e(n)$ is zero [43,44].

$$y(n) = \sum_{k=1}^{L} b_n(n)x(n-k) \tag{13}$$

The structure shows in Figure 7 that, based on the fact that the SCG signal is measured in the frequency band 1–20 Hz. It feeds the signal of the frequency band of interest to the adaptive filter to make the adaptive filter much more efficient and effective [23,45]. For this, a finite impulse response filter (FIR) or an infinite impulse response filter (IIR) can be used [44,46]. As a result, an accelerometer can be used for measurement. Apart from the fundamental adaptive filter configuration, the frequency band is 5–45 Hz for adaptive filter with delay Figure 7. Additionally, the performance to detect AO peaks is analyzed with FIR and IRR filters.

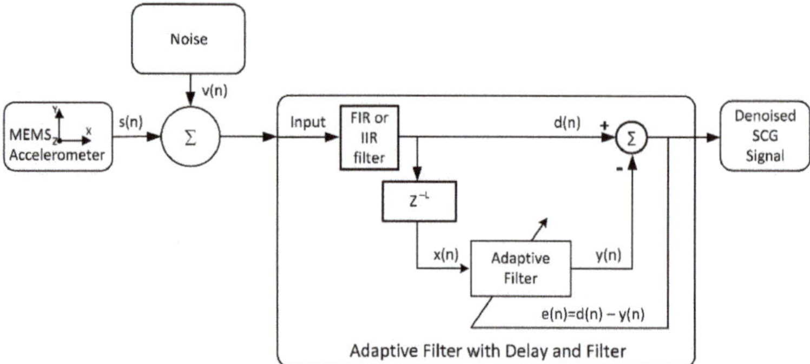

Figure 7. Adaptive filter configuration. Where: s(n)—input signal, x(n)—delayed signal, d(n)—desired signal, e(n)—adaptation error, y(n)—denoised signal, v(n) noise signal.

Processing data from the simulated SCG are presented in the Table 4. The second table shows the measurement results with an adjusted 410 ms duration window, which indicates the shortest interval between the AO peaks [47]. This duration is taken from the data in Table 1 to avoid misinterpretation of seismocardiogram signal artifacts when aortic valve closure is accepted for aortic opening. A mitral valve opening peak may also be faulty. The minimum peak interval can also be called a blind zone or a silence mode, when signal peaks are not evaluated with a level higher than the lowest signal root mean square (RMS) value [48,49]. For the first adapter filter (AF), the convergence parameter or the adaptation step mu is equal to 0.00023, and for the second, this parameter is 0.00019.

Table 4. Modeled SCG signal processing data.

Filter Order	1 Adaptive Filter	2 Adaptive Filter
Filter Order	FIR (100)/892	IIR (5)/892
mu AF step	2.3105×10^{-4}	1.9370×10^{-4}
Heart Rate, beats/min	67	69
RMS, m/s^2	0.2145	0.2635
SNR, dB	-6.3288	-6.7995
RMSE, m/s^2	0.1447	0.3401
Peaks number	63	67
Peaks Interval Mean, ms	887.032	858.313
Peaks Interval STD	446.838	237.043
Adaptation duration, s	1.237476	1.185759
Adaptation time, s	1.200769	1.193749

Knowing that the signal noise ratio (SNR) [50,51] of the SCG signal at the adaptive filters input is -14.009 dB, the processing efficiencies of these two filters can be evaluated. The processing benefit of the first adaptive filter 7.6802 dB is achieved in 1.2375 s, while the 7.2095 dB processing benefit of the second adaptive filter is achieved in 1.1857 s. The root mean square error (RMSE) of the modeled seismocardiogram signal for the first adaptive filter is 0.1447 (m/s^2), and for the second it is 0.3401 (m/s^2). Hence, the modeled seismocardiogram signal processing is less stable using the second adaptive filter.

3. Results and Discussions

Based on the data described above, signal processing is performed in the form of a mathematical model of the SCG signal and adaptive filter algorithms. The signal processing is carried out in an experimental way, measured following the theoretical SCG model during desk work when slow moving or in a resting state.

The severity of the generalized seismocardiogram was primarily analyzed by filters of limited and unlimited impulsive reactions. For this purpose, there were limited pulse and indefinite impulsive reaction filters, based on the available initial data suggesting that it is appropriate to examine the signal from the frequency band 1 to 40 Hz.

Ascertainment of whether the limits of the frequency band in question are correctly selected and whether efficient use of the signal processing system resources is performed by a fast Furrier transformation (FFT).

The FFT is performed for the investigated signal when a frequency distribution spectrum is generated which contains the maximum signal powers located in the adjusted frequency band from 0 to 50 Hz.

Similarly, the processing of further SCG signals and its results are focused on the seismocardiogram measurement in order to determine the working condition of the human heart using a nonelectrical measurement method. This choice is motivated by the fact that it is possible to perform an evaluation of a heart condition without disturbing the activity of the patient. The accelerometer, in response to mechanical movements, captures changes in acceleration and converts them into electrical signals. Heart rate is one of the main objects of measurement and is simply determined to give a minimal indication of the state of health.

In order to achieve more reliable data, FIR and IIR filters are applied, which are serially connected with signal delay before the adaptive filter. The decision to use one of the FIR or IIR filters at the beginning will facilitate the adaptive filter's ability to effectively eliminate suddenly changed signal components related to movement or activity. This responds directly to the purpose of the study and research group to measure the seismocardiogram while the patient performs daily activities such as driving. Based on this, the heart rhythms of the possible scenarios are calculated simultaneously with the evaluation of the time of SCG signal processing using adaptive filter. All seismocardiogram signal processing was performed with MATLAB 2021a software installed on an Intel (R) Core (TM) i5-7200U CPU @2.50 GHz 2.70 GHz with 16 GB of RAM and 109 GB of free space on a 500 GB SSD.

The analyzed resting SCG signal consists of 61,000 samples, which were recorded at a sampling rate of 490 Hz on an MPU9250 accelerometer connected to an ESP32 WROOM microcontroller via an I2C data bus. The measured data analysis was performed with MATLAB software.

After representative calculation of the signals of all three coordinate axes of the accelerometer was performed, a steady-state SCG signal with an SNR of -12.5946 dB was measured.

Figures 8 and 9 shows signal processing results, where (a) signal represents the measured signal in the input.

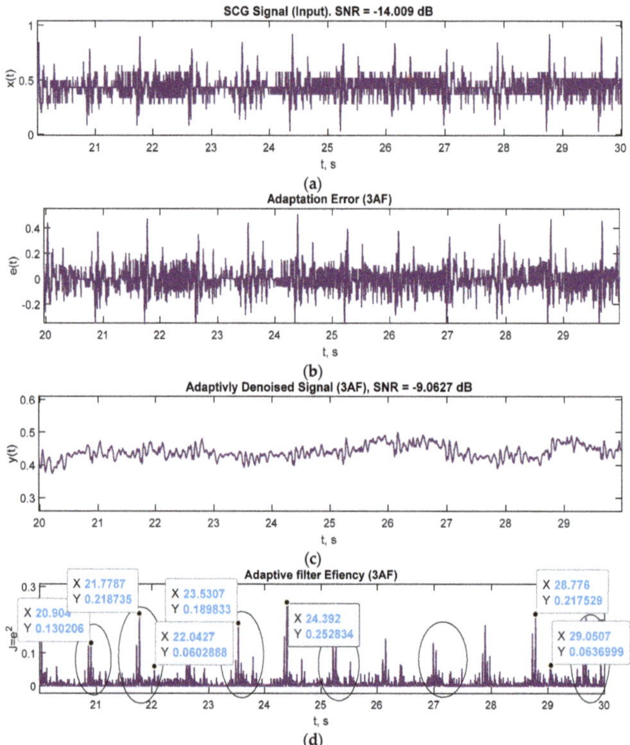

Figure 8. Processing seismocardiogram signal in the first adaptive filter with sample delay 5 and the FIR filter in the input. (**a**) SCG signal on input. (**b**) Adaptation error of the first adaptive filter. (**c**) Adaptively filtered signal. (**d**) The efficiency of the filter.

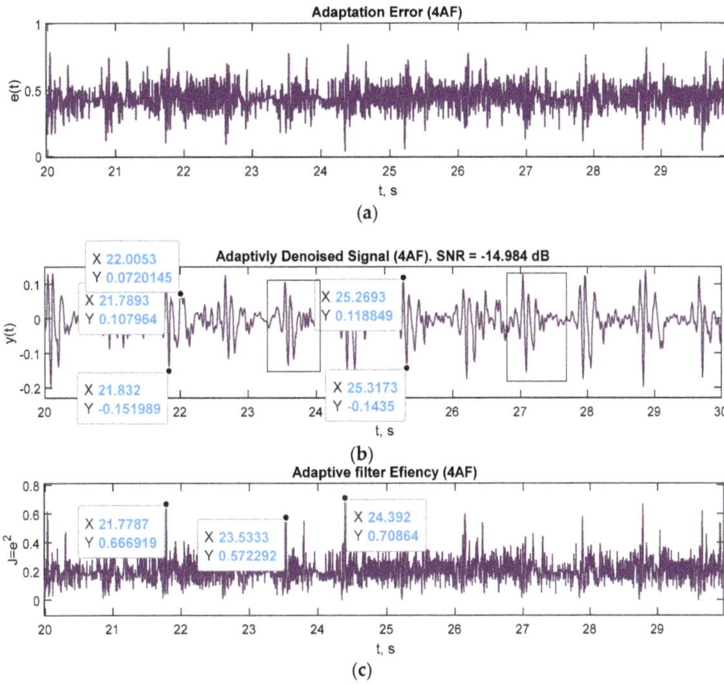

Figure 9. Processing SCG signal in the second adaptive filter with sample delay order 5 and IIR filter in the input. (**a**) Adaptation error of the second adaptive filter. (**b**) Adaptively filtrated signal. (**c**) The efficiency of the adaptive filter.

Figures 8 and 9 show two cases of the measured SCG signal processing data when two different LMS adaptive filter algorithms were adjusted based on the mathematical model simulation data. Both Figures 8 and 9 allow received values of the acceleration amplitudes and fiducial points, which make it possible to perform a deeper analysis of received data, calculate heart rate and heart rate variability.

Following the theoretical seismocardiogram processing data, the same adjustments use adaptive filters for real seismocardiogram signal processing. These processing data are tabled in Table 5.

Table 5. Rest SCG signal processing data.

Filter Order	1 Adaptive Filter	2 Adaptive Filter
Filter Order	FIR (100)/892	IIR (5)/892
mu AF step	3.5503×10^{-4}	5.1336×10^{-4}
Heart Rate, beats/min	107	98
RMS, m/s^2	0.5503	0.0523
SNR, dB	-5.1047	-7.8809
RMSE, m/s^2	0.0717	0.0945
Peaks number	106	97
Peaks Interval Mean, ms	556.406	611.619
Peaks Interval STD	137.600	171.220
Peaks Variation	18,933.805	29,316.363
Adaptation time, s	1.200769	1.193749

The 7.49 dB processing benefit of the first adaptive filter was achieved in 1.2 s, while the 4.7137 dB processing benefit of the second adaptive filter was achieved in 1.194 s.

Correlation analysis of the two signals showed that for the filter with a 5 s delay and an IIR filter at the input, the processed steady-state SCG signal level is sufficient and, in some cases, correlates with the SCG-modeled signal without noise signal. Therefore, the adaptive filters must be combined individually (Table 6) to achieve the desired processing results. The negative SNR value in both cases −5.1 and −7.88 dB indicate that denoised signal has a significant level of noise.

Table 6. Rest SCG signal processing data with individually adjusted adaptive filters.

Filter Order	3 Adaptive Filter	4 Adaptive Filter
Filter Order	200	50
mu AF step	2.1226×10^{-3}	1.4774×10^{-1}
Heart Rate, beats/min	95	88
RMS, m/s^2	0.4373	0.0478
SNR, dB	−9.0627	−14.9807
RMSE, m/s^2	0.0503	0.0984
Peaks number	61	57
Peaks Interval Mean, ms	623.885	664.281
Peaks Interval STD	229.843	132.981
Peaks Variation	52,827.737	17,683.920
Processing time, s	1.091161	1.110726

Both adaptive filters have been adjusted by reducing filter orders, respectively, from 892 to 200 and 50. Figure 10 represents the seismocardiogram signal in the second adaptive filter with marked AO and AC artifacts.

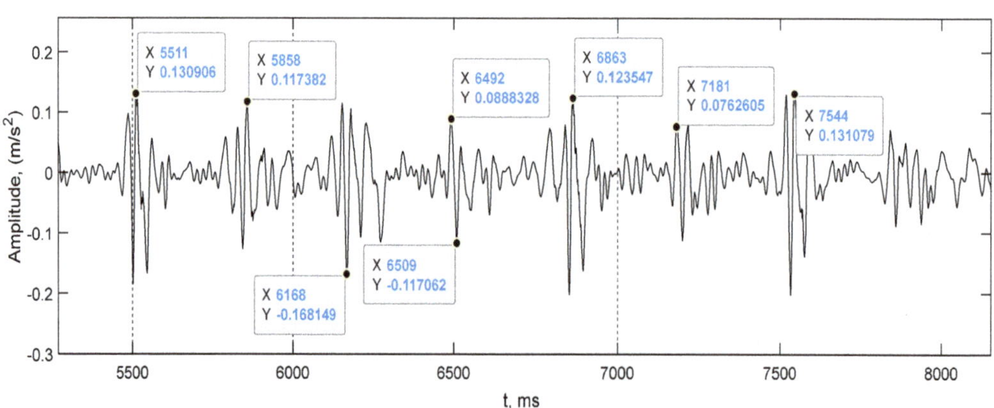

Figure 10. Denoised SCG signal in the second adaptive filter output after individual adjustment.

Figure 11 presents the same signal power frequency density, where one can see the extremums of signal amplitude and artifacts sequences. White spaces indicates each HR cycle signal peaks maximum power in dB.

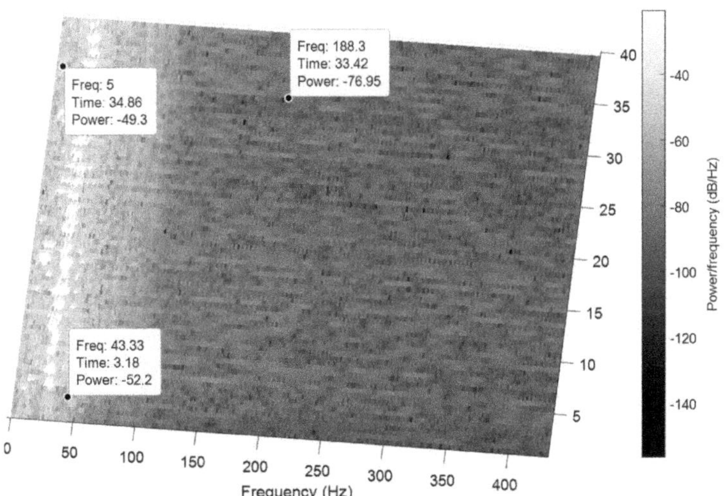

Figure 11. Spectrogram of the denoised SCG signal in the second adaptive filter output after individual adjustment.

4. Conclusions

A simpler mathematical model was developed and three algorithms were created for a more detailed explanation and understanding of the process and behavior of the model. Using this algorithm, the processing program can be written in several programming languages, not only in MATLAB. This seismocardiogram model can be used to find optimal parameters (fifth-order delay before adaptive filter, 892 filter order) for adaptive filters that can perform real cardio-mechanical vibration signal processing for estimating AO peaks. As a result, the heart rate was calculated. However, filter order 200 and 50 were evaluated again, so the processing duration decreased in both filters, respectively, from 1.20 to 1.09 and from 1.19 to 1.11 s. In each case, the signal processing and the necessary calculations were performed with programs written in MATLAB.

The proposed system and methodology can be useful to increase such research efficiency. Furthermore, future works will relate to improvements in processing methods by including machine learning to estimate systolic and diastolic intervals and heart rate variability during daily activity.

Author Contributions: Conceptualization, G.U., A.V. and D.N.; methodology, D.A., M.Z. and M.P.; software, G.U. and J.K.; validation, D.A., A.V., M.Z., M.P. and M.F.; visualization, G.U., D.N. and A.V.; investigation, D.N., G.U., D.A., A.V. and M.Z.; resources, M.Z., M.F., J.K. and M.P.; data curation, D.N., M.Z. and G.U.; writing—original draft preparation, G.U., A.V., D.A. and M.P.; writing—review and editing, D.N., D.A., J.K. and M.F.; supervision, A.V.; funding acquisition, M.Z. and D.A. All authors have read and agreed to the published version of the manuscript.

Funding: This research received no external funding.

Institutional Review Board Statement: Not applicable.

Informed Consent Statement: Not applicable.

Data Availability Statement: Not applicable.

Conflicts of Interest: The authors declare no conflict of interest.

References

1. Virani, S.S.; Alonso, A.; Aparicio, H.J.; Benjamin, E.J.; Bittencourt, M.S.; Callaway, C.W.; Carson, A.P.; Chamberlain, A.M.; Cheng, S.; Delling, F.N.; et al. Heart Disease and Stroke Statistics—2021 Update. *Circulation* **2021**, *143*, E254–E743. [CrossRef] [PubMed]
2. Bhatnagar, P.; Wickramasinghe, K.; Wilkins, E.; Townsend, N. Trends in the epidemiology of cardiovascular disease in the UK. *Heart* **2016**, *102*, 1945–1952. [CrossRef] [PubMed]
3. Saini, S.K.; Gupta, R. Artificial intelligence methods for analysis of electrocardiogram signals for cardiac abnormalities: State-of-the-art and future challenges. *Artif. Intell. Rev.* **2022**, *55*, 1519–1565. [CrossRef]
4. Maršánová, L.; Ronzhina, M.; Smíšek, R.; Vítek, M.; Němcová, A.; Smital, L.; Nováková, M. ECG features and methods for automatic classification of ventricular premature and ischemic heartbeats: A comprehensive experimental study. *Sci. Rep.* **2017**, *7*, 11239. [CrossRef]
5. Miramontes, R.; Aquino, R.; Flores, A.; Rodríguez, G.; Anguiano, R.; Ríos, A.; Edwards, A. PlaIMoS: A Remote Mobile Healthcare Platform to Monitor Cardiovascular and Respiratory Variables. *Sensors* **2017**, *17*, 176. [CrossRef] [PubMed]
6. D'Mello; Skoric; Xu; Roche; Lortie; Gagnon; Plant Real-Time Cardiac Beat Detection and Heart Rate Monitoring from Combined Seismocardiography and Gyrocardiography. *Sensors* **2019**, *19*, 3472. [CrossRef] [PubMed]
7. Mehrang, S.; Jafari Tadi, M.; Lahdenoja, O.; Kaisti, M.; Vasankari, T.; Kiviniemi, T.; Airaksinen, J.; Pankaala, M.; Koivisto, T. Machine Learning Based Classification of Myocardial Infarction Conditions Using Smartphone-Derived Seismo- and Gyrocardiography. In Proceedings of the Computing in Cardiology; IEEE Computer Society, Maastricht, The Netherlands, 23–26 September 2018; Volume 45.
8. Andreozzi, E.; Centracchio, J.; Punzo, V.; Esposito, D.; Polley, C.; Gargiulo, G.D.; Bifulco, P. Respiration Monitoring via Forcecardiography Sensors. *Sensors* **2021**, *21*, 3996. [CrossRef]
9. Di Rienzo, M.; Vaini, E.; Castiglioni, P.; Merati, G.; Meriggi, P.; Parati, G.; Faini, A.; Rizzo, F. Wearable seismocardiography: Towards a beat-by-beat assessment of cardiac mechanics in ambulant subjects. *Auton. Neurosci. Basic Clin.* **2013**, *178*, 50–59. [CrossRef]
10. Polley, C.; Jayarathna, T.; Gunawardana, U.; Naik, G.; Hamilton, T.; Andreozzi, E.; Bifulco, P.; Esposito, D.; Centracchio, J.; Gargiulo, G. Wearable Bluetooth Triage Healthcare Monitoring System. *Sensors* **2021**, *21*, 7586. [CrossRef]
11. Sahoo, P.K.; Thakkar, H.K.; Lin, W.Y.; Chang, P.C.; Lee, M.Y. On the design of an efficient cardiac health monitoring system through combined analysis of ECG and SCG signals. *Sensors* **2018**, *18*, 379. [CrossRef]
12. Nguyen, T.-N.; Nguyen, T.-H. Deep Learning Framework with ECG Feature-Based Kernels for Heart Disease Classification. *Elektron. Elektrotechnika* **2021**, *27*, 48–59. [CrossRef]
13. Taebi, A.; Solar, B.; Bomar, A.; Sandler, R.; Mansy, H. Recent Advances in Seismocardiography. *Vibration* **2019**, *2*, 64–86. [CrossRef] [PubMed]
14. Jain, P.K.; Tiwari, A.K. Heart monitoring systems-A review. *Comput. Biol. Med.* **2014**, *54*, 1–13. [CrossRef]
15. Aboltins, A.; Pikulins, D.; Grizans, J.; Tjukovs, S. Piscivorous Bird Deterrent Device Based on a Direct Digital Synthesis of Acoustic Signals. *Elektron. Elektrotechnika* **2021**, *27*, 42–48. [CrossRef]
16. Conn, N.J.; Schwarz, K.Q.; Borkholder, D.A. In-home cardiovascular monitoring system for heart failure: Comparative study. *JMIR mHealth uHealth* **2019**, *7*, e12419. [CrossRef]
17. Sahoo, P.K.; Thakkar, H.K.; Lee, M.Y. A cardiac early warning system with multi channel SCG and ECG monitoring for mobile health. *Sensors* **2017**, *17*, 711. [CrossRef] [PubMed]
18. Leitão, F.; Moreira, E.; Alves, F.; Lourenço, M.; Azevedo, O.; Gaspar, J.; Rocha, L.A. High-Resolution Seismocardiogram Acquisition and Analysis System. *Sensors* **2018**, *18*, 3441. [CrossRef]
19. Andreozzi, E.; Fratini, A.; Esposito, D.; Naik, G.; Polley, C.; Gargiulo, G.D.; Bifulco, P. Forcecardiography: A Novel Technique to Measure Heart Mechanical Vibrations onto the Chest Wall. *Sensors* **2020**, *20*, 3885. [CrossRef]
20. Andreozzi, E.; Gargiulo, G.D.; Esposito, D.; Bifulco, P. A Novel Broadband Forcecardiography Sensor for Simultaneous Monitoring of Respiration, Infrasonic Cardiac Vibrations and Heart Sounds. *Front. Physiol.* **2021**, *12*, 725716. [CrossRef]
21. Holcik, J.; Moudr, J. Mathematical Model of Seismocardiogram. In *World Congress on Medical Physics and Biomedical Engineering 2006*; Springer: Berlin/Heidelberg, Germany, 2007; Volume 14, pp. 3415–3418.
22. Casas, B.; Lantz, J.; Viola, F.; Cedersund, G.; Bolger, A.F.; Carlhäll, C.J.; Karlsson, M.; Ebbers, T. Bridging the gap between measurements and modelling: A cardiovascular functional avatar. *Sci. Rep.* **2017**, *7*, 6214. [CrossRef]
23. Sørensen, K.; Schmidt, S.E.; Jensen, A.S.; Søgaard, P.; Struijk, J.J. Definition of Fiducial Points in the Normal Seismocardiogram. *Sci. Rep.* **2018**, *8*, 15455. [CrossRef]
24. Mohammed, Z.; Elfadel, I.; Rasras, M. Monolithic Multi Degree of Freedom (MDoF) Capacitive MEMS Accelerometers. *Micromachines* **2018**, *9*, 602. [CrossRef]
25. Centracchio, J.; Andreozzi, E.; Esposito, D.; Gargiulo, G.D.; Bifulco, P. Detection of Aortic Valve Opening and Estimation of Pre-Ejection Period in Forcecardiography Recordings. *Bioengineering* **2022**, *9*, 89. [CrossRef]
26. Andreozzi, E.; Centracchio, J.; Esposito, D.; Bifulco, P. A Comparison of Heart Pulsations Provided by Forcecardiography and Double Integration of Seismocardiogram. *Bioengineering* **2022**, *9*, 167. [CrossRef]
27. Shi, W.; Chew, M.-S. Mathematical and physical models of a total artificial heart. In Proceedings of the 2009 IEEE International Conference on Control and Automation, Christchurch, New Zealand, 9–11 December 2009; pp. 637–642. [CrossRef]

28. Guidoboni, G.; Sala, L.; Enayati, M.; Member, S.; Sacco, R.; Szopos, M.; Keller, J.M.; Fellow, L.; Popescu, M.; Member, S.; et al. Cardiovascular Function and Ballistocardiogram: A Relationship Interpreted via Mathematical Modeling. *IEEE Trans. Biomed. Eng.* **2019**, *66*. [CrossRef]

29. Htet, Z.L.; Aye, T.P.P.; Singhavilai, T.; Naiyanetr, P. Hemodynamics during Rotary Blood Pump support with speed synchronization in heart failure condition: A modelling study. In Proceedings of the 2015 37th Annual International Conference of the IEEE Engineering in Medicine and Biology Society (EMBC), Milan, Italy, 25–29 August 2015; pp. 3307–3310. [CrossRef]

30. Pockevicius, V.; Markevicius, V.; Cepenas, M.; Andriukaitis, D.; Navikas, D. Blood Glucose Level Estimation Using Interdigital Electrodes. *Electron. Electr. Eng.* **2013**, *19*. [CrossRef]

31. Abdolrazaghi, M.; Navidbakhsh, M.; Hassani, K. Mathematical modelling and electrical analog equivalent of the human cardiovascular system. *Cardiovasc. Eng.* **2010**, *10*, 45–51. [CrossRef]

32. Yang, X.; Leandro, J.S.; Cordeiro, T.D.; Lima, A.M.N. An Inverse Problem Approach for Parameter Estimation of Cardiovascular System Models. In Proceedings of the 2021 43rd Annual International Conference of the IEEE Engineering in Medicine & Biology Society, Virtual Conference, 1–5 November 2021; pp. 5642–5645. [CrossRef]

33. Jain, K.; Patra, A.; Maka, S. Modeling of the Human Cardiovascular System for Detection of Atherosclerosis. *IFAC PapersOnLine* **2018**, *51*, 545–550. [CrossRef]

34. Zia, J.; Kimball, J.; Hersek, S.; Inan, O.T. Modeling Consistent Dynamics of Cardiogenic Vibrations in Low-Dimensional Subspace. *IEEE J. Biomed. Health Informatics* **2020**, *24*, 1887–1898. [CrossRef]

35. Uskovas, G.; Valinevicius, A.; Zilys, M.; Navikas, D.; Frivaldsky, M.; Prauzek, M.; Konecny, J.; Andriukaitis, D. Driver Cardiovascular Disease Detection Using Seismocardiogram. *Electronics* **2022**, *11*, 484. [CrossRef]

36. Prauzek, M.; Konecny, J. Optimizing of Q-Learning Day/Night Energy Strategy for Solar Harvesting Environmental Wireless Sensor Networks Nodes. *Elektron. Elektrotechnika* **2021**, *27*, 50–56. [CrossRef]

37. Skovierova, H.; Pavelek, M.; Okajcekova, T.; Palesova, J.; Strnadel, J.; Spanik, P.; Halašová, E.; Frivaldsky, M.; Foresta, F. The Biocompatibility of Wireless Power Charging System on Human Neural Cells. *Appl. Sci.* **2021**, *11*, 3611. [CrossRef]

38. Hrbac, R.; Kolar, V.; Bartlomiejczyk, M.; Mlcak, T.; Orsag, P.; Vanc, J. A Development of a Capacitive Voltage Divider for High Voltage Measurement as Part of a Combined Current and Voltage Sensor. *Elektron. Elektrotechnika* **2020**, *26*, 25–31. [CrossRef]

39. Surgailis, T.; Valinevicius, A.; Markevicius, V.; Navikas, D.; Andriukaitis, D. Avoiding Forward Car Collision using Stereo Vision System. *Electron. Electr. Eng.* **2012**, *18*. [CrossRef]

40. Soni, N.; Malekian, R.; Andriukaitis, D.; Navikas, D. Internet of Vehicles Based Approach for Road Safety Applications Using Sensor Technologies. *Wirel. Pers. Commun.* **2019**, *105*, 1257–1284. [CrossRef]

41. Ieremeiev, O.; Lukin, V.; Okarma, K.; Egiazarian, K. Full-Reference Quality Metric Based on Neural Network to Assess the Visual Quality of Remote Sensing Images. *Remote Sens.* **2020**, *12*, 2349. [CrossRef]

42. Zhang, D.; Tian, Q. A Novel Fuzzy Optimized CNN-RNN Method for Facial Expression Recognition. *Elektron. Elektrotechnika* **2021**, *27*, 67–74. [CrossRef]

43. Semmlow, J.L.; Griffel, B. *Biosignal and Medical Image Processing MATLAB-Based Application*, 3rd ed.; Taylor & Francis Group: New York, NY, USA, 2014; ISBN 9781466567368.

44. Humaidi, A.J.; Ibraheem, I.K.; Ajel, A.R. A novel adaptive LMS algorithm with genetic search capabilities for system identification of adaptive FIR and IIR filters. *Information* **2019**, *10*, 176. [CrossRef]

45. Mora, N.; Cocconcelli, F.; Matrella, G.; Ciampolini, P. Detection and Analysis of Heartbeats in Seismocardiogram Signals. *Sensors* **2020**, *20*, 1670. [CrossRef]

46. Sotner, R.; Domansky, O.; Jerabek, J.; Herencsar, N.; Petrzela, J.; Andriukaitis, D. Integer-and Fractional-Order Integral and Derivative Two-Port Summations: Practical Design Considerations. *Appl. Sci.* **2019**, *10*, 54. [CrossRef]

47. Choudhary, T.; Bhuyan, M.K.; Sharma, L.N. Orthogonal subspace projection based framework to extract heart cycles from SCG signal. *Biomed. Signal Process. Control* **2019**, *50*, 45–51. [CrossRef]

48. Sadhukhan, D.; Mitra, M. R-peak detection algorithm for ECG using double difference and RR interval processing peer-review under responsibility of C3IT. *Procedia Technol.* **2012**, *4*, 873–877. [CrossRef]

49. Paterova, T.; Prauzek, M. Estimating Harvestable Solar Energy from Atmospheric Pressure Using Deep Learning. *Elektron. Elektrotechnika* **2021**, *27*, 18–25. [CrossRef]

50. Peksinski, J.; Zeglinski, G.; Mikolajczak, G.; Kornatowski, E. Estimation of BER Bit Error Rate Using Digital Smoothing Filters. *Elektron. Elektrotechnika* **2021**, *27*, 75–83. [CrossRef]

51. Peric, Z.H.; Denic, B.D.; Savic, M.S.; Vucic, N.J.; Simic, N.B. Binary Quantization Analysis of Neural Networks Weights on MNIST Dataset. *Elektron. Elektrotechnika* **2021**, *27*, 55–61. [CrossRef]

 *electronics*

MDPI

Article

Energy Flow Analysis of Excavator System Based on Typical Working Condition Load

Deying Su, Liang Hou, Shaojie Wang *, Xiangjian Bu and Xiaosong Xia

Department of Mechanical and Electrical Engineering, Xiamen University, Xiamen 361000, China; 19920190154058@stu.xmu.edu.cn (D.S.); hliang@xmu.edu.cn (L.H.); bxj@xmu.edu.cn (X.B.); 19920200156022@stu.xmu.edu.cn (X.X.)
* Correspondence: wsj@xmu.edu.cn

Abstract: Accurate energy flow results are the premise of excavator energy-saving control research. Only through an accurate energy flow analysis based on operating data can a practical excavator energy-saving control scheme be proposed. In order to obtain the excavator's accurate energy flow, the excavator components' performance and operating data requirements are obtained, and the experimental schemes are designed to collect it under typical working conditions. The typical working condition load is reconstructed based on wavelet decomposition, harmonic function, and theoretical weighting methods. This paper analyzes the excavator system's energy flow under the typical working condition load. In operation conditions, the output energy of the engine only accounts for 50.21% of the engine's fuel energy, and the actuation and the swing system account for 9.33% and 4%, respectively. In transportation conditions, the output energy of the engine only accounts for 49.80% of the engine's fuel energy, and the torque converter efficiency loss and excavator driving energy account for 15.09% and 17.98%, respectively. The research results show that the energy flow analysis method based on typical working condition load can accurately obtain each excavator component's energy margin, which provides a basis for designing energy-saving schemes and control strategies.

Keywords: excavator; typical working condition; load; operating data; energy flow

Citation: Su, D.; Hou, L.; Wang, S.; Bu, X.; Xia, X. Energy Flow Analysis of Excavator System Based on Typical Working Condition Load. *Electronics* **2022**, *11*, 1987. https://doi.org/10.3390/electronics11131987

Academic Editors: Darius Andriukaitis, Yongjun Pan and Peter Brida

Received: 15 May 2022
Accepted: 23 June 2022
Published: 24 June 2022

Publisher's Note: MDPI stays neutral with regard to jurisdictional claims in published maps and institutional affiliations.

1. Introduction

Excavators are commonly used in farmland water conservancy, urban greening and construction projects and are the construction machinery with the largest fuel consumption [1–3]. The engine output energy is converted into hydraulic energy or vehicle driving energy during the excavator's working process. Due to the significant mass attribute and the characteristics of multi-mechanism linkage, the excavator energy consumption is much higher than the actual working load. With the aggravation of the global energy crisis, air quality problems, and the tightening of construction machinery emission policies and regulations in various countries, many scholars have undertaken a large number of studies on the energy-saving control of excavators.

Yang et al. improved excavator engine efficiency through engine deactivation technology [4]. Researchers and major manufacturers worldwide have carried out a series of studies on hydraulic system control methods such as a load sensing (LS) system [5], negative flow system (NFS) [6], positive flow system (PFS) [7], and independent metering valve (IMV) [8], which effectively improved the excavator system efficiency. Kim and Zhao et al. researched boom gravitational potential energy and swing system energy recovery, respectively [9–11]. Xiao and Kwon researched the gas–electric hybrid excavator [12,13], and Shen studied the hybrid hydraulic excavator based on the accumulator [14,15]. Zimmerman and Paolo et al. established a mathematical model of the excavator system on simulation platforms MATLAB-Simulink and AMESim, respectively [16–18]. Based on the mathematical model, they realized the energy flow analysis of the excavator system and designed a hybrid excavator scheme [19,20]. In the above studies, simple signals such as constants, steps, and ramps are often used to simulate the working load, which are difficult to accurately represent the actual working condition, which will inevitably lead to deviations in the results of energy flow analysis and the effect of energy-saving schemes. In response to this problem, An et al. proposed an excavator energy flow analysis method based on operating data [21], but the study did not involve the construction of typical working condition load and could not be further applied to mathematical models. Therefore, it is urgent to build up the typical working condition load of the excavator, realize the accurate analysis of the energy flow, obtain the energy margin of the excavator system, and provide accurate simulation load for the mathematical model.

Zhai et al. compiled a load spectrum for hydraulic pump fatigue analysis based on the mixed distribution method [22,23], which focuses on the frequency of load action and cannot be applied to the energy flow analysis of excavators. Chang et al. use the theoretical weighted method to construct the typical working condition load of the loader [24], which ignored the non-stationary characteristics of the random term of the load. Based on the wavelet decomposition method, Wang obtained the swing condition load's random term and trend term, and the swing condition load function composed of the random term and the trend term function [25,26]. This study does not involve the load's reconstruction of the excavator's complete working condition, and it is not easy to achieve parametric characterization of the operation load of the main pump or other components. Based on the above analysis, this paper combines the wavelet decomposition method and the theoretical weighting method to construct the excavator's typical working condition load. The energy flow of the excavator system is analyzed based on typical working condition load, and an accurate energy margin is provided for the excavator system's energy-saving scheme.

The rest of the paper is organized as follows: Section 2 conducts a theoretical energy flow analysis of the excavator, Section 3 obtains component performance and operating data, Section 4 constructs the typical working condition load and analyzes the energy flow of the excavator, and Section 5 is a discussion of the results and gives conclusions.

2. Theoretical Energy Flow Analysis of Excavator System

The research objective of this paper is a wheel excavator, and its work system is shown in Figure 1, which consists of an engine, transmission system, and hydraulic system. The transmission system includes a hydraulic torque converter, front and rear drive axles, transmission shaft, etc. The hydraulic system consists of the main pump, steering pump, main control valve (MCV), actuators and swing motor, etc. Transportation and operation conditions are typical conditions of wheel excavators. The operation conditions are divided into five stages: dig preparation, digging, lifting, unloading, and swinging [27], as shown in Figure 2. Under transportation conditions, engine energy is converted into driving energy. In operation conditions, the engine leads the hydraulic pump to work through the coupling, and the main control valve distributes the flow rate of the hydraulic pump to the actuator, such as the boom, stick, bucket, and swing motor. There is no steering action during the wheel excavator's operation, and the steering pump's output flow rate is distributed to the swing system and the bucket mechanism by the main control valve. Except for the mode of transportation, wheel excavators are no different from crawler excavators, which will be referred to as excavators.

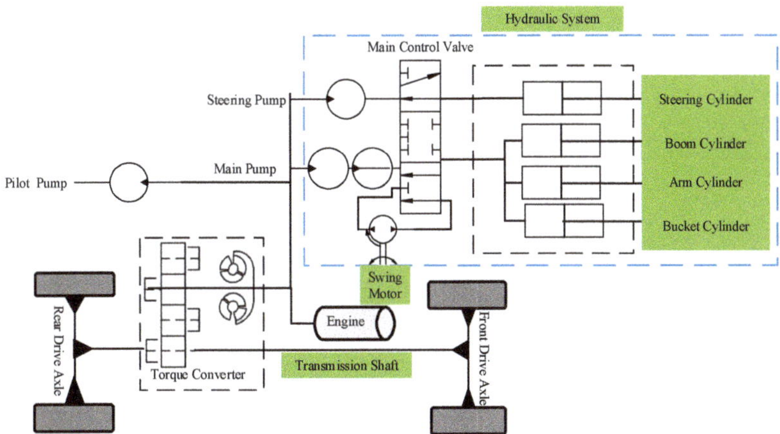

Figure 1. Excavator work system.

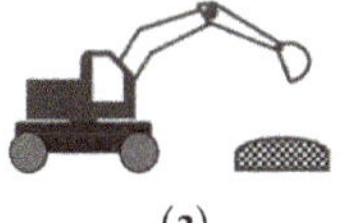

(a) (b) (c) (d) (e)

Figure 2. Operation stages of excavator. (**a**) Dig preparation; (**b**) digging; (**c**) lifting; (**d**) unloading; (**e**) swinging.

2.1. Engine Theoretical Energy Flow

The engine's energy transfer process includes heat loss, cylinder loss, mechanical friction loss, generator consumption, etc. The remaining output energy is transferred to the work system, and the theoretical energy flow is shown in Figure 3. The output energy is related to the output torque and speed, obtained by Equation (1). The heat loss is associated with the coolant temperature change and specific heat capacity, expressed by Equation (2). Cylinder loss, mechanical friction loss, and alternator consumption are expressed as other consumption by Equation (3). Finally, the theoretical energy flow of the engine is calculated by Equations (1)–(3).

$$P_{E\_outputenergy} = \frac{n \times M}{9549} \tag{1}$$

$$P_{E\_heatloss} = C_e \rho_e q_e \Delta \theta_e \tag{2}$$

$$P_{E\_otherconsumption} = mQ - P_{E\_outputenergy} - P_{E\_heatloss} \tag{3}$$

where m is the engine fuel quality, kg/s; Q is the fuel calorific value, J/kg; n is the engine output speed, rpm; M is the engine output torque, Nm; C_e is the coolant specific heat capacity, J/(kg °C); ρ_e is the coolant density, kg/m^3; q_e is the coolant flow rate, m^3/s; $\Delta \theta_e$ is the coolant temperature difference, °C.

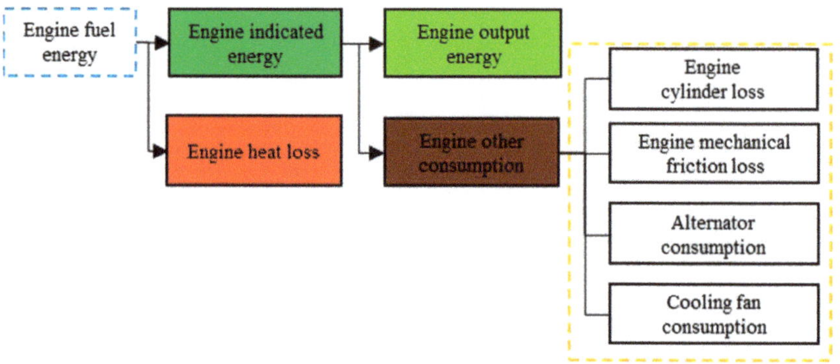

Figure 3. Engine energy flow.

2.2. Hydraulic Pump Theoretical Energy Flow

The engine's output energy is transmitted to the main pump, steering pump, and auxiliary components such as the cooling pump and the oil pump. The energy loss of the hydraulic pumps includes the efficiency loss and the hydraulic system's heat loss. The hydraulic pumps' energy flow is shown in Figure 4. According to the calculation method of hydraulic pump energy and the hydraulic system heat loss, the hydraulic pumps' energy flow is expressed by Equations (4)–(7).

$$P_{C\&O\_consumption} = P_{E\_outputenergy} - P_{P\_efficiencyloss} - P_{Hy\_heatloss} - P_{P\_outputenergy} \quad (4)$$

$$P_{P\_efficiencyloss} = \sum_{i=1}^{3} (1 - \eta_{hi}) P_{Pumpi} \quad (5)$$

$$P_{Pumpi} = \frac{P_i q_i}{60} \quad (6)$$

$$P_{Hy\_heatloss} = C_h \rho_h q_h \Delta \theta_h \quad (7)$$

where η_{hi} is the main pump 1, main pump 2, and steering pump efficiency; P_i is the main pump 1, main pump 2, and steering pump output pressure, bar; q_i is the main pump 1, main pump 2, and steering pump output flow rate, L/min; C_h, ρ_h, $\Delta\theta_h$ is the hydraulic oil specific heat capacity, density, and temperature difference, respectively, J/(kg·°C), kg/m^3, °C; q_h is the hydraulic oil flow rate, m^3/s.

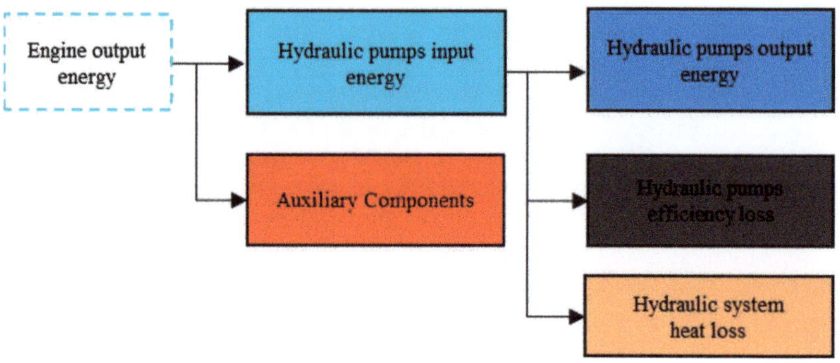

Figure 4. Excavator pump energy flow.

2.3. Operation System Theoretical Energy Flow

The energy flow of the excavator operation system is illustrated in Figure 5. The output energy of the hydraulic pump is transmitted to the actuation system and the swing motor through the MCV. The actuation system energy includes the boom energy, the stick energy, and the bucket energy. The operation system's loss includes the MCV loss, the hydraulic circuit loss, the cylinder loss, and overflow loss during the swing. Finally, the energy flow of the operation system is expressed by Equations (8)–(11).

$$P_{Actuator} = P_{boom} + P_{arm} + P_{bucket} \tag{8}$$

$$P_{System\_loss} = P_{Contralvalve\_inputenergy} - P_{Actuatorsystem} - P_{Swingmotor} \tag{9}$$

$$P_{Swingmotor} = P_{motorin} q_{motorin} \tag{10}$$

$$P_{Cylinder} = P_{in} q_{in} - P_{out} q_{out} \tag{11}$$

where $P_{motorin}$ is the motor inlet pressure, bar; $q_{motorin}$ is the motor inlet flow, L/min; P_{in} is the cylinder inlet pressure, bar; q_{in} is the cylinder inlet flow, L/min; P_{out} is the cylinder outlet pressure, bar; q_{out} is the cylinder outlet flow, L/min.

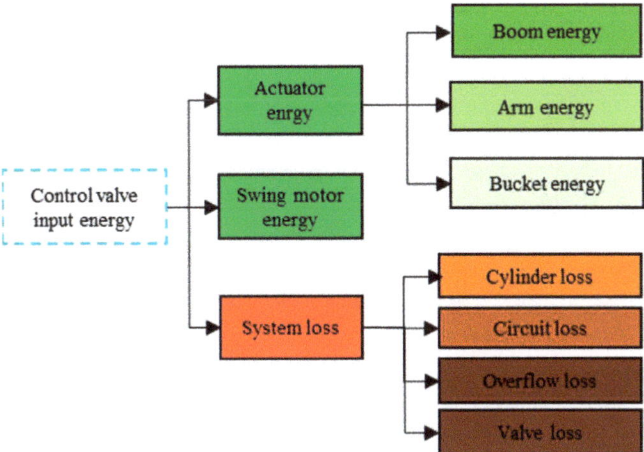

Figure 5. Excavator operation system energy flow.

2.4. Excavator System Theoretical Energy Flow under Transportation Condition

In the excavator transportation condition, the engine's output energy is sequentially transmitted to the transmission shaft, the drive axle, and the wheel reducer through the hydraulic torque converter, and finally, drive the excavator. The actuation system does not work during this period, and the main pump consumes the engine energy at the lowest output flow rate. The excavator energy flow under the transportation condition is shown in Figure 6. Depending on the calculation method of the theoretical energy flow and the tractive force [28], the energy flow in the transportation condition is expressed by Equations (12)–(16).

$$P_{Transmissionsystem} = P_{E\_outputenergy} - P_{Mainpunp\_loss} - P_{Steeringpump\_consumption} \tag{12}$$

$$P_{Excavator\_drivingenergy} = Fv \tag{13}$$

$$P_{T\_loss} = (1 - \eta_T) P_{Transmissionsystem} \tag{14}$$

$$F = \frac{M_{Out} \cdot i_T \cdot i_0 \cdot i_g \cdot \eta_T \cdot \eta_0 \cdot \eta_g}{1000 r_d} \tag{15}$$

$$v = 0.377 \frac{n_{Out} \cdot r_d}{i_T \cdot i_0 \cdot i_g} \tag{16}$$

where M_{Out}, n_{Out} are the output torque and speed of the torque converter, respectively; r_d is the radius of the wheel; η_T is the torque converter efficiency, which varies with speed and gear positions, and is measured by bench experiments; i_T, i_0, and i_g are the torque converter gear ratio, the total transmission ratio of drive shaft and front axle, and transmission ratio of the wheel reducer, respectively; η_0, η_g are the total efficiency of drive shaft and front axle, and the wheel reducer efficiency, which are taken as 0.9 and 0.8, respectively.

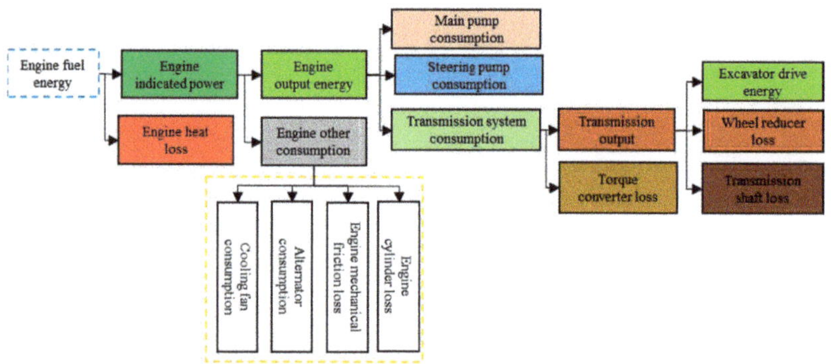

Figure 6. Excavator transportation condition energy flow.

Based on the above research, the energy flow analysis of the excavator system needs to obtain the component performance and operating data shown in Table 1. Operating data construct the typical working condition load, and the energy flow analysis is realized based on the typical working condition load and component performance parameters.

Table 1. Data requirements of components for the excavator energy flow analysis.

Serial Number	Components' Data	Serial Number	Components' Data
1	Engine fuel rate	16	Swing motor outlet pressure
2	Engine output torque	17	Swing motor inlet flow rate
3	Engine output speed	18	Swing motor outlet flow rate
4	Coolant temperature	19	Boom A chamber pressure
5	Coolant flow rate	20	Boom B chamber pressure
6	Main pump 1 pressure	21	Arm A chamber pressure
7	Main pump 1 flow rate	22	Arm B chamber pressure
8	Main pump 2 pressure	23	Bucket A chamber pressure
9	Main pump 2 flow rate	24	Bucket B chamber pressure
10	Steering pump pressure	25	Boom displacement
11	Steering pump flow rate	26	Arm displacement
12	Main pump 1 efficiency	27	Bucket displacement
13	Main pump 2 efficiency	28	Torque converter efficiency
14	Steering pump efficiency	29	Torque converter output speed
15	Swing motor inlet pressure	30	Torque converter output torque

3. Excavator Data Acquisition

3.1. Operating Data Acquisition

The data requirements of the excavator's components are obtained through the theoretical energy flow analysis. This section describes the operating data acquisition. The data types to be collected for components include speed, torque, temperature, pressure, flow rate, displacement, etc., as shown in Figure 7. The experiment requires that the operating data be collected and recorded stably and synchronously. The acquisition rate of the Dewe data acquisition system is 200 KS/s/ch, which can simultaneously collect analogue signals and digital signals. It has USB, CAN, GPS, video, and other data acquisition interfaces, and the data storage space is 500 GB, which can meet the requirements of this data acquisition.

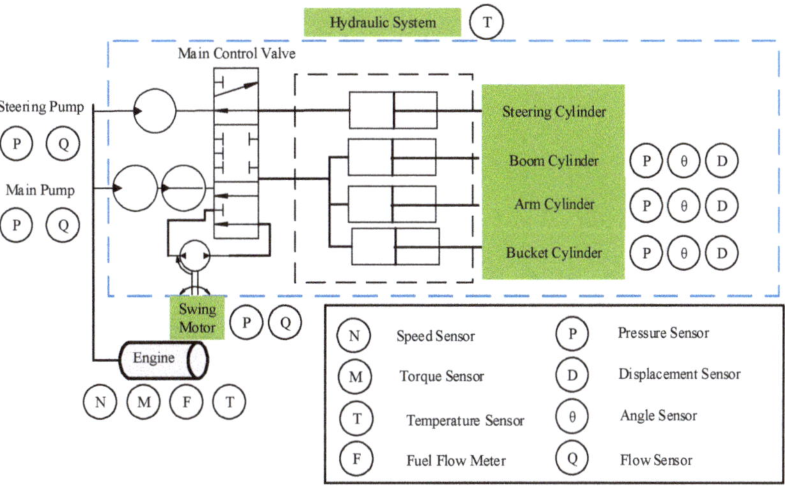

Figure 7. Data required for each component of excavator.

The excavator operation environment is very harsh, the flow sensor is expensive and easily damaged, and it is not easy to collect the flow rate data of the actuator through the flow sensor. In order to solve this problem, the cylinder pressure and displacement operating data are collected, and the cylinder's inlet and outlet flow rate is calculated through the cylinder dynamic relationship of Equations (17) and (18) [16]. The schematic of the cylinder parameters is shown in Figure 8, and the displacement and pressure sensor arrangement is shown in Figure 9.

$$\dot{p}_A = \frac{1}{C_{HA}} \cdot (Q_A - Q_{Li} - A_A \cdot \dot{x}) \tag{17}$$

$$\dot{p}_B = \frac{1}{C_{HB}} \cdot (-Q_B + Q_{Li} + A_B \cdot \dot{x}) \tag{18}$$

where P_A, P_B are the cylinder chamber A and B pressure, bar; Q_A, Q_B are the cylinder chamber A and B flow rate, L/min; A_A, A_B are the cylinder bore and rod side annular area, m^2; x is the cylinder displacement, m; Q_{Li} is the leakage flow rate from chamber A to B; C_{HA}, C_{HB} are the hydraulic volume of chambers A and B, calculated by Equations (19)–(22).

$$Q_{Li} = k_{Li} \cdot (p_A - p_B) \tag{19}$$

$$C_{HA} = \frac{V_A}{K_{FL}}, \; C_{HB} = \frac{V_B}{K_{FL}} \tag{20}$$

$$V_A = \left(\left[\frac{h}{2} + x \right] \cdot A_A + V_{LA} \right) \tag{21}$$

$$V_B = \left(\left[\frac{h}{2} - x \right] \cdot A_B + V_{LB} \right) \tag{22}$$

where k_{Li} is the flow rate leakage coefficient of chamber A to B; K_{FL} is fluid bulk modulus V_{LA}, V_{LB} are the chamber A and B dead zone volume.

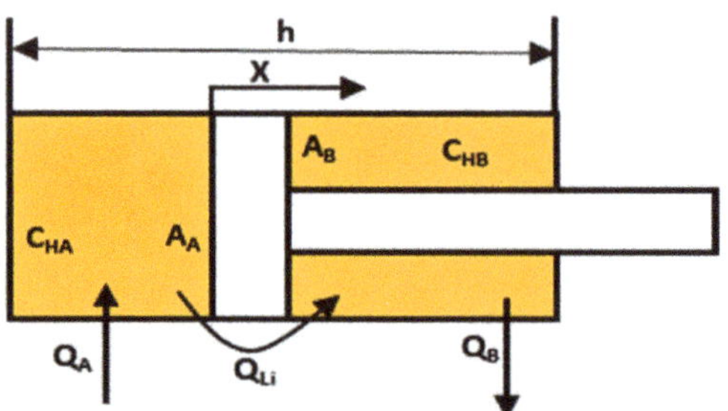

Figure 8. Schematic diagram of cylinder parameters.

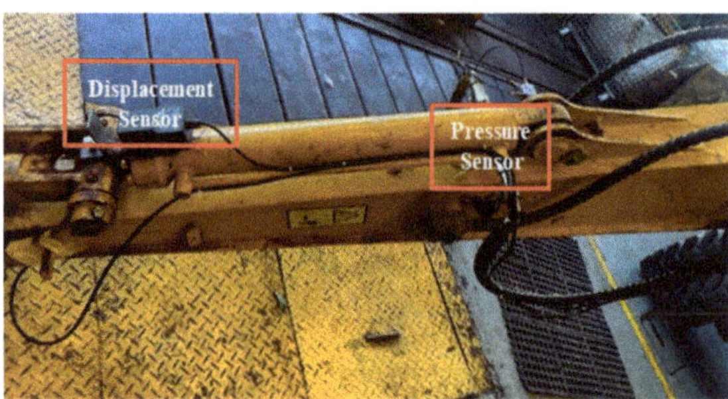

Figure 9. Installation position of cylinder displacement and pressure sensor.

In order to obtain sufficient operating data and ensure the consistency of each experiment, the experimental process and requirements are as follows:

1. Prepare two soil pits No. 1 and No. 2 at the experimental site, to ensure that the soil in the pits has a similar degree of looseness.

2. The excavator is operated by the same operator, its working speed is set to 1200 rpm.

3. During the operation, dig the original soil in the No. 1 pit and unloading it into the No. 2 pit.

4. Ensure that the duration of each cycle's operation and the duration of the same operation stages are similar, and the loading direction is changed after every 15 buckets.

5. Record the total number of buckets in the experiment for 15 min. The experimental process and the experimental site are shown in Figure 10.

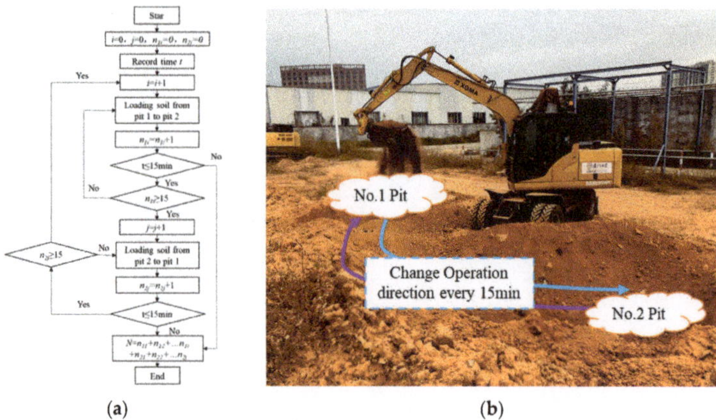

(a)   (b)

Figure 10. Experimental process and test site. (**a**) Experimental process; (**b**) test site.

The experimental site for transportation conditions is a hard cement floor with a total length of 400 m. During the experiment, the engine speed was set at 2200 rpm, the gear was at the highest gear position, and the excavator was used for 30 cycles (12 km) without a load. Compared with the operation condition, the actuation system and the swing motor do not work in the transportation condition, and the data of the engine, torque converter, steering pump, and main pump are mainly collected. The collected data for transportation conditions are shown in Table 2.

Table 2. Data collection under transportation condition.

Serial Number	Components' Data	Serial Number	Components' Data
1	Main pump 1 outlet pressure	6	Main pump 2 outlet flow rate
2	Main pump 2 outlet pressure	7	Torque converter output speed
3	Steering pump outlet pressure	8	Torque converter output torque
4	Steering pump outlet flow rate	9	Engine output speed
5	Main pump 1 outlet flow rate	10	Engine output torque

3.2. Performance Data Acquisition

The excavator system theoretical energy analysis shows that the accurate energy flow analysis needs the components' performance data support. In this section, the component bench experiment to measure the hydraulic pumps and torque converter efficiencies is described. The main pump of the excavator is a variable displacement pump, and the steering pump is a fixed pump. Fixed pump efficiencies vary with loads, while variable displacement pump efficiencies vary with the loads and displacement ratios. The operating speed of the main pump and steering pump were set to 1200 rpm and their efficiencies were tested according to Standard JB/T7043. The test results are shown in Figure 11.

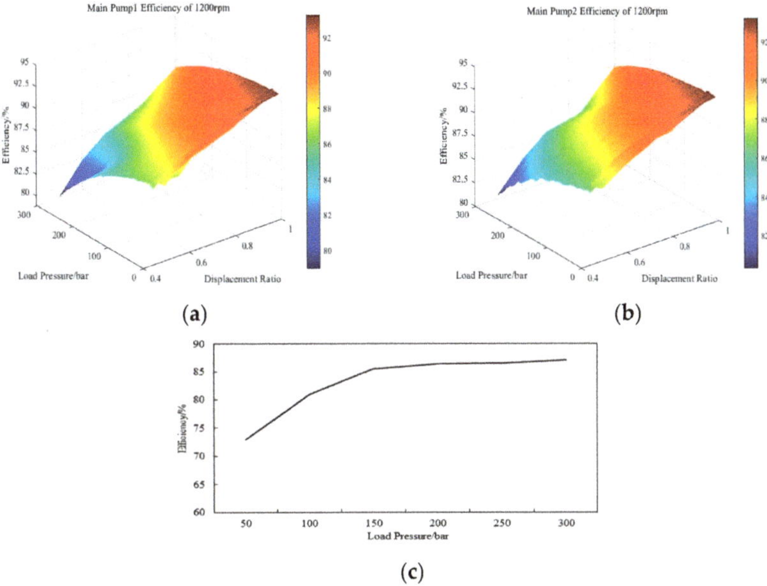

Figure 11. Efficiency results of hydraulic pumps at 1200 rpm. (**a**) Pump 1 displacement ratio-load-efficiency results; (**b**) pump 2 displacement ratio-load-efficiency results; (**c**) steering pump load-efficiency results.

The torque converter's efficiency test experiments were carried out. The experimental bench is shown in Figure 12a. The experimental steps are as follows:

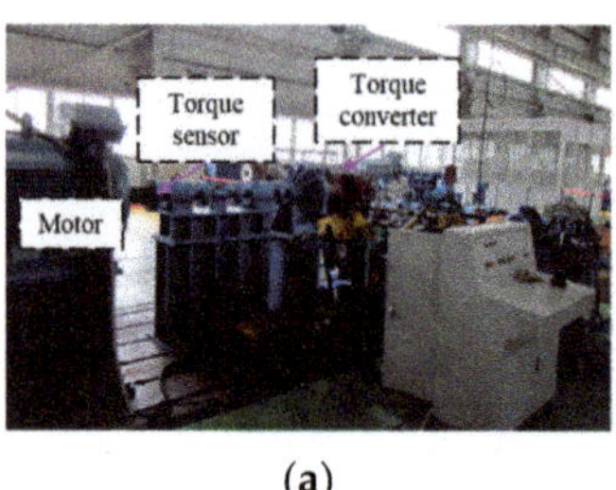

(a)

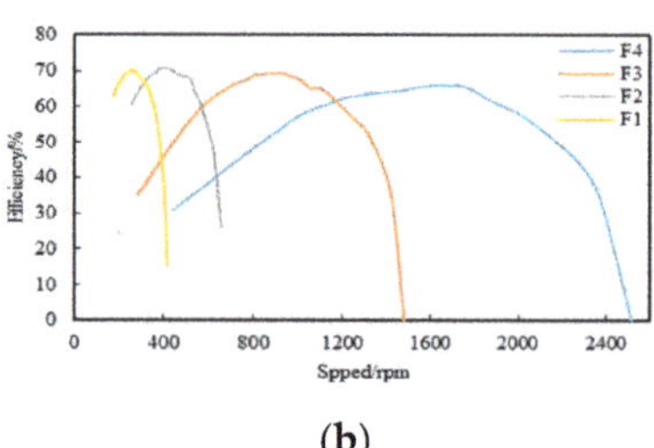

(b)

Figure 12. Torque converter efficiencies test. (**a**) The torque converter efficiencies test bench; (**b**) the torque converter's efficiencies under different gear positions.

1. The motor simulates the output of the engine, keeps the input speed constant, and the experimental speed is set to 2200 r/min under the transport condition.

2. In the no-load state, increase the input speed to the set value. After the speed is stable, load successively at the output, reduce the output speed, and keep the input speed constant. After the loaded speed is fully stable, record the data.

3. The same experiment was repeated 5 times, and the average values of the 5 experiments were used as the experimental results.

According to the above experimental method, the efficiencies of the torque converter under different gear positions were measured, and the results are shown in Figure 12b.

4. System Energy Flow Analysis

4.1. Excavator Typical Working Condition Load Reconstruction

The wavelet decomposition method decomposes the excavator operation load into random and trend terms. The random term is processed by data segmentation, noise reduction, and singular value removal, and its stationarity is verified by the round-robin method. The random term's power spectrum is analyzed, and its parametric characterization is realized using the harmonic function method. The trend term is a non-stationary random signal. After the data segmentation, its mean square value is measured, and the load trend term is reconstructed by the weighted theory method. The reconstructed random and trend terms are combined, and the reconstruction of the typical condition load is complete. The process is shown in Figure 13.

As pump 2's load as an example, it is firstly decomposed into a random term and trend term through wavelet decomposition, then filtered and singular values are removed. The results are shown in Figure 14.

Its statistical characteristics are representative only when the load's random term is a stationary ergodic process of various states. The load random term of 60 cycles' operation data are divided into 10 subsample sequences of equal length. The mean square of each series is calculated and compared to the overall mean square of the random term. The round-robin method is used to carry out the ergodic process test of the stationary states [10], and the test results are shown in Table 3.

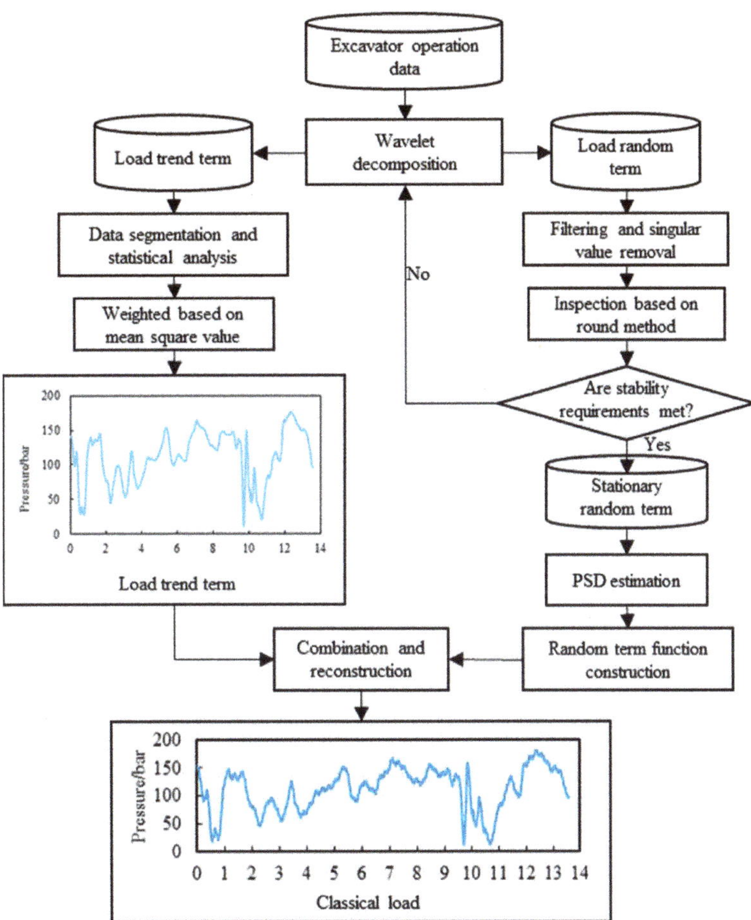

Figure 13. The typical working condition load reconstruction.

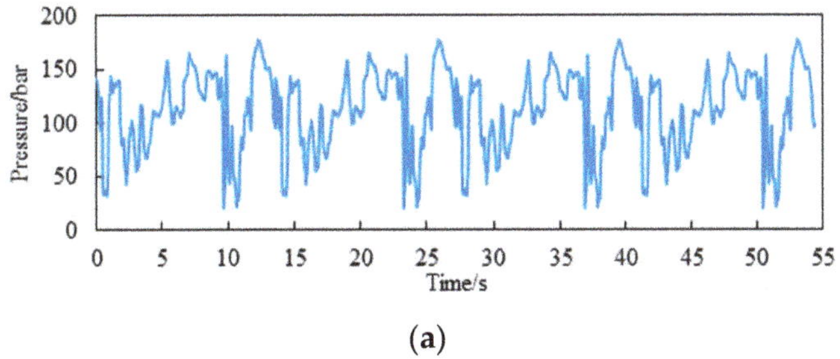

(**a**)

Figure 14. *Cont.*

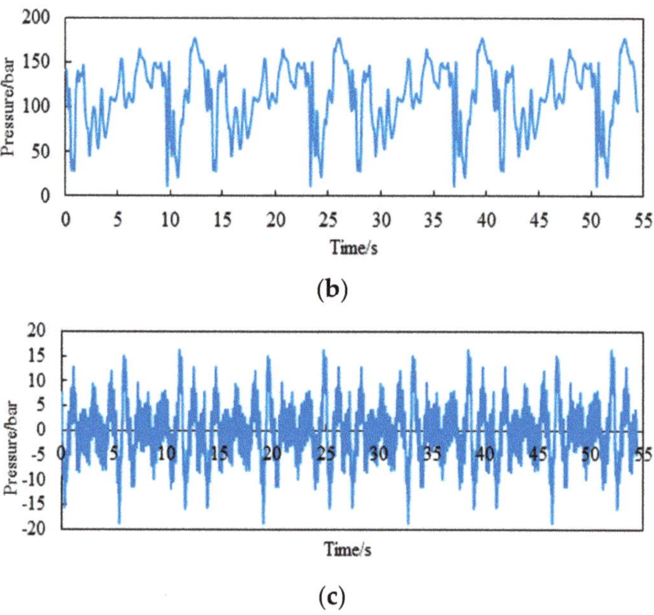

(b)

(c)

Figure 14. Wavelet decomposition results of main pump 2 in operation condition. (**a**) Original data; (**b**) trend term data; (**c**) random term data.

Table 3. Rounds statistics of load random term.

Serial Number	Mean Square Value	Rounds Statistics	Serial Number	Mean Square Value	Rounds
1	10.127	-	6	8.688	-
2	11.559	+	7	8.581	-
3	12.089	+	8	9.666	-
4	9.337	-	9	10.810	+
5	12.127	+	10	9.395	-
Total mean square value		10.316	Round numbers		6

When the number of subsamples is $n = 10$, under the significance level $\alpha = 0.05$, the number of rounds should be [3,8]. The number of statistical rounds of the load's random term is 6, so the assumption of stationarity is acceptable. If the samples of a random process are stationary and the experimental condition for obtaining each sample are basically the same, the stationary random process can be treated as an ergodic process.

Using the Welch method to estimate the power spectrum of the load's random term, the power spectral density function curve is shown in Figure 15. It can be seen from the figure that the load energy of the excavator's main pump 2 is concentrated in the range of 0 to 9 Hz, and there are two peak frequencies of 3.08 Hz and 4.02 Hz, which are consistent with the significant inertia of the excavator and its cyclic operation characteristics. The random load's power density spectrums are divided into several intervals according to the intermediate frequencies (the intermediate frequencies are $\omega_1, \omega_2, \dots, \omega_n$), and the harmonic function approximation replaces the original random term at the discrete frequencies. Each harmonic component must satisfy the energy equivalence condition of Equation (23).

$$\frac{A_n^2}{2} = 2 \int_{\omega_{n-1}}^{\omega_n} P(\omega) d(\omega) \tag{23}$$

where A_n^2 is the amplitude of the n_{th} harmonic function; $P(\omega)$ is the power spectral density of the random term.

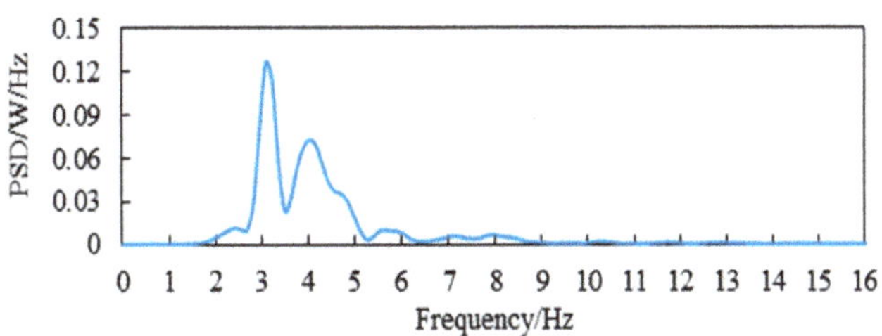

Figure 15. Power spectral density of load random term.

Combined with the results of the power spectral density of the load random term, the parameterization of the load random term is expressed as Equation (24). The parameterized expression of the load random term function is completed in the frequency range of 1 to 9 Hz. The accuracy of the reconstructed data can be evaluated by Formula (25), and the evaluation result $R = 0.952$, which proves that the reconstruction method has high accuracy. Figure 16 shows the comparison results of the reconstructed load's random term and the original load's random term.

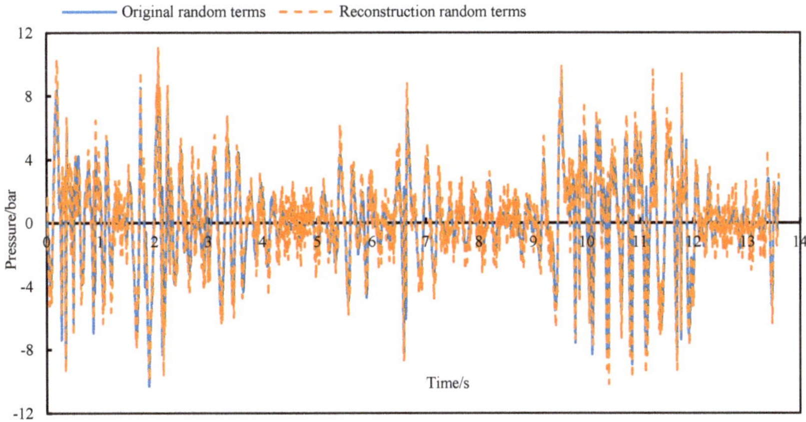

Figure 16. The comparison result between the reconstructed and original load random term.

$$f(t) = 2\sqrt{\int_{\omega_0}^{\omega_1} P(\omega)d\omega}\cos\left(\frac{\omega_1}{2}t + \varphi_1\right) + 2\sqrt{\int_{\omega_1}^{\omega_2} P(\omega)d\omega}\cos\left(\frac{\omega_1 + \omega_2}{2}t + \varphi_2\right) + \ldots 2\sqrt{\int_{\omega_{n-1}}^{\omega_n} P(\omega)d\omega}\cos\left(\frac{\omega_{n-1} + \omega_n}{2}t + \varphi_n\right) \tag{24}$$

$$R = 1 - \frac{\sum_{i=1}^{T}\frac{|P_{Ri} - P_{Oi}|}{P_{Oi}}}{T} \tag{25}$$

where P_{Ri} is the load random term after reconstruction; P_{Oi} is the original load random term; T is the total time of the sample.

Compared with the random term, the trend term cannot be parameterized by the harmonic function. Therefore, the trend term samples of 60 cycles' operation data are divided into 10 groups of subsamples, each group of subsamples has 6 operation segments of data, and each segment of operation segment data is divided into 5 operation stages. Based on the method from the literature [24], the divided operation sample data are theoretically weighted and reconstructed by Equations (26) and (27).

$$X_i = \sum_{j=1}^{10} \left(\frac{R_{ij}}{\sum\limits_{j=1}^{10} R_{ij}} x_{ij} \right) (i = 1, \ldots, 6;\ j = 1, \ldots, 10) \tag{26}$$

where i is the sequence number of the operation segment; j is the sequence number of the subsample; R_{ij} is the mean square value of the i_{th} operation segment of the j_{th} subsample; x_{ij} is the time domain waveform data of the i_{th} operation segment of the j_{th} subsample; X_i is the weighted data of the i_{th} operation segment.

$$y_i = \sum_{i=1}^{6} \left(\frac{R_{ik}}{\sum\limits_{k=1}^{6} R_{ik}} x_{ik} \right) (x_{ik} \in X_i)(i = 1, \ldots, 5;\ k = 1, \ldots, 6) \tag{27}$$

where k is the sequence number of the operation cycle; R_{ik} is the mean square value of the i_{th} operation stage of the k_{th} cycle; x_{ik} is the waveform data of the i_{th} operation stage of the k_{th} cycle; y_i is the weighted data of the i_{th} operation stage.

The result of the load's trend term after weighted reconstruction is shown in Figure 17. The operation conditions of the excavator include 5 stages of dig preparation, digging, lifting, unloading, and swinging. The boom and the swing system work in the stages of dig preparation, lifting, and swinging, so the load amplitude ranges in the three stages are close and smooth. During the digging stage, the bucket, boom, and stick work together, the load of the main pump is affected by the working resistance, and the load fluctuates greatly. During the dumping stage, where only the bucket works, there is a shock caused by load transients.

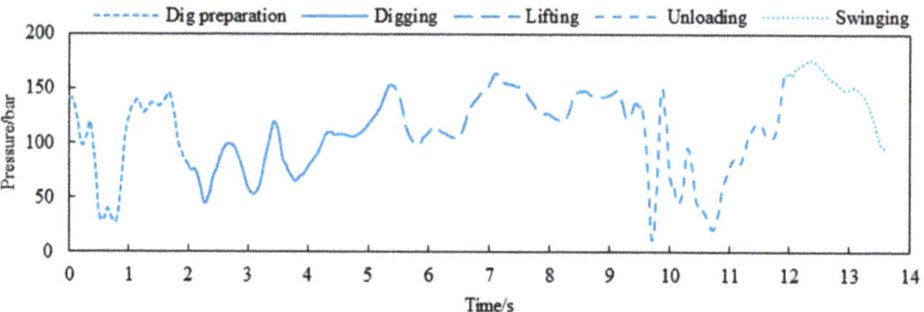

Figure 17. Reconstruction results of trend terms of typical load.

The load's random term is combined with the load's trend term to obtain the reconstructed load. The accuracy of the reconstructed data is evaluated by Equation (25), and the evaluation result R is 0.965, indicating that the accuracy of the reconstructed load is high, which proves that the load reconstruction method for typical working conditions is reasonable. The comparison result between the reconstructed load and the original load is shown in Figure 18.

Similarly, under transportation conditions, the output torque is decomposed into random and trend terms. The output torque is reconstructed by the harmonic function and the theoretical weighting method, and its evaluation result R is 0.973. The comparison results of the reconstructed output torque and the original torque are shown in Figure 19. Finally, the same method is used to reconstruct the typical working condition load of the main pump 1, boom, swing motor, engine, and other components. The energy flow analysis of the excavator under operation and transportation conditions is carried out based on typical working condition load.

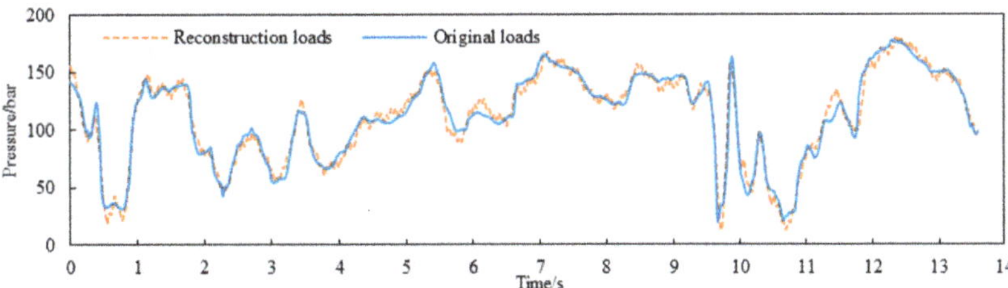

Figure 18. Comparison between reconstructed load and original data.

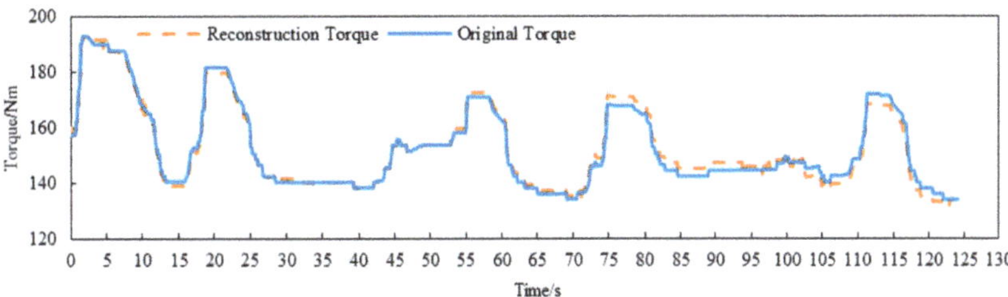

Figure 19. The comparison between reconstructed and original load.

4.2. Energy Flow Analysis Results

The results of the engine energy flow analysis are shown in Figure 20. The output energy of the engine only accounts for 50.21% of the engine's fuel energy, the heat loss accounts for 18.85% of the engine's fuel energy, and other consumption of the engine accounts for 30.94% of the engine's fuel energy.

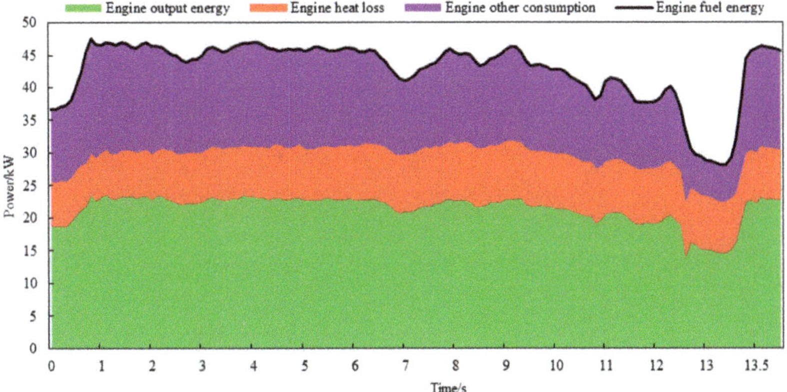

Figure 20. The engine energy flow.

The hydraulic pump energy flow analysis results are shown in Figure 21. The hydraulic pumps' efficiency loss, the hydraulic system's heat loss, the auxiliary components' consumption, and the hydraulic pumps' output energy account for 12.00%, 12.77%, 19.31%, and 55.92% of the engine's output energy, respectively.

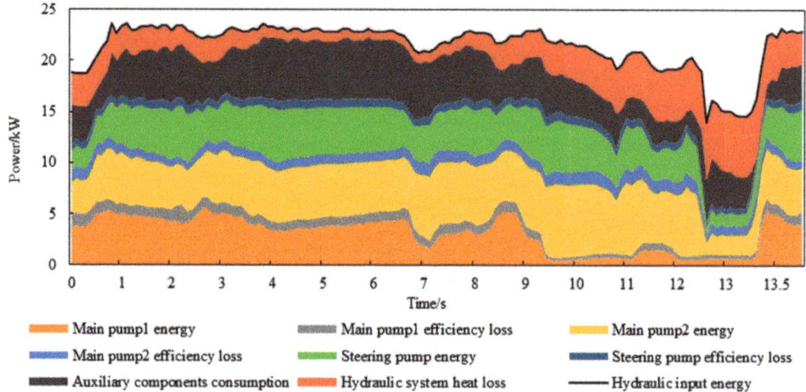

Figure 21. The hydraulic pump energy flow.

The energy flow analysis results of the operation system are shown in Figure 22. Motor consumption accounts for 14.23% of the system input energy. The boom, stick, and bucket work accounted for 11.09%, 3.46%, and 18.67% of the system input energy, respectively, totaling 33.22%. The loss in the hydraulic system circuit (cylinder loss, circuit loss, overflow loss, and main control valve loss) account for 52.55% of the system input energy. At the same time, in process of lowering the boom and stick, the gravitational potential energy released by itself is equivalent to 6.07% of the system input energy.

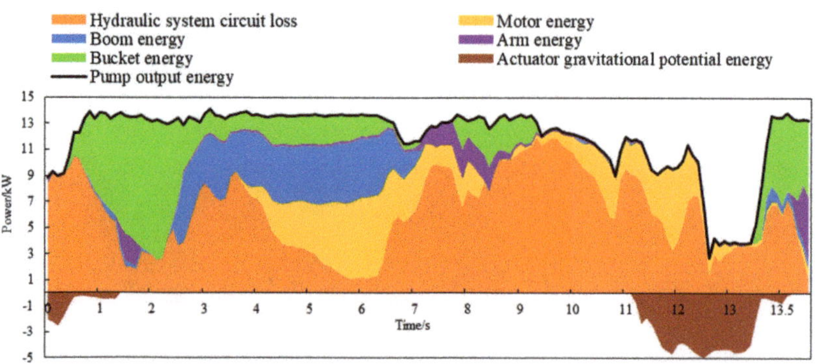

Figure 22. The operation system energy flow.

From each component's energy flow results, the proportion of component energy in the engine's fuel energy is further analyzed, and the energy flow in the excavator's one operation cycle is obtained. It can be seen from the analysis that the engine heat loss, engine other loss, and the engine's output energy account for 18.85%, 30.94%, and 50.21% of the engine's fuel energy, respectively. The hydraulic pump's efficiency loss, the hydraulic system's heat loss, the auxiliary components' consumption, and the hydraulic pump's output energy account for 6.03%, 6.41%, 9.69%, and 28.08% of the engine's fuel energy, respectively. The boom, stick, bucket, and swing motor energy consumption accounted for 3.11%, 0.98%, 5.24%, and 4.00% of the engine's fuel energy, respectively, totaling 9.33%. The hydraulic circuit system consumes 14.75% of the engine's fuel energy. In addition, during the lowering of the boom or stick, its gravitational potential energy is equivalent to 1.70% of the engine's fuel energy. The results are shown in Figure 23.

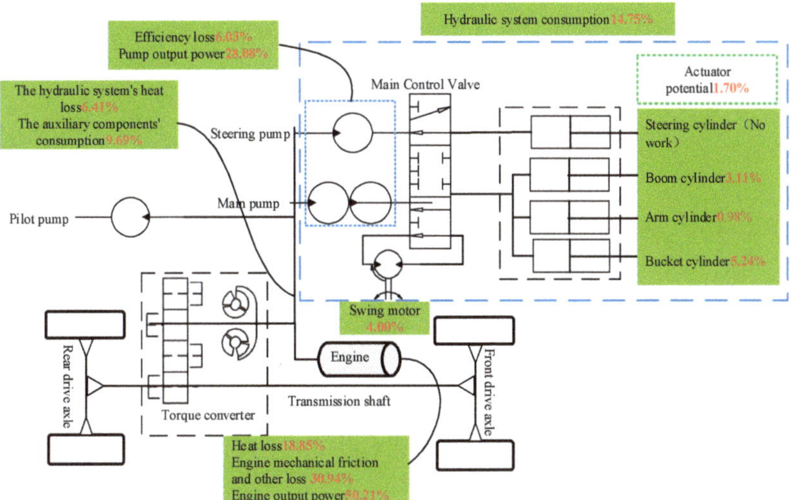

Figure 23. Excavator operation condition energy flow.

Under transportation conditions, the engine output energy, engine heat loss, and other loss account for 49.80%, 23.73%, and 26.47% of the engine's fuel energy. The main pump and steering pump consume 8.66% and 1.25% of the engine's fuel energy. The efficiency loss of the torque converter, the loss of the drive axle and wheel reducer, and the driving energy in transportation conditions account for 15.09%, 6.82%, and 17.98% of the engine fuel energy, respectively. The results are shown in Figure 24.

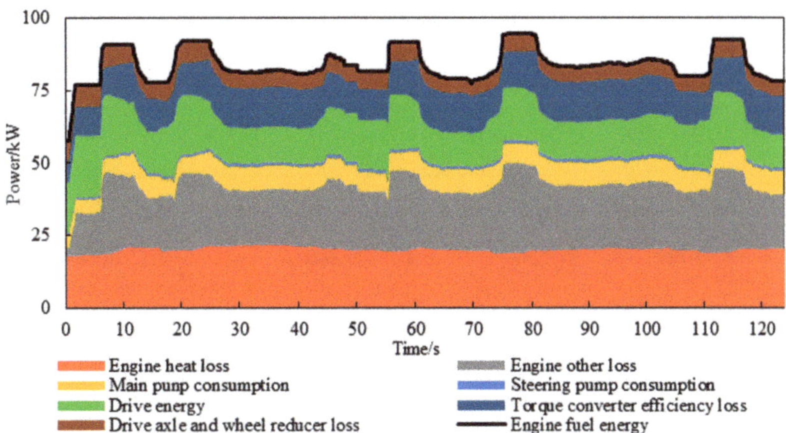

Figure 24. Transportation condition energy flow.

4.3. Experimental Verification

Ten experiments were carried out under operation and transportation conditions, the energy flow results based on the experimental data were obtained, and the average value was calculated. The comparison of energy flow analysis results based on typical working condition load and experimental data are shown in Tables 4–8. Under operation conditions, the maximum error between the two is 4.80%, and under transportation conditions, the maximum error between the two is 4.23%. The error of the energy flow analysis results of the two working conditions is less than 5%, which proves that the energy flow analysis method based on typical working conditions is reasonable.

Table 4. Engine energy flow results comparison under operation conditions.

Results	Fuel Energy	Heat Loss	Other Loss	Output Energy
Under typical condition load	46.26 kW	18.85%	30.94%	50.21%
Experimental results	48.53 kW	19.80%	30.99%	49.21%
Deviation	4.68%	4.80%	0.16%	2.03%

Table 5. Hydraulic pump energy flow results comparison under operation conditions.

Results	Output Energy	Hyraulic System Heat Loss and Auxiliary Components' Consumption	Efficiency Loss
Under typical condition load	28.08%	16.10%	6.03%
Experimental results	27.58%	15.70%	5.93%
Deviation	1.81%	2.55%	1.69%

Table 6. Operation system energy flow results comparison under operation conditions.

Results	Hydraulic System Consumption	Actuation System	Swing Motor	Regeneration of Actuators
Under typical condition load	14.75%	9.33%	4.00%	1.70%
Experimental results	13.98%	8.93%	3.82%	1.78%
Deviation	0.72%	4.48%	4.71%	4.49%

Table 7. Engine energy flow results comparison under transportation conditions.

Results	Fuel Energy	Heat Loss	Other Loss
Under typical condition load	84.05 kW	23.73%	26.47%
Experimental results	85.88 kW	24.05%	27.11%
Deviation	2.13%	1.33%	2.36%

Table 8. The transportation condition energy flow comparison results.

Results	Main Pump Consumption	Steering Pump Consumption	Torque Converter Efficiency Loss	Drive Axle and Wheel Reducer	Driving Energy
Under typical condition load	8.66%	1.25%	15.09%	6.82%	17.98%
Experimental results	8.36%	1.22%	14.95%	7.06%	17.25%
Deviation	3.59%	2.46%	0.94%	3.40%	4.23%

Whether it is operation or transportation conditions, the energy utilization rate of excavators is relatively low. Under the operation condition, the recoverable potential energy and the overflow loss of the swing process account for a large proportion. Reasonable energy recovery methods may be considered to recover it. Under transportation conditions, the main pump in the non-working state has a significant loss. Reducing the minimum output flow of the main pump may be considered, such as the use of a positive flow control pump.

5. Conclusions

In order to obtain accurate excavator energy flow results, this paper analyzes the excavator system's theoretical energy flow and obtains the excavator experimental data. The typical working condition load is reconstructed and the accurate analysis of the excavator energy flow is realized based on it. The work can be summarized as follows:

1. Data requirements are obtained according to the results of the system theoretical energy flow analysis, the experimental schemes are designed, and the excavator data are collected. The typical working condition load is reconstructed based on wavelet decomposition, harmonic function, and theoretical weighting methods, for which the accuracy evaluation value R is 0.965. The reconstructed load is close to the excavator's actual working condition load.

2. Under operation conditions, the output energy of the engine only accounts for 50.21% of the engine's fuel energy, and the energy consumption of the actuation and the swing system accounts for 9.33% and 4%, respectively. Under transportation conditions, the output energy of the engine only accounts for 49.80% of the engine's fuel energy, and the torque converter efficiency loss and excavator driving energy account for 15.09% and 17.98%, respectively. In both working conditions, the effective utilization rate of engine energy is low.

3. The energy flow analysis results provide energy margins for designing energy-saving schemes and control strategies. Energy recovery and reducing the main pump's minimum output flow rate can be considered for excavators' energy-saving control.

Author Contributions: Data collection experiments, D.S., X.B. and X.X.; Investigation, D.S.; Methodology, D.S., L.H. and S.W.; Writing—original draft, D.S.; Writing—review and editing, D.S., L.H. and S.W. All authors have read and agreed to the published version of the manuscript.

Funding: This project was supported by the National Key R&D Program of China (Grant No. 2020YFB1709901, No. 2020YFB1709904), National Natural Science Foundation of China (Grant No. 51975495, 51905460) and Guangdong Basic and Applied Basic Research Foundation (Grant No. 2021A1515012286).

Institutional Review Board Statement: The study did not require ethical approval.

Informed Consent Statement: The study did not involve humans.

Data Availability Statement: The study did not report any data.

Conflicts of Interest: The authors declare no conflict of interest.

References

1. Trani, M.L.; Bossi, B.; Gangolells, M.; Casals, M. Predicting fuel energy consumption during earthworks. *J. Clean. Prod.* **2016**, *112*, 3798–3809. [CrossRef]
2. Abekawa, T.; Tanikawa, Y.; Hirosawa, A. *Introduction of Komatsu Genuine Hydraulic Oil KOMHYDRO HE*; Komatsu Technical Report; Yumpu: Komatsu, Japan, 2010; Volume 56.
3. Ohira, S.; Suehiro, M.; Ota, K.; Kawamura, K. Use of emission rights for construction machinery to help prevent global warming. *Hitachi Rev.* **2013**, *62*, 123–130.
4. Yang, J.; Quan, L.; Yang, Y. Excavator energy-saving efficiency based on diesel engine cylinder deactivation technology. *Chin. J. Mech. Eng.* **2012**, *25*, 897–904. [CrossRef]
5. Xu, D.F. Analysis and Computing of Load Sensing System in Hydraulic Excavator. *J. Tongji Univ.* **2001**, *29*, 1097–1100.
6. Gao, F.; Pan, S.X. Experimental study on model of negative control of hydraulic system. *J. Mech. Eng.* **2005**, *41*, 107–111. [CrossRef]
7. Ge, L.; Dong, Z.; Quan, L.; Li, Y. Potential energy regeneration method and its engineering applications in large-scale excavators. *Energy Convers. Manag.* **2019**, *195*, 1309–1318. [CrossRef]
8. Choi, K.; Seo, J.; Nam, Y.; Kim, K.U. Energy-saving in excavators with application of independent metering valve. *J. Mech. Sci. Technol.* **2015**, *29*, 387–395. [CrossRef]
9. Zhang, J.; Li, K.; Guan, L.; Zhu, T. Research on Positive Control System with Synthetic Pilot Pressure of Excavator. *China Mech. Eng.* **2012**, *23*, 1537–1541. [CrossRef]
10. Kim, Y.; Kim, P.; Murrenhoff, H. Boom potential energy regeneration scheme for hydraulic excavators. In Proceedings of the ASME/BATH 2016 Symposium on Fluid power and Motion Control, Bath, UK, 7–9 September 2016.
11. Zhao, P.; Chen, Y.; Zhou, H. Simulation analysis of potential energy recovery system of hydraulic hybrid excavator. *Int. J. Precis. Eng. Manuf.* **2017**, *18*, 1575–1589. [CrossRef]
12. Xiao, Q.; Wang, Q.; Zhang, Y. Control strategies of power system in hybrid hydraulic excavator. *Autom. Constr.* **2008**, *4*, 361–367. [CrossRef]

13. Kwon, T.S.; Lee, S.W.; Sul, S.K.; Park, C.G.; Kim, N.I.; Kang, B.I.; Hong, M.S. Power Control Alg-orithm for Hybrid Excavator with Supercapacitor. *IEEE Trans. Ndustry Appl.* **2010**, *46*, 1447–1455. [CrossRef]
14. Deppen, T.O.; Alleyne, A.G.; Stelson, K.; Meyer, J. A model predictive control approach for a parallel hydraulic hybrid powertrain. In Proceedings of the American Control Conference, San Francisco, CA, USA, 29 June–1 July 2011; IEEE: Piscataway, NJ, USA; pp. 2713–2718.
15. Shen, W.; Jiang, J.; Su, X.; Karimi, H.R. Energy-saving analysis of hydraulic hybrid excavator based oncommon pressure rail. *Sci. World J.* **2013**, *2013*, 560694. [CrossRef] [PubMed]
16. Zimmerman, J.D.; Pelosi, M.; Williamson, C.A. Energy consumption of an LS excavator hydraulic system. In Proceedings of the ASME International Mechanical Engineering Congress and Exposition, Seattle, WA, USA, 11–15 November 2007; pp. 117–126.
17. Casoli, P.; Riccò, L.; Campanini, F.; Lettini, A.; Dolcin, C. Mathematical model of a hydraulic excavator forfuel consumption predictions. In Proceedings of the ASME/BATH 2015 symposium on Fluid Power and Motion Control, Chicago, IL, USA, 12–14 October 2015.
18. Altare, G.; Padovani, D.; Nervegna, N. A Commercial Excavator: Analysis, Modelling and Simulation of theHydraulic Circuit. In Proceedings of the SAE 2012 Commercial Vehicle Engineering congress, Rosemont, IL, USA, 13–14 September 2012.
19. Casoli, P.; Riccò, L.; Campanini, F.; Bedotti, A. Hydraulic hybrid excavator—Mathematical model validationand energy analysis. *Energies* **2016**, *9*, 1002. [CrossRef]
20. Bedotti, A.; Campanini, F.; Pastori, M.; Riccò, L.; Casoli, P. Energy saving solutions for a hydraulic excavator. *Energy Procedia* **2017**, *126*, 1099–1106. [CrossRef]
21. An, K.; Kang, H.; An, Y.; Park, J.; Lee, J. Methodology of excavator system energy flow-down. *Energies* **2020**, *13*, 951. [CrossRef]
22. Zhai, X.; Wang, J.; Chen, J. Parameter Estimation Method of Mixture Distribution for Construction Machinery. *Math. Probl. Eng.* **2018**, *2018*, 3124048. [CrossRef]
23. Zhai, X.; Zhang, X.; Jiang, Z.; Li, Y.; Zhang, Q.; Wang, J. Load spectrum compiling for wheel loader semi-axle based on mixed distribution. *Trans. Chin. Soc. Agric. Eng. (Trans. CSAE)* **2018**, *34*, 78–84.
24. Chang, L.; Xu, L.; Lü, M.; Liu, Y.; Zhao, Y. Driving cycle construction of loader based on typical working condition test. *Trans. Chin. Soc. Agric. Eng. (Trans. CSAE)* **2018**, *34*, 63–69.
25. Jiang, T.; Fu, Z.Y.; Wang, A.L. Parameterized load model under non-stationary random cyclic conditions. *J. Northeast. Univ. (Nat. Sci.)* **2016**, *37*, 845–850.
26. Chen, Y.; Wang, A.; Li, X. Load Classification Design of Excavator Multiway Valve. *J. Xi'an Jiaotong Univ.* **2020**, *54*, 100–108.
27. Peng, B. Research on Stage-Based Power Matching Control of Hydraulic Excavator. Ph.D. Thesis, ZheJiang University, Hangzhou, China, 2016.
28. Wang, S.J.; Hou, L.; Huang, H.T.; Zhu, Q. Application of optimal shift control strategy on human intelligent fuzy control. *J. Xiamen Univ. (Nat. Sci.)* **2016**, *55*, 131–136. (In Chinese)

electronics

MDPI

Article

Performance Analysis on Low-Power Energy Harvesting Wireless Sensors Eco-Friendly Networks with a Novel Relay Selection Scheme

Hoang-Sy Nguyen [1], Lukas Sevcik [2] and Hoang-Phuong Van [3,*]

[1] Becamex Business School, Eastern International University,
Thu Dau Mot City 820000, Binh Duong Province, Vietnam; sy.nguyen@eiu.edu.vn
[2] Faculty of Electrical Engineering and Information Technology, University of Zilina, 01026 Zilina, Slovakia;
lukas.sevcik@uniza.sk
[3] Institute of Engineering and Technology, Thu Dau Mot University,
Thu Dau Mot City 820000, Binh Duong Province, Vietnam
* Correspondence: phuongvh@tdmu.edu.vn

Abstract: Simultaneous wireless information and power transfer (SWIPT) has been utilized widely in wireless sensor networks (WSNs) to design systems that can be sustained by harvesting energy from the surrounding areas. In this study, we investigated the performance of the low-power energy harvesting (LPEH) WSN. We equipped each relay with a battery that consisted of an on/off (1/0) decision scheme according to the Markov property. In this context, an optimal loop interference relay selection was proposed and investigated. Moreover, the crucial role of the log-normal distribution method in characterizing the LPEH WSN's constraints was proven and emphasized. System performance was evaluated in terms of the overall ergodic outage probability (OP) both analytically and numerically with Monte Carlo simulation. The system had the lowest overall ergodic OP, thus, performed the best with an energy harvesting time switch of 0.175. Following the increase in the signal-to-noise ratio (SNR), the system without a direct link performed the worst. Furthermore, as more relays were deployed, the better the system performed. Finally, results showed that more than 80% of the data rates can be obtained under the household condition, without the need for extra bandwidth and power supply.

Keywords: eco-friendly networks; wireless sensor network; energy efficiency; low-power networks; Markov model; log-normal fading

Citation: Nguyen, H.-S.; Sevcik, L.; Van, H.-P. Performance Analysis on Low-Power Energy Harvesting Wireless Sensors Eco-Friendly Networks with a Novel Relay Selection Scheme. *Electronics* **2022**, *11*, 1978. https://doi.org/10.3390/electronics11131978

Academic Editor: Nurul I. Sarkar

Received: 21 May 2022
Accepted: 20 June 2022
Published: 24 June 2022

Publisher's Note: MDPI stays neutral with regard to jurisdictional claims in published maps and institutional affiliations.

1. Introduction

Following the advancement of the fifth and upcoming sixth generation communication, simultaneous wireless information and power transfer (SWIPT) holds its place as a promising technology for self-sustaining communication systems. In particular, it can offer notable system performance gains, indicated by improvements in spectral efficiency (SE), power consumption, interference management, and transmission delay, due to its nature of facilitating simultaneous information and energy transmission [1–3]. For the SWIPT cooperative relaying system, there are two major modes in use: the half-duplex (HD) and the full-duplex (FD). Specifically, in HD mode, the relay is equipped with one antenna that functions so that data retransmission from the source is carried out on dedicated and orthogonal channels [4]. Meanwhile, in FD mode, a relay is equipped with two antennas to facilitate data reception and transmission in the same time frame and bandwidth. Moreover, it is worth noting that the FD wireless network is praised for its ability to double the SE, considerably improving the network throughput in comparison with the HD network.

The combination of the FD operating mode in EH relaying networks has been a debatable topic regarding the possibility of sustainability when operating FD mode and performing the relaying task. However, with the advancement of the antenna, battery technologies, and signal processing capacity, it is reasonable to re-assess the FD-EH relaying networks for the promising benefits they can offer, as in [5–7]. Additionally, in [8] Li et al. showed that the utilization of energy storage can bring in zero diversity gain when the signal-to-noise ratio (SNR) level is high. Remarkably, Zhong et al. in [5] extensively discussed the constraints of FD-EH relaying networks and proved that system performance in three different transmission modes can be significantly enhanced.

To describe the outdoor wireless channel, Rayleigh, Rician, and Nakagami-m fading channels have been widely accepted among network designers [9]. However, for indoor settings, due to the shadowing effects caused by walls, obstructions, and human movements, log-normal is a better choice [10–13]. Thus, log-normal is a more suitable candidate for smart homes and industrial Internet of Things (IIoTs) applications. In addition, it is reasonable to use the current advancement in the relaying network technology to further boost the system performance of the household networks. In fact, this has been proven to provide higher channel capacity, a wider network coverage range, and a lesser shadowing effect [11]. Moreover, equipping more relays in a cooperative relaying system provides a greater degree of freedom (DoF), enabling the system to combine more independent fading signals from several relays in operation, hence resulting in higher system performance [14]. The above studies have provided a background to develop a low-power energy harvesting (LPEH) wireless sensor network (WSN), which can counteract the high signal attenuation and noise in household environments [15–17].

Although the energy harvested from radio frequency (RF) is relatively low in comparison to other sources such as thermal, solar, etc., it is sufficient to operate several devices and sensors as noted by Assogba et al. [18]. Specifically, authors in [19] attempted to measure the free ambient RF and found that by densifying the Wi-Fi access points, it is possible to not only power the indoor devices (smoke detectors, wearable electronics, etc.) but to also extend their lifetime via trickle charging. Furthermore, in regard to upgrading indoor electrical grids, a Power Line Communication (PLC) has been investigated; wireless devices can exploit the existing electricity cables to improve data transmission and energy efficiency [20]. PLC enables establishing communication with remote or highly attenuated nodes that are underground, in buildings with metal walls, obstructed, or in hospitals where the electromagnetic interference is not allowed. In addition, PLC can work as either a standalone system or as a facilitator of Wi-Fi extension in a large in-house environment [21]. In the past, PLC operated in narrowband with a range of 3–500 kHz and a low data rate, which is undesirable for modern demands. Thus, Ref. [22] has investigated the possibilities of deploying broadband PLC with the range of 1.8–2.5 MHz and a high data rate to serve multimedia services in household environments. Several PLC standards have been developed [23–25]. Application wise, PLC has constituted to the smart grid (SG), which is the foundation for smart city applications including traffic and lighting control [26,27], lighting systems in urban areas [28], irrigation, and other smart city applications. [29].

Regarding the security of the physical layer, the available techniques for wireless networks are applicable for PLC as well. For example, a case study of how to utilize the jamming technique to guard against eavesdroppers for cooperative PLC networks was presented in [30]. In the household setup, it is highly recommended to utilize log-normal fading channels because other fading channels cannot describe efficiently the fading effects caused by walls, human movement, and indoor obstructions [31]. Several papers have extensively analyzed the performance of relay-aided EH wireless networks embedded in the PLC using different key parameters. Research has proven this combination is promising for the future of smart grids and indoor wireless communication [32–36], which is undeniably powered by the development of LPEH WSN.

As can be concluded from the reviewed literature, WSN cannot perform well in indoor environments due to the extensive shadowing effect caused by indoor obstructions. To create an efficient communication system for indoor setups, it is of great interest to deploy the WSN together with PLC and EH modules. However, the effect of the interactions between the three technologies on system performance has not been well studied in the existing literature. Therefore, in this manuscript, we assess the performance of a relay selection (RS) scheme, the optimal loop interference RS scheme in the context of LPEH WSN. Notably, the relays are equipped with batteries and operate with the FD-AF protocol. The shadowing effect of the network is characterized by log-normal fading channels.

2. System Models

Figure 1 below depicts an LPEH WSN that includes a source (S), an in-between cluster (C) of K cooperative relays (R_i), $(1 \leq i \leq K)$, and a destination (D). As investigated in [37,38], this setup is typical and appropriate for studying the impact of RS schemes operating in wireless networks. Additionally, the coverage extension scenario is assumed so that communication between (S) and (D) can be achieved only with the help of the intermediate relays [39,40]. Moreover, the solution can help overcome the deep shadowing effects caused by the movement of humans or obstacles in the surrounding area, making it more suitable for the studied LPEH WSN, [37].

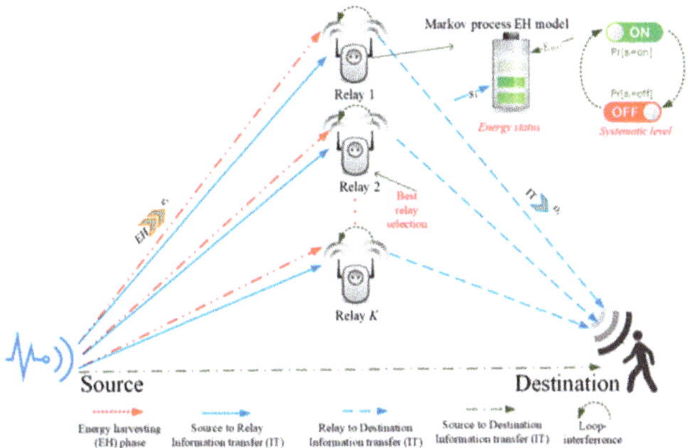

Figure 1. System models.

In addition, the carrier and symbol synchronization are assumed to be ideal. Each terminal knows beforehand its own channel state information (CSI). (S) is powered by a stable power source P_s and every (R) is powered by a battery P_R together with an EH module. The additive white Gaussian noise (AWGN), n_j, $(j \in \{R, D\})$ is with zero mean and variance N_0 at the (R_i) and (D), respectively.

In fact, the distances from (S) to (R_i), (R_i) to (D), and (S) to (D) are denoted with d_{SR_i}, d_{R_iD} and d_{SD}. Their corresponding channel coefficients are $l_{s,r}$, $l_{r,d}$, and $l_{s,d}$. As in Laourine et al. [10] and Mellios et al. [41], (S) and (D) are HD and (R)s are FD. The FD setup causes the loop interference channel $l_{r,r}$ to the system. During the communication, which is split into time slots, the i-th relay $R_i(R_i \in C)$ is selected as per a (RS) scheme to help establishing the information transmission. Within a signal block, the narrowband transmits signals at (S) and is denoted as $s(t)$, (R_i), and has zero mean. The statistical mean operation is denoted with $E[|s(t)|^2] = 1$.

We consider random variables (RVs) that are independently and identically distributed (i.i.d) as per log-normal distribution as $l_{s,r}^2$, $l_{r,d}^2$, and $l_{s,d}^2$. The RVs are associated with parameters $\mathcal{LN}\left(2\omega_{l_{s,r}}, 4\Omega_{l_{s,r}}^2\right)$, $\mathcal{LN}\left(2\omega_{l_{r,d}}, 4\Omega_{l_{r,d}}^2\right)$, and $\mathcal{LN}\left(2\omega_{l_{s,d}}, 4\Omega_{l_{s,d}}^2\right)$, respectively. Furthermore, the i.i.d loop interference channel following log-normal distribution $l_{r,r}^2$ is with parameter $\mathcal{LN}\left(2\omega_{l_{r,r}}, 4\Omega_{l_{r,r}}^2\right)$. The loop interference strength given by this parameter is important to characterize the performance of the FD system.

2.1. Standalone Direct Link

To serve as a foundation for further study, a direct transmission protocol, in which (S) transmits information directly to (D) without any assisting (R) in the LPEH WSNs is assumed and investigated.

Definition 1. *In the context of the direct transmission protocol, the direct (S)-(D) link is the unique option. Thus, it will use all the time slots in the signal block for data transmission.*

As in [42], we employ a block fading channel assuming that the service process is stationary, and that the buffer receives data at a constant rate. From the analytical point of view, the effective capacity after normalization is the same as the conventional Shannon ergodic capacity under the condition of no delay constraint.

Thus, employing the direct transmission protocol, the overall (S)-(D) capacity can be obtained with zero mean, circularly symmetric

$$C_{s,d} = W log_2\left(1 + \frac{P_s}{N_0}\frac{|l_{s,d}|^2}{d_{s,d}^m}\right),\tag{1}$$

where m is the path loss exponent, and W is the frequency bandwidth.

As mentioned above, the ergodic OP is an indicator for evaluations of system performance. This is the probability that instantaneous capacity falls below a given threshold's bits per channel use (BPCU) being R_0, and $Pr[C_{s,d} < R_0]$. Here, the probability density function (PDF) and the cumulative distribution function (CDF) of the RV X in log-normal distribution [43] are, respectively, calculated by

$$F_X(z) = 1 - \mathcal{Q}\left(\frac{\frac{10}{\ln(10)}\ln(z) - 2\omega_X}{2\Omega_X}\right),\tag{2}$$

and

$$f_X(z) = \frac{10/\ln(10)}{z\sqrt{8\pi\Omega_X^2}}e^{\left(-\frac{\left(\frac{10}{\ln(10)}\ln(z) - 2\omega_X\right)^2}{8\Omega_X^2}\right)},\tag{3}$$

where $\mathcal{Q}(\cdot)$ is the Gaussian $\mathcal{Q} -$ function, $\mathcal{Q}(x) = \int_x^\infty \frac{1}{\sqrt{2\pi}}e^{\left(-\frac{t^2}{2}\right)}\mathrm{dt}$.

By utilizing the CDF of the log-normally distributed RV $|l_{s,d}|^2$ in (1), the ergodic OP for the direct transmission protocol can be expressed as

$$EOP_{s,d} = \Pr\left(|l_{s,d}|^2 < \left(2^{\frac{R_0}{W}} - 1\right)\frac{N_0 d_{s,d}^m}{P_s}\right) = 1 - \mathcal{Q}\left(\frac{\xi\ln(a) - 2\omega_{s,d}}{2\Omega_{s,d}}\right),\tag{4}$$

where $\xi = \frac{10}{\ln(10)}$ is a scaling constant [44], $a = \frac{\gamma_0 N_0}{P_s d_{s,d}^{-m}}$, and $\gamma_0 = 2^{R_0/W} - 1$

Remark 1. *The direct link of the relay-aided systems offers better data rate gain in noisier condition, that is, when the SINR and data rates are low. It is also worth noting that relay-aided links are highly beneficial for transmission over long distance. In normal conditions, the direct link is sufficient to establish a successful communication.*

2.2. Relay-Aided Cooperative Protocol

As previously mentioned, the relay-aided cooperative protocol is deployed to improve the LPEH WSN performance.

Definition 2. *In the context of the relay-aided cooperative protocol, all relay links can alternatively transmit the data in place of the direct link. Specifically, if the direct link is severely attenuated, a relay will be chosen among K_s from the cluster (C) to realize the data transmission alternatively. This is performed on a condition that the relay has a sufficient amount of energy to conduct the task. This process is shown below in Figure 2.*

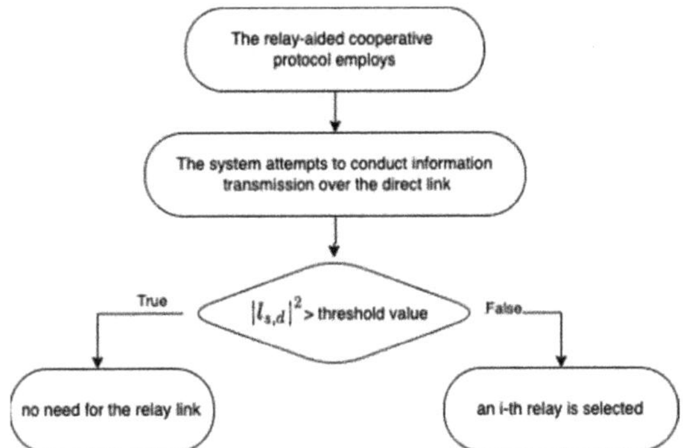

Figure 2. Criteria for relay selection.

By this approach, more time slots are saved, and overhead information is reduced, because only the needed relay link is activated. Moreover, in LPEH WSN, we can receive the maximal diversity gain equaling the relay nodes that are available in the network, as proven in [4].

As described in [39], one transmission cycle from (S) to (D) realized within T time is split into three slots as per the TSR protocol. The first time slot τT, where τ is the EH time factor, $(0 \leq \tau \leq 1)$, is spent by (R) to harvest the energy from the signal that (S) broadcasts. The remaining $(1 - \tau)T$ is halved with one half for information of $(S) - (R_i)$ and the other for $(R_i) - (D)$.

Within τT, given the energy harvested the i-th (R) has harvested, $E_H = \frac{\eta \tau T P_s |l_{s,r_i}|^2}{d_{s,r_i}^m}$, the power that the relay transmits can be described as

$$P_{R_i} = \frac{E_H}{(1 - \tau)T} = \frac{\eta \tau P_s |l_{s,r_i}|^2}{(1 - \tau)d_{s,r_i}^m}, \tag{5}$$

where the EH efficiency η, $0 \leq \eta \leq 1$, shows the circuitry's characteristics.

As mentioned in [5] as a relay can recognize its own signal in the FD multi-relay scenario, the interference cancellation can be applied to itself.

For $(1 - \tau)T$, the signal after the interference cancellation at the i-th (R) can be expressed as

$$y_r(t) = \sqrt{\frac{P_s}{d_{s,r_i}^m}} l_{s,r_i} s(t) + \widehat{l_{r,r}} \widehat{r}(t) + n_r, \tag{6}$$

where the information signal $s(t)$ is normalized as $E[|s(t)|^2] = 1$. The imperfect interference cancelation leaves out the residual loop interference $\widehat{l_{r,r}}$, and $E|\widehat{r}(t)|^2 = P_{R_i}$.

In an FD-AF system, after the signal is base-band processed at (R_i) as per (6), it is amplified and then sent to (D). Therefore, (D) receives the signal of

$$y_d(t) = \sqrt{\frac{P_s P_{R_i}}{d_{s,r_i}^m d_{r_i,d}^m}} l_{s,r_i} l_{r_i,d} G s(t) + \sqrt{\frac{P_{R_i}}{d_{r_i,d}^m}} l_{r_i,d} l_{r,r} G r(t) + \sqrt{\frac{P_{R_i}}{d_{r_i,d}^m}} l_{r_i,d} G n_r + n_d, \tag{7}$$

where the relay gain, G, denotes the instantaneous received power normalization, within which process the relay transmission is allowed with maximum power, $P_s \to \infty$, according to [45,46]

$$G = \frac{1}{\sqrt{\frac{P_s}{d_{s,r_i}^m}|l_{s,r_i}|^2 + |\widehat{l_{s,r_i}}|^2 P_{R_i} + N_0}} \approx \frac{1}{\sqrt{\frac{P_s}{d_{s,r_i}^m}|l_{s,r_i}|^2 + |\widehat{l_{s,r_i}}|^2 P_{R_i}}}. \tag{8}$$

Then, (8) is substituted into (7) and modified so that the end-to-end SNR of the i-th (R) at (D) can be obtained

$$\gamma_{s,R_i,d} = \frac{\frac{P_s \gamma_{S,R_i}}{P_{R_i} \gamma_{R,R_i}} P_{R_i} \gamma_{R_i,D}}{\frac{P_s \gamma_{S,R_i}}{P_{R_i} \gamma_{R,R_i}} + P_{R_i} \gamma_{R_i,D} + 1}, \tag{9}$$

where $\gamma_{S,R_i} = |l_{s,r_i}|^2 d_{s,r_i}^{-m}$, $\gamma_{R_i,D} = |l_{r_i,d}|^2 d_{r_i,d}^{-m}$, and $\gamma_{R,R_i} = |l_{r,r}|^2$.

Additionally, the instantaneous capacity of the FD-AF-TSR system can be calculated

$$C_{s,R_i,d} = (1 - \tau) \log_2 (1 + \gamma_{s,R_i,d}). \tag{10}$$

From this, the RS scheme for the in-studied LPEH WSN optimal loop interference relay selection scheme is derived. In particular, the RS scheme first chooses (R) having the best end-to-end link, then updates the SINR in the first branch. Thus, the selected k relay, with $P_s \to \infty$, has the condition of

$$k \approx \arg \max_i \min \left\{ \frac{P_s \gamma_{S,R_i}}{P_{R_i} \gamma_{R,R_i}}, P_{R_i} \gamma_{R_i,D} \right\}. \tag{11}$$

It should be noted that the optimal loop interference RS scheme is established only if the full CSI is known.

2.3. Energy Storage Modeling at Relay

The stationary stochastic process in Tutuncuoglu et al. [46], states that the minimum energy needed to activate R_k and the harvested energy within the k-th signal block as en_k. Firstly, the R_k is assumed without energy storage, thus, operates only with en_k. The R_k runs out of energy if $en_k \leq P_{R_k}$. The probability that this event occurs is named energy-exhausted probability and can be formulated as follows.

$$EOP_k = Pr[en_k < P_{R_k}] = \int_0^{P_{R_k}} f_{en_k}(x)dx, \tag{12}$$

where $f_{en_k}(x)$ is used to signify the PDF of the stationary stochastic process e_k.

In practice, however, the R_k is installed with a battery with the capacity of en_{max}. The k-th signal block transmission consumes an amount of ou_k energy. As per the condition in Definition 2, the ou_k can be expressed with the stationary random function as

$$Pr(ou_k) = \begin{cases} EOP_{s,d}/\mathcal{K}, & \text{if } Pr(ou_k = P_{R_k}) \\ 1 - EOP_{s,d}/\mathcal{K}, & \text{if } Pr(ou_k = 0), \\ 0, & \text{otherwise} \end{cases} \tag{13}$$

where $\mathcal{K}$ denotes that every (R_k) can be activated with an equal probability. Notably, the variables en_k and ou_k are stationary and independent.

Additionally, to formulate the energy buffer status according to the Markov stochastic process, we use the denotation s_k, which stands for the initial energy amount the (R_k) stores before starting the k-th signal block transmission.

Subsequently, we can combine (12), (13), and the original PDF of en_k to obtain the stationary PDF of s_k being $f_{s_k}(x)$. Theoretically, all the stochastic characteristics of the EH module can be described with $f_{s_k}(x)$. However, because it is not necessary to include all these energy buffer statuses in the system described here, a simplified model is proposed. Specifically, according to Definition 2, the number of statuses is reduced to two processing steps and described in Lemma 1.

Lemma 1. *The s_k and the possible energy that can be consumed at (R_k) are compared to make the decision on whether the transmission should be realized on the direct or the relaying link. This is shown below in Figure 3.*

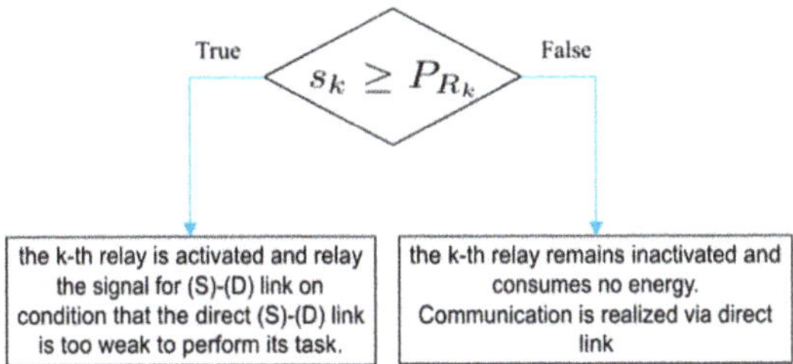

Figure 3. Criteria for choosing the direct or relaying link for information transmission.

These two statuses are therefore used to characterize the flow of the energy that has been harvested. Accordingly, we proposed an on–off $(1/0)$ model with denotation s'_k, which is described in Figure 1. Then, we used the stationary PDF of s_k, $f_{s_k}(x)$ to formulate the stationary PDF of s'_k as follows.

$$
Pr(s'_i) = \begin{cases} \int_{P_{R_k}}^{en_{max}} f_{s_k}(x)dx, & \text{if } Pr(s'_i = 1) \\ \int_{0}^{P_{R_k}} f_{s_k}(x)dx, & \text{otherwise} \end{cases}.
\tag{14}
$$

Remarkably, (14) is not the same as the original PDF $f_{en_k}(x)$ of the harvested energy in (13).

Subsequently, an on–off $(1/0)$ (on/off status) model to describe the harvested energy flow has been formulated and described in the following Remark 2. For more details on the derivation, see [47].

Remark 2. *The relaying link in LPEH WSN can be activated if the direct link cannot perform the information transmission task and the available energy stored at the relay is greater than P_{R_k}. Additionally, thanks to the two parameters P_{R_k} and OP_k, which characterize the EH module employed in R_k, we can capture the stochastic property of the energy harvested beforehand with no loss.*

3. Performance Analysis

3.1. Formulating the Problem

In practice, the relaying link over R_k cannot always be established due to deep fading or energy exhausting when the surrounding energy resources fluctuate. This event is named an overall outage event and can be caused by three reasons as follows.

At least one of the data packets cannot be successfully delivered during the information transmission from (S) to (D). This causes the direct link's end-to-end SNR to be below the threshold value of the system.

The energy harvested by R_k is not sufficient, so it cannot be activated to realize the information transmission via the relaying link. This is known as energy-exhausted probability.

The relay R_k is activated but cannot transfer all the data packets successfully with a sufficient amount of energy from the EH process. This also causes the relaying link's end-to-end SNR to be below the threshold value of the system.

Having acknowledged the above constraints, for the herein study, we assume that R_k possesses a sufficient amount of energy to realize the information transmission. Subsequently, the overall ergodic OP of the FD-AF-TSR LPEH WSN described in Definition 2 can be formulated as follows

$$EOP_{oc} = EOP_{s,d} \times \prod_{k=1}^{K} \left[EOP_k + (1 - OEP_k) \times EOP_{s,R_k,d} \right]. \tag{15}$$

3.2. Analysis of the Probability of Overall Ergodic Outage

With regard to (10), the ergodic OP for S–R_k–D link can be fully defined as

$$OP_{s,R_k,d} = Pr \left\{ \frac{\frac{P_s \gamma_{S,R_k}}{P_{R_k} \gamma_{R,R_k}} P_{R_k} \gamma_{R_k,D}}{\frac{P_s \gamma_{S,R_k}}{P_{R_k} \gamma_{R,R_k}} + P_{R_k} \gamma_{R_k,D} + 1} < \gamma_1 \right\}, \tag{16}$$

where $\gamma_1 = 2^{R_0/(1-\tau)W} - 1$.

Proposition 1. *The SNR in (11) is assumed to be high and identical. Thus, we can formulate the corresponding overall ergodic OP for the high SNR range under the RS scheme as*

$$EOP_{oc}(\gamma_1) = \left[1 - Q\left(\frac{\xi \ln(a) - 2\omega_{l_{s,d}}}{2\Omega_{l_{s,d}}} \right) \right] \times EOP_k + (1 - EOP_k) \times EOP_{s,R_k,d}(\gamma_1)^K, \tag{17}$$

where $EOP_{s,R_k,d}(\gamma_1) = 1 - Q\left(\frac{\xi \ln\left(\frac{\eta\tau}{1-\tau}\gamma_1\right) + 2\omega_{l_{r,r}}}{2\Omega_{l_{r,r}}} \right) \times Q\left(\frac{\xi \ln(c) - 2\left(\omega_{l_{s,r}} + \omega_{l_{r,d}}\right)}{\sqrt{2}\left(\Omega_{l_{s,r}} + \Omega_{l_{r,d}}\right)} \right), a = \frac{(1-\tau)}{\eta\tau P_s}\gamma_1.$

Proof of Proposition 1. It is worth noting that the γ_{R_k} in (11) is crucial for the RS process. As we combine (5) and (6), we can formulate the SNR at (R) as follows

$$\gamma_{S,R_k} = \frac{P_s \gamma_{S,R_k}}{P_{R_k}} = \frac{1-\tau}{\eta\tau} \times \frac{1}{\gamma_{R,R_k}}. \tag{18}$$

Similarly, the SNR at (D) is given by

$$\gamma_{R_k,D} = P_{R_k} \gamma_{R_k,D} = \frac{\eta\tau P_s}{(1-\tau)} \gamma_{S,R_k} \gamma_{R_k,D}. \tag{19}$$

Subsequently, we combine (18) and (19), then substitute them into (11) to obtain the ergodic OP in (16), considering R_k, as follows

$$EOP_{s,R_k,d} = Pr \left\{ \min\left\{ \frac{1-\tau}{\eta\tau Y}, \frac{\eta\tau P_s X}{(1-\tau)} \right\} < \gamma_1 \right\}, \tag{20}$$

where $X = \gamma_{S,R_k} \gamma_{R_k,D}$ and $Y = \gamma_{R,R_i}$.

It should be noted that all the aforementioned channel means and variances are i.i.d. as per the log–normal distribution. Assuming that X and Y are two independent RVs, the ergodic OP of the system with the optimal RS scheme in (11) can be expressed as

$$EOP_{s,R_k,d} = 1 - \overline{F_Y}\left(\frac{\eta\tau}{1-\tau}\gamma_1 \right) \overline{F_X}\left(\frac{(1-\tau)}{\eta\tau P_s}\gamma_1 \right), \tag{21}$$

where $\overline{F_Y}(\cdot)$ and $\overline{F_X}(\cdot)$ are the complementary CDFs of Y and X. Because RV Y is distributed in a log–normal manner, its complementary CDF $\overline{F_Y}(\cdot)$ can be obtained with ease as follows:

$$\overline{F_Y}\left(\frac{\eta\tau}{1-\tau}\gamma_1 \right) = Q\left(\frac{\xi \ln\left(\frac{\eta\tau}{1-\tau}\gamma_1\right) + 2\omega_{l_{r,r}}}{2\Omega_{l_{r,r}}} \right). \tag{22}$$

Moreover, the RV X is a product of two RVs which are as well log–normally distributed, its complementary CDF can be obtained as

$$\overline{F_X}\left(\frac{(1-\tau)}{\eta \tau P_s}\gamma_1\right) = \mathcal{Q}\left(\frac{\xi \ln\left(\frac{(1-\tau)}{\eta \tau P_s}\gamma_1\right) - 2\left(\omega_{l_{s,r}} + \omega_{l_{r,d}}\right)}{\sqrt{2}\left(\Omega_{l_{s,r}} + \Omega_{l_{r,d}}\right)}\right). \tag{23}$$

As we substitute (22) and (23) into (21), the ergodic OP can be obtained as

$$OP_{s,R_k,d} = 1 - \mathcal{Q}\left(\frac{\xi \ln\left(\frac{\eta \tau}{1-\tau}\gamma_1\right) + 2\omega_{l_{r,r}}}{2\Omega_{l_{r,r}}}\right) \times \mathcal{Q}\left(\frac{\xi \ln\left(\frac{(1-\tau)}{\eta \tau P_s}\gamma_1\right) - 2\left(\omega_{l_{s,r}} + \omega_{l_{r,d}}\right)}{\sqrt{2}\left(\Omega_{l_{s,r}} + \Omega_{l_{r,d}}\right)}\right), \tag{24}$$

Consequently, (24) and (4) are substituted into (15) to obtain the overall ergodic OP in the optimal RS scheme, as given in (17). $\square$

4. Results and Discussion

In this section, the Monte Carlo numerical simulation was conducted to investigate the overall ergodic OP of the in-studied system with an optimal RS scheme. In summary, the analytic results were obtained by evaluating the log–normal distribution following [45]. The system parameters for the simulations are listed below in Table 1. The results are calculated based on Proposition 1 in (17), by means of MatLab simulation.

Table 1. Simulation parameters.

Primary Parameters	Description	Values
OP_k	energy-exhausted probability	10^{-1}
W	frequency bandwidth	2 (W)
R_0	transmission rate threshold	2 (bps/Hz)
P_s	traditional stabilized power source	5 (dB)
N_0	overall AWGNs	1
H	EH efficiency	1
T	EH time fraction	0.2
m	path-loss exponent	2
$d_{s,r}$	(S)-(R_k) distance	5 (m)
$d_{r,d}$	(R_k)-(D) distance	5 (m)
$d_{s,d}$	(S)-(D) distance	10 (m)
$\Omega_{l_{s,r}}$	(S)-(R_k) channel mean	4 (dB)
$\Omega_{l_{r,d}}$	(R_k)-(D) channel mean	4 (dB)
$\Omega_{l_{s,d}}$	(S)-(D) channel mean	4 (dB)

Figure 4 illustrates the overall ergodic OP versus the EH time switch τ. Two values of loop interference channel variance being $\Omega_{l_{r,r}} = 2$(dB) and 4 (dB) are used together with the constant power source of $P_s = 5$ (dB). It is possible to observe that the two curves reach their minimum at $\tau = (0.175)$, at which the system performs the best and the coding gap is the largest. As τ increases to more than 0.3, the overall ergodic OP curves exponentially decrease, corresponding to the extreme case where too much time is dedicated to EH activity, resulting in insufficient time for data transmission purpose. The principle for this behavior is that, starting from 0, when we increase the power source, the message from the source has higher probability to be successfully decoded, leading to a more successful DF process. Nevertheless, when the power source passes the threshold value, the probability that the relay can decode the message from the source is so high that it reduces the relay power and the quality of the second hop. Therefore, the existence of an optimal τ must be considered when designing such a network.

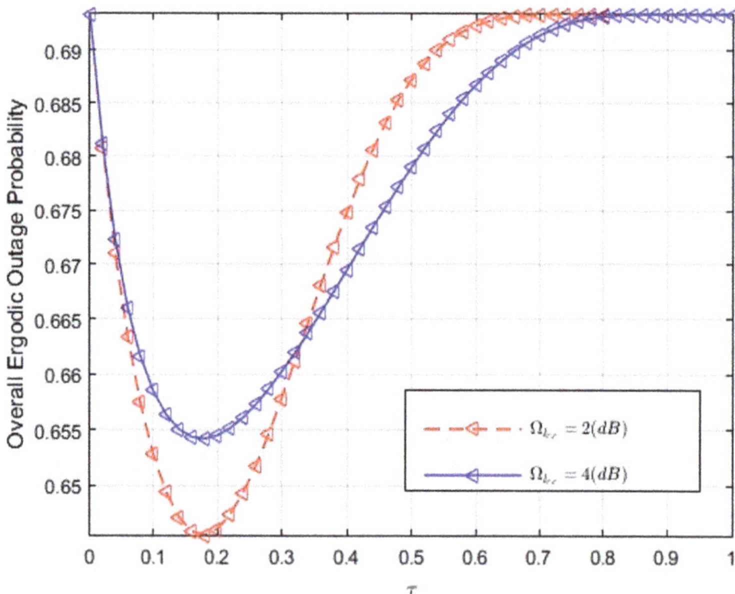

Figure 4. Relation between the overall ergodic OP and the EH time switch τ, with two $\Omega_{l_{r,r}}$ values.

Figure 5 shows the relation between the overall ergodic OP of the LPEH WSN and the SNR. For comparison, we use three values of 0, 0.5, and 1 for the energy-exhausted probability OP of best RS. The $OP_k = 0$ when the LPEH WSN operates with the cooperative relays, $OP_k = 1$ with direct link, and $OP_k = 0.5$ for both or only the relay link (red color). It can be noted that as SNR increases from -20 to 30 dB, the $OP_k = 0.5$ without a direct link delivers the highest overall ergodic OP, thus, the worst system performs. The remaining curves are relatively closed to each other and sharply approach 0 overall ergodic OP when the SNR increases to 30 (dB). Theory and simulation agree well with each other and suggest the importance of implementing the EH cooperative relays in boosting the performance of the LPEH WSN. Additionally, it is remarkable that the data rate has been significantly improved, with over 80% data rates achieved under the PLC condition, requiring no extra bandwidth or power supply.

Figure 6 plots the overall ergodic OP versus the SNR when changing the number of relays, K, from 1 to 3 and 5. The three curves are simulated with $OP_k = 10^{-1}$. It is quite intuitive that the more intermediate relays are installed, the lower the overall ergodic OP curve, leading to better system performance. Remarkably, changing the number of relays results in performance curves with similar shapes as changing the OP_k value. So far, it can be concluded that the utilization of intermediate EH relays is beneficial for LPEH WSNs with the best RS, and the higher the number of relays, the better the system performs.

Using log–normal fading channels to characterize the shadowing effects, the herein system can characterize real indoor scenarios. However, being too specific in LPEH WSN means that it is hard for the proposed model to be scaled up without the expertise for both software and hardware and how to implement them together. Moreover, since the studies on the smart grid are just in their early stages, in-depth help is not easily accessible in the existing literature. Other shortcomings of the model are mentioned in the next section as we discuss possible future research to improve the model in question.

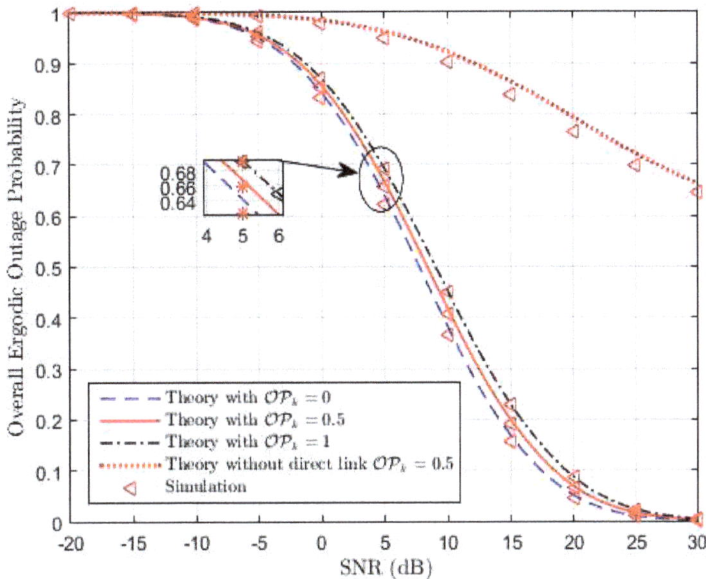

Figure 5. Relation between the overall ergodic OP and the SNR, with three OP_k values.

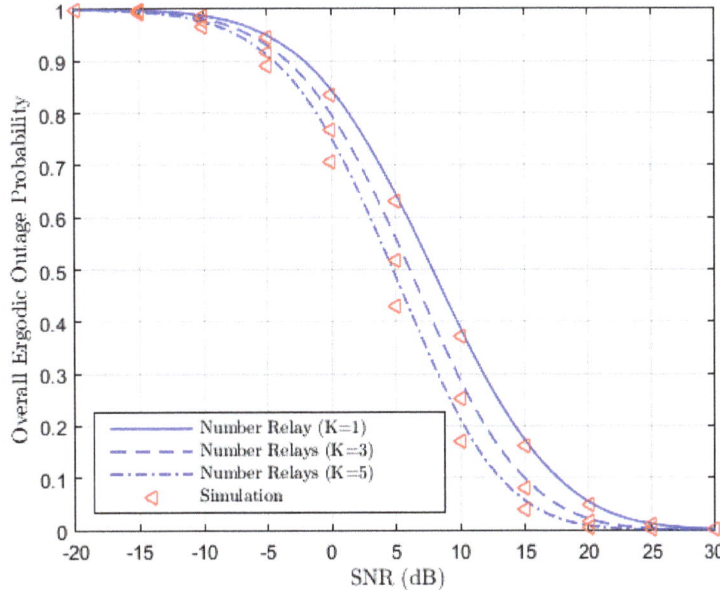

Figure 6. Relation between the overall ergodic OP and the SNR, with three K values.

5. Conclusions

In general, the paper presents the performance analysis in terms of the overall ergodic OP of the optimal RS scheme in the context of LPEH WSN. Since the simulations correlate well with the theory, the expressions that we derived show potential to be applicable for future studies. It is also proven that the log–normal fading channel is appropriate for modelling such indoor scenarios. Future studies, which would complement the shortcomings of this study, can consider different RS schemes and compare them with the studies in this paper to find the best one for the LPEH WSN setup. In addition, experimental studies can be conducted to compare the results in this study with reality. With regard to eco-friendly networks, in the next studies, we can compare the technical requirements for the coexistence of PLC systems and other communication technologies. We can evaluate how the resource should be allocated in a system with a multicarrier and improve the signal processing and channel coding abilities of the system. Furthermore, we can compare the protocols that are used for LPEH WSN with other protocols that have been utilized for WSN and the dynamics of the physical layers to facilitate further implementation of PLC in the near future.

Author Contributions: Conceptualization, H.-S.N. and H.-P.V.; methodology, formal analysis, H.-P.V. and L.S.; writing—original draft preparation, L.S.; writing—review and editing, H.-P.V. and H.-S.N.; visualization, H.-S.N.; supervision, H.-S.N. All authors have read and agreed to the published version of the manuscript.

Funding: This work has been supported by the Slovak Scientific Grant Agency (VEGA), Project No. 1/0588/22 "Research of a location-aware system for achievement of QoE in 5G and B5G networks" and under the project of Operational Programme Integrated Infrastructure: Independent research and development of technological kits based on wearable electronics products, as tools for raising hygienic standards in a society exposed to the virus causing the COVID-19 disease, ITMS2014+ code 313011ASK8. The project is co-funded by the European Regional Development Fund. This research is also funded by Thu Dau Mot University under grant number DT.20.2-020.

Data Availability Statement: The data presented in this study are available on request from the corresponding author.

Conflicts of Interest: The authors declare no conflict of interest.

References

1. Perera, T.D.P.; Jayakody, D.N.K.; Sharma, S.K.; Chatzinotas, S.; Li, J. Simultaneous wireless information and power transfer (SWIPT): Recent advances and future challenges. *IEEE Commun. Surv. Tutor.* **2018**, *20*, 264–302. [CrossRef]
2. Sidhu, R.K.; Ubhi, J.S.; Aggarwal, A. A survey study of different RF energy sources for RF energy harvesting. In Proceedings of the 2019 IEEE International Conference on Automation, Computational and Technology Management (ICACTM), London, UK, 24–26 April 2019. [CrossRef]
3. Nguyen, H.-S.; Ly, T.; Nguyen, T.-S.; Huynh, V.; Nguyen, T.-L.; Voznak, M. Outage performance analysis and SWIPT optimization in energy harvesting wireless sensor network deploying NOMA. *Sensors* **2019**, *19*, 613. [CrossRef]
4. Laneman, J.; Tse, D.; Wornell, G. Cooperative diversity in wireless networks: Efficient protocols and outage behavior. *IEEE Trans. Inf. Theory* **2004**, *50*, 3062–3080. [CrossRef]
5. Zhong, C.; Suraweera, H.A.; Zheng, G.; Krikidis, I.; Zhang, Z. Wireless information and power transfer with full duplex relaying. *IEEE Trans. Commun.* **2014**, *62*, 3447–3461. [CrossRef]
6. Liu, H.; Kim, K.J.; Kwak, K.S.; Poor, H.V. Power splitting-based SWIPT with decode-and-forward full-duplex relaying. *IEEE Trans. Wirel. Commun.* **2016**, *15*, 7561–7577. [CrossRef]
7. Zeng, Y.; Zhang, R. Full-duplex wireless-powered relay with self-energy recycling. *IEEE Wirel. Commun. Lett.* **2015**, *4*, 201–204. [CrossRef]
8. Li, T.; Fan, P.; Letaief, K.B. Outage probability of energy harvesting relay-aided cooperative networks over rayleigh fading channel. *IEEE Trans. Veh. Technol.* **2016**, *65*, 972–978. [CrossRef]
9. Zhang, Z.; Chai, X.; Long, K.; Vasilakos, A.V.; Hanzo, L. Full duplex techniques for 5g networks: Self-interference cancellation, protocol design, and relay selection. *IEEE Commun. Mag.* **2015**, *53*, 128–137. [CrossRef]
10. Laourine, A.; Stephenne, A.; Affes, S. Estimating the ergodic capacity of log-normal channels. *IEEE Commun. Lett.* **2007**, *11*, 568–570. [CrossRef]
11. Zhang, Z.; Zhang, W.; Tellambura, C. Cooperative OFDM channel estimation in the presence of frequency offsets. *IEEE Trans. Veh. Technol.* **2009**, *58*, 3447–3459. [CrossRef]
12. Di Renzo, M.; Graziosi, F.; Santucci, F. Performance of Cooperative Multi-Hop Wireless Systems over Log-Normal Fading Channels. In Proceedings of the IEEE GLOBECOM 2008—2008 IEEE Global Telecommunications Conference, New Orleans, LA, USA, 2008; pp. 1–6. [CrossRef]
13. Zhu, B.; Cheng, J.; Yan, J.; Wang, J.; Wu, L.; Wang, Y. A new technique for analyzing asymptotic outage performance of diversity over lognormal fading channels. In Proceedings of the IEEE 2017 15th Canadian Workshop on Information Theory (CWIT), Quebec City, QC, Canada, 11–14 June 2017. [CrossRef]

14. Foschini, G.J. Layered space-time architecture for wireless communication in a fading environment when using multi-element antennas. *Bell Labs Tech. J.* **2002**, *1*, 41–59. [CrossRef]

15. Rubin, I.; Lin, Y.-Y.; Kofman, D. Relay-aided networking for power line communications. In Proceedings of the IEEE Information Theory and Applications Workshop (ITA), San Diego, CA, USA, 9–14 February 2014. [CrossRef]

16. Rabie, K.M.; Adebisi, B.; Yousif, E.H.G.; Gacanin, H.; Tonello, A.M. A comparison between orthogonal and non-orthogonal multiple access in cooperative relaying power line communication systems. *IEEE Access* **2017**, *5*, 10118–10129. [CrossRef]

17. Pu, H.; Liu, X.; Zhang, S.; Xu, D. Adaptive cooperative non-orthogonal multiple access-based power line communication. *IEEE Access* **2019**, *7*, 73856–73869. [CrossRef]

18. Assogba, O.; Mbodji, A.K.; Diallo, A.K. Efficiency in RF energy harvesting systems: A comprehensive review. In Proceedings of the IEEE International Conference on Natural and Engineering Sciences for Sahel's Sustainable Development Impact of Big Data Application on Society and Environment (IBASE-BF), Ouagadougou, Burkina Faso, 4–6 February 2020. [CrossRef]

19. Georgiou, O.; Mimis, K.; Halls, D.; Thompson, W.H.; Gibbins, D. How many wi-fi aps does it take to light a lightbulb? *IEEE Access* **2016**, *4*, 3732–3746. [CrossRef]

20. Wu, X. Reliable Indoor Power Line Communication Systems: Via Application of Advanced Relaying Processing. Ph.D. Thesis, Curtin University, Perth, Australia, 5 November 2016.

21. Galli, S.; Latchman, H.; Oksman, V.; Prasad, G.; Yonge, L. Multimedia PLC systems. In *Power Line Communications: Principles, Standards and Applications from Multimedia to Smart Grid*; John Wiley and Sons: Hoboken, NJ, USA, 2016; pp. 473–508. [CrossRef]

22. Berger, L.T.; Schwager, A.; Pagani, P.; Schneider, D.M. MIMO power line communications. *IEEE Commun. Surv. Tutor.* **2015**, *17*, 106–124. [CrossRef]

23. Ben-Yehezkel, Y.; Gazit, R.; Haidine, A. Performance evaluation of medium access control mechanisms in high-speed narrowband PLC for smart grid applications. In Proceedings of the IEEE International Symposium on Power Line Communications and Its Applications (ISPLC), Beijing, China, 27–30 March 2012; pp. 94–101. [CrossRef]

24. *IEEE Standard for Broadband over Power Line Networks: Medium Access Control and Physical Layer Specifications*; IEEE Std 1901-2020 (Revision of IEEE Std 1901-2010); IEEE: Piscataway, NJ, USA, 2021; pp. 1–1622. [CrossRef]

25. HomePlug Green PHY Specification Release Version 1.1, HomePlug Alliance Standard, January 2012. Available online: http://www.homeplug.org/tech/whitepapers/HomePlug_Green_PHY_whitepaper_121003.pdf (accessed on 1 March 2020).

26. Calderoni, L.; Maio, D.; Rovis, S. Deploying a network of smart cameras for traffic monitoring on a 'city kernel. *Expert Syst. Appl.* **2014**, *41*, 502–507. [CrossRef]

27. Martínez-Rodríguez-Osorio, R.; Calvo-Ramon, M.; Fernández-Otero, M.Á.; Navarrete, L.C. Smart control system for LEDs traffic-lights based on PLC. In Proceedings of the 6th WSEAS International Conference on Power Systems, Lisbon, Portugal, 22–24 September 2006; pp. 256–260.

28. Lighting PLC Applications. Available online: https://www.smart-energy.com/regional-news/europe-uk/street-lighting-problems-and-solutions/ (accessed on 1 March 2020).

29. Litra, G. *The Smart City Opportunity for Utilities*; Technical Reports; Scottmadden Management Consultants: Atlanta, GA, USA, 2017. Available online: https://www.scottmadden.com/insight/the-smart-city-opportunity-for-utilities/ (accessed on 1 January 2020).

30. Salem, A.; Hamdi, K.A.; Alsusa, E. Physical layer security over correlated log-normal cooperative power line communication channels. *IEEE Access* **2017**, *5*, 13909–13921. [CrossRef]

31. Renzo, M.D.; Graziosi, F.; Santucci, F. A comprehensive framework for performance analysis of cooperative multi-hop wireless systems over log-normal fading channels. *IEEE Trans. Commun.* **2010**, *58*, 531–544. [CrossRef]

32. Cheng, X.; Cao, R.; Yang, L. Relay-aided amplify-and-forward powerline communications. *IEEE Trans. Smart Grid* **2013**, *4*, 265–272. [CrossRef]

33. Ezzine, S.; Abdelkefi, F.; Cances, J.P.; Meghdadi, V.; Bouallegue, A. Evaluation of PLC Channel Capacity and ABER Performances for OFDM-Based Two-Hop Relaying Transmission. *Wirel. Commun. Mob. Comput.* **2017**, *2017*, 4827274. [CrossRef]

34. Rabien, K.M.; Adebisi, B.; Salem, A. Improving energy efficiency in dual-hop cooperative plc relaying systems. In Proceedings of the 2016 International Symposium on Power Line Communications and its Applications (ISPLC), Bottrop, Germany, 20–23 March 2016; pp. 196–200. [CrossRef]

35. Rabie, K.M.; Adebisi, B.; Gacanin, H.; Yarkan, S. Energy-per-bit performance analysis of relay-assisted power line communication systems. *IEEE Trans. Green Commun. Netw.* **2018**, *2*, 60–368. [CrossRef]

36. Ramesh, R.; Gurugopinath, S.; Muhaidat, S. Outage performance of relay-assisted noma over power line communications. In Proceedings of the 2020 IEEE 31st Annual International Symposium on Personal, Indoor and Mobile Radio Communications, London, UK, 31 August–3 September 2020; pp. 1–6. [CrossRef]

37. Da Costa, D.; Aissa, S. Performance analysis of relay selection techniques with clustered fixed-gain relays. *IEEE Signal Process. Lett.* **2010**, *17*, 201–204. [CrossRef]

38. Wang, H.; Ma, S.; Ng, T.-S.; Poor, H.V. A general analytical approach for opportunistic cooperative systems with spatially random relays. *IEEE Trans. Wirel. Commun.* **2011**, *10*, 4122–4129. [CrossRef]

39. Nguyen, H.-S.; Nguyen, T.-S.; Voznak, M. Relay selection for SWIPT: Performance analysis of optimization problems and the trade-off between ergodic capacity and energy harvesting. *AEU-Int. J. Electron. Commun.* **2018**, *85*, 59–67. [CrossRef]

40. Riihonen, T.; Werner, S.; Wichman, R. Mitigation of loopback self-interference in full-duplex MIMO relays. *IEEE Trans. Signal Process.* **2011**, *59*, 5983–5993. [CrossRef]

41. Mellios, E.; Goulianos, A.; Dumanli, S.; Hilton, G.; Piechocki, R.; Craddock, I. Off-body channel measurements at 2.4 GHz and 868 MHz in an indoor environment. In Proceedings of the 9th International Conference on Body Area Networks, ICST, London, UK, 29 September–1 October 2014. [CrossRef]

42. Wu, D.; Negi, R. Effective capacity: A wireless link model for support of quality of service. *IEEE Trans. Wirel. Commun.* **2003**, *2*, 630–643. [CrossRef]

43. Zimmermann, M.; Dostert, K. A multipath model for the powerline channel. *IEEE Trans. Commun.* **2002**, *50*, 553–559. [CrossRef]

44. Rabie, K.M.; Adebisi, B.; Tonello, A.M.; Nauryzbayev, G. For more energy-efficient dual-hop DF relaying power-line communication systems. *IEEE Syst. J.* **2018**, *12*, 2005–2016. [CrossRef]

45. Mehta, N.B.; Wu, J.; Molisch, A.F.; Zhang, J. Approximating a sum of random variables with a lognormal. *IEEE Trans. Wirel. Commun.* **2007**, *6*, 2690–2699. [CrossRef]

46. Tutuncuoglu, K.; Ozel, O.; Yener, A.; Ulukus, S. The binary energy harvesting channel with a unit-sized battery. *IEEE Trans. Inf. Theory* **2017**, *63*, 4240–4256. [CrossRef]

47. Gorlatova, M.; Wallwater, A.; Zussman, G. Networking low-power energy harvesting devices: Measurements and algorithms. *IEEE Trans. Mob. Comput.* **2013**, *12*, 1853–1865. [CrossRef]

 *electronics*

Article

Performance Enhancement of Functional Delay and Sum Beamforming for Spherical Microphone Arrays

Yang Zhao [1], **Zhigang Chu** [1,*] **and Linyong Li** [2,3]

[1]   College of Mechanical and Vehicle Engineering, Chongqing University, Chongqing 400044, China; zy970330@cqu.edu.cn
[2]   Key Laboratory of Environmental Protection of Guangdong Power Grid Co., Ltd., Guangzhou 510080, China; lilyong@foxmail.com
[3]   Electric Power Research Institute of Guangdong Power Grid Co., Ltd., Guangzhou 510080, China
*   Correspondence: zgchu@cqu.edu.cn

Citation: Zhao, Y.; Chu, Z.; Li, L. Performance Enhancement of Functional Delay and Sum Beamforming for Spherical Microphone Arrays. *Electronics* **2022**, *11*, 1132. https://doi.org/10.3390/electronics11071132

Academic Editor: João Soares

Received: 3 March 2022
Accepted: 31 March 2022
Published: 2 April 2022

Abstract: Functional delay and sum (FDAS) beamforming for spherical microphone arrays can achieve 360° panoramic acoustic source identification, thus having broad application prospects for identifying interior noise sources. However, its acoustic imaging suffers from severe sidelobe contamination under a low signal-to-noise ratio (SNR), which deteriorates the sound source identification performance. In order to overcome this issue, the cross-spectral matrix (CSM) of the measured sound pressure signal is reconstructed with diagonal reconstruction (DRec), robust principal component analysis (RPCA), and probabilistic factor analysis (PFA). Correspondingly, three enhanced FDAS methods, namely EFDAS-DRec, EFDAS-RPCA, and EFDAS-PFA, are established. Simulations show that the three methods can significantly enhance the sound source identification performance of FDAS under low SNRs. Compared with FDAS at SNR = 0 dB and the number of snapshots = 1000, the average maximum sidelobe levels of EFDAS-DRec, EFDAS-RPCA, and EFDAS-PFA are reduced by 6.4 dB, 21.6 dB, and 53.1 dB, respectively, and the mainlobes of sound sources are shrunk by 43.5%, 69.0%, and 80.0%, respectively. Moreover, when the number of snapshots is sufficient, the three EFDAS methods can improve both the quantification accuracy and the weak source localization capability. Among the three EFDAS methods, EFDAS-DRec has the highest quantification accuracy, and EFDAS-PFA has the best localization ability for weak sources. The effectiveness of the established methods and the correctness of the simulation conclusions are verified by the acoustic source identification experiment in an ordinary room, and the findings provide a more advanced test and analysis tool for noise source identification in low-SNR cabin environments.

Keywords: spherical microphone array; sound source identification; functional delay and sum beamforming; denoising; performance enhancement

1. Introduction

The construction of an environmentally friendly society is one of the development themes in the world today, and many related research works have been carried out in different fields [1,2]. Noise pollution control is also one of the important research contents, and its foremost work is noise source identification. In recent years, beamforming sound source identification technology based on microphone arrays has attracted much attention by virtue of its qualities of fast measurement and high localization accuracy [3]. Planar arrays and spherical arrays are two popular forms of microphone arrays used in beamforming. Planar arrays can identify the sound sources in the half space in front of the array, but their performance in cabin environments is poor due to their inability to suppress the interference sources behind the array. Spherical arrays can panoramically record the sound field information, thus exhibiting broad application prospects for identifying interior noise sources of high-speed trains, automobiles, and airplanes [4,5]. Nevertheless, sound source identification in cabin environments currently faces the problem of low signal-to-noise ratio (SNR) caused by reverberation.

The classical beamforming algorithm that matches the spherical arrays is spherical harmonics beamforming (SHB), which exploits the orthogonality of spherical harmonics to create the beamforming output, with maximum values in the directions of arrival. SHB is computationally efficient; however, it suffers from poor spatial resolution and severe sidelobe contamination [6]. To improve the acoustic source identification performance of beamforming with spherical microphone arrays, filter and sum (FAS) [7], deconvolution [8,9], and functional delay and sum (FDAS) [10,11] are successively proposed. FAS shows computational efficiency comparable to SHB and stronger sidelobe suppression than SHB, but it does not show improved spatial resolution. Deconvolution outperforms SHB in both sidelobe suppression and spatial resolution [12], and it can enhance resistance to noise interference by using the auto-spectrum-removed cross-spectral matrix (CSM) as input, which is at the cost of the computational efficiency or quantification accuracy [13]. FDAS not only maintains low computational complexity, but also improves spatial resolution and sidelobe suppression. However, FDAS is less resistant to noise interference, and the amplitude of sidelobes and the width of mainlobes increase again in the presence of strong noise interference. FDAS must take the complete CSM as input, so the noise interference cannot be suppressed by removing the auto-spectral elements. In view of this, it is meaningful to investigate noise reduction methods that are suitable for FDAS, which can help to enhance the sound source identification performance of FDAS in low-SNR cabin environments.

In recent years, the denoising methods based on the reconstructed CSM have been proposed, which can suppress noise interference while maintaining the completeness of the CSM. The CSM reconstruction methods, such as diagonal reconstruction (DRec), robust principal component analysis (RPCA), and probabilistic factor analysis (PFA), have been extensively studied in planar microphone arrays [14,15], but their applications in spherical arrays have not been reported. DRec, reported by Dougherty [16], minimizes the sum of the diagonal elements of the measured CSM under the constraint that the denoised CSM is positive semidefinite, and is solved by linear programming. Next, Hald et al. [17] applied the CVX toolbox in MATLAB to DRec to accelerate the solving process and analyzed the effect of snapshot number on noise reduction. RPCA, an early image processing method, was introduced into CSM denoising by Dinsenmeyer et al. [18]; this method exploits the low-rankness of the signal matrix and the sparsity of the noise matrix to reconstruct the signal matrix. The noise reduction of RPCA is better than DRec, but it is influenced by the regularization parameter. To overcome this problem, in the same literature, Dinsenmeyer et al. proposed the PFA based on a probabilistic framework [18]. No regularization parameter is required for PFA to obtain excellent noise reduction. In the literature [19], a benchmarking of the above methods was performed using planar array-based beamforming, and the result showed that PFA and RPCA provided the best noise reduction, significantly better than DRec. Since the above CSM reconstruction methods can maintain the completeness of CSM while suppressing the noise interference, they are particularly suitable for the noise reduction of spherical array-based FDAS.

In this paper, DRec, RPCA, and PFA are introduced into the spherical array-based FDAS to improve its sound source identification performance in low-SNR cabin environments, that is, to suppress sidelobes, shrink mainlobes, and maintain the strong identification ability for weak sources. Correspondingly, three enhancement methods, namely EFDAS-DRec, EFDAS-RPCA, and EFDAS-PFA, are established. The remainder of this paper is organized as follows: In Section 2, the overview of DRec, RPCA, and PFA, as well as the theory of EFDAS, are described. Next, in Section 3, the sound source identification performance of the three EFDAS methods is numerically simulated and compared. Subsequently, in Section 4, the simulation results are verified using an acoustic experiment in an ordinary room. Finally, the conclusions are drawn in Section 5.

2. Theory of EFDAS

FDAS beamforming employs the pressure signals collected by the spherical microphone array to construct an output function, thus identifying the sound sources. The function strengthens the output in the directions of the sound sources and suppresses the output in the directions of non-sources. EFDAS enhances the anti-noise interference capability of FDAS by denoising the measured CSM. In this section, the signal model of solid spherical microphone arrays, the denoised CSM, and the output of EFDAS are described.

2.1. Signal Model of Solid Spherical Microphone Arrays

Figure 1 depicts the acoustic pressure signal model of a solid spherical microphone array and the corresponding coordinate system with the origin at the array center. An arbitrary point in 3D space can be described with the spherical coordinate (r, Ω), where r denotes the distance from the point to the origin, $\Omega \equiv (\theta, \varphi)$ represents the direction of the point with θ, and φ being the elevation and azimuth angles, respectively. The symbols "●" indicate the microphones arranged on the surface of the solid spherical array, and the coordinate of the qth microphone is $(a, \Omega_{Mq})(q = 1, 2, \ldots, Q)$, where a is the array radius, Ω_{Mq} denotes the direction of the qth microphone, and Q is the total number of microphones. The symbols "★" indicate sound sources, and the coordinate of the ith source is $(r_{Si}, \Omega_{Si})(i = 1, 2, \ldots, I)$, where r_{Si} denotes the distance from the ith source to the array center, Ω_{Si} denotes the direction of the ith source, and I is the total number of sound sources. The corresponding source intensity of the ith source is defined as $s(r_{Si}, \Omega_{Si})$. The source intensity vector constructed by all sound sources is formulated as $s = [s(r_{S1}, \Omega_{S1})], \ldots, s(r_{SI}, \Omega_{SI})^{\mathrm{T}} \in \mathbb{C}^{I \times 1}$, where the superscript "T" represents the transposition.

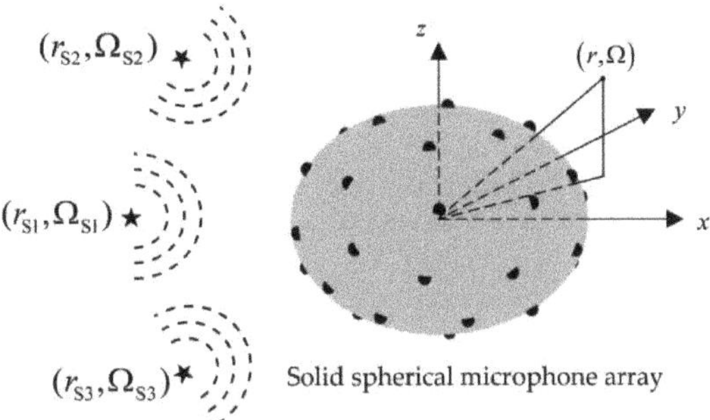

Figure 1. Sound pressure signal model of a solid spherical array.

The sound pressure vector $p^{\star} \in \mathbb{C}^{Q \times 1}$ measured by the array can be written as

$$p^{\star} = p_A + n = Hs + n \tag{1}$$

where $p_A \in \mathbb{C}^{Q \times 1}$ is the sound pressure vector generated by the sound sources, $n \in \mathbb{C}^{Q \times 1}$ is the independent stochastic noise, and $H \in \mathbb{C}^{Q \times I}$ is the propagation matrix from I sources to Q microphones. Here, SNR is defined as SNR $= 20 \log_{10}(\|p_A\|_2 / \|n\|_2)$, where "$\| \cdot \|_2$" denotes the 2 norm. In H, the element $H_{q,i}$ represents the sound pressure generated at the qth microphone by a unit intensity sound source located at the ith source and is formulated as

$$H_{q,i} = \sum_{n=0}^{\infty} \sum_{m=-n}^{n} R_n(kr_{Si}, ka)(Y_n^m(\Omega_{Si}))^* Y_n^m(\Omega_{Mq}) \tag{2}$$

where $Y_n^m(\Omega_{Si})$ and $Y_n^m(\Omega_{Mq})$ denote spherical harmonics in the directions of Ω_{Si} and Ω_{Mq}, respectively, both n and m are the degrees of spherical harmonics, the superscript "*" denotes the conjugation, $k = 2\pi f / c$ is the wavenumber corresponding to frequency f and the speed of sound c, and $R_n(kr_{Si}, ka)$ is the radial function. For spherical wave, $R_n(kr_{Si}, ka)$ is expressed as [7,8]

$$R_n(kr_{Si}, ka) = -4\pi j h_n^{(2)}(kr_{Si})\left(j_n(ka) - \frac{j_n'(ka)h_n^{(2)}(ka)}{h_n^{(2)'}(ka)}\right) \tag{3}$$

where $h_n^{(2)}(\cdot)$ and $h_n^{(2)'}(\cdot)$ are the spherical Hankel function of the second kind and its derivative, respectively, $j_n(\cdot)$ and $j_n'(\cdot)$ are the spherical Bessel function of the first kind and its derivative, respectively, and $j = \sqrt{-1}$ is the imaginary unit.

The measured sound pressure is further processed with FFT according to the Welch method [20], and the averaged CSM $C \in \mathbb{C}^{Q \times Q}$ over multiple snapshots is obtained as follows

$$C = \mathbb{E}(p^{\star}p^{\star H}) = \mathbb{E}(p_A p_A^H) + \mathbb{E}(nn^H) + \mathbb{E}(p_A n^H + np_A^H) \tag{4}$$

where "$\mathbb{E}(\cdot)$" is the expectation operator and the superscript "H" represents the Hermitian transpose. The averaged CSM consists of the signal CSM $C_A = \mathbb{E}(p_A p_A^H) \in \mathbb{C}^{Q \times Q}$, the noise CSM $C_N = \mathbb{E}(nn^H) \in \mathbb{C}^{Q \times Q}$, and additional crossed terms. Since noise signal is considered stochastic and independent of the source signals, the off-diagonal elements of C_N and the crossed terms tend to zero as the number of snapshots tends to infinity. Therefore, the noise is mainly concentrated on the diagonal of C in the case of enough snapshots.

2.2. Reconstruction of the Measured CSM

Since FDAS requires the complete CSM C as input, its performance is affected by the noise on the diagonal of C, resulting in a significant degradation of acoustic imaging clarity in the presence of strong noise interference. In this subsection, the noise on the diagonal of C is removed by reconstructing the measured CSM with DRec, RPCA, and PFA, thus promoting the anti-noise interference capability of FDAS.

2.2.1. DRec

As explained previously, with a sufficient number of snapshots engaged in the averaging, the noise is mainly concentrated on the diagonal of C. The idea of DRec is to remove the noise auto-spectrum from the diagonal of C as much as possible, i.e., minimizing the sum of auto-spectrum in C while maintaining the denoised CSM positive semidefinite. Therefore, a diagonal matrix with unknown non-positive diagonal elements d_q is added to C to cancel the noise auto-spectrum. The unknown diagonal elements d_q are arranged in the vector $d \in \mathbb{R}^{Q \times 1}$, and then the solution problem of DRec can be formulated as [17]

$$\min \left(\sum_{q=1}^{Q} d_q\right) \text{ subject to } C + \text{Diag}(d) \succcurlyeq 0 \tag{5}$$

where $\text{Diag}(d) \in \mathbb{R}^{Q \times Q}$ represents a diagonal matrix whose diagonal entries are the elements in vector d, and the symbol "$\succcurlyeq 0$" represents positive semidefinite for a matrix. Equation (5) can be quickly solved using the CVX toolbox in MATLAB, and thus the denoised CSM $\hat{C}_A \in \mathbb{C}^{Q \times Q}$ can be estimated as

$$\hat{C}_A = C + \text{Diag}(d) \tag{6}$$

2.2.2. RPCA

RPCA utilizes the low-rankness of the signal CSM and the sparsity of the noise CSM to denoise the measured CSM. The rank of the signal CSM is determined by the number of incoherent sources. Considering that the number of incoherent main sources to be identified is usually smaller than the number of microphones, the signal CSM can be assumed to have low rank. Additionally, the noise CSM can be considered as a sparse diagonal matrix since the off-diagonal elements of the noise CSM tend to zero after multiple averaging. RPCA aims at recovering the low-rank signal matrix from noise-contaminated measurements, and its optimization problem can be expressed as [18]

$$\min \|C_A\|_* + \lambda \|\text{vec}(C - C_A)\|_1 \tag{7}$$

where "$\|\cdot\|_*$" and "$\|\cdot\|_1$" denote the nuclear norm (sum of the eigenvalues) and the 1 norm respectively, "vec" stands for converting a matrix into a vector, and λ is the regularization parameter. Wright et al. [21] recommended $\lambda = Q^{-0.5}$ and employed an accelerated proximal gradient algorithm to solve the above optimization problem, resulting in the denoised CSM being estimated as $\hat{C}_A$. In our case, with $Q = 36$, $\lambda = 0.167$ is adopted.

2.2.3. PFA

PFA is a probabilistic framework-based inference method, whose goal is to fit the measurements with the following model

$$p_j^{\star} = Ls_{Ej} + n_j \,, \, j = 1, \ldots, N_s \tag{8}$$

where $j = 1, \ldots, N_s$ is the serial number of snapshots, $s_E \in \mathbb{C}^{\kappa \times 1}$ is the source intensity vector of $\kappa(\kappa < Q)$ equivalent sound sources, and $L \in \mathbb{C}^{Q \times \kappa}$ is a mixing matrix. $S_E \in \mathbb{C}^{\kappa \times \kappa}$ is defined as the CSM of s_E. The number of equivalent sound sources cannot be less than the number of true sound sources; otherwise, it is difficult to describe the sound field comprehensively. Since PFA still takes advantage of the fact that the noise is mainly concentrated on the diagonal of CSM after multiple averaging, the measured CSM is further modeled as follows:

$$C = \frac{1}{N_s} \sum_{j=1}^{N_s} (Ls_{Ej} + n_j)(Ls_{Ej} + n_j)^{\mathrm{H}} \approx L(\frac{1}{N_s} \sum_{j=1}^{N_s} s_{Ej}s_{Ej}^{\mathrm{H}})L^{\mathrm{H}} + \frac{1}{N_s} \sum_{j=1}^{N_s} n_j n_j^{\mathrm{H}} = LS_E L^{\mathrm{H}} + C_N \tag{9}$$

In Equation (9), except that the measured CSM C is known, L, s_E, and n are considered as random variables. The prior probability density functions (PDFs) assigned to the unknown variables are illustrated in Table 1.

Table 1. Prior PDFs assigned to the unknown variables of PFA.

Priors	Hyper-Priors
$s_E \sim \mathcal{CN}(0, \gamma^2 I_\kappa)$	$\gamma^2 \sim \mathcal{IG}(a_\gamma, b_\gamma)$
$n \sim \mathcal{CN}(0, \text{Diag}(\sigma^2))$	$\sigma^2 \sim \mathcal{IG}(a_\sigma, b_\sigma)$
$L \sim \mathcal{CN}(0, I_{Q\kappa}/\kappa)$	

$\mathcal{CN}$ and $\mathcal{IG}$ denote the complex Gaussian PDF and the inverse gamma PDF (pair of conjugate distribution), respectively, γ^2 and $\sigma^2 \in \mathbb{R}^{Q \times 1}$ are the variance of equivalent sources and noise, $I_\kappa \in \mathbb{R}^{\kappa \times \kappa}$ and $I_{Q\kappa} \in \mathbb{R}^{Q\kappa \times Q\kappa}$ are two identity matrices with different dimensions, and the remaining hyperparameters ($a_\gamma = b_\gamma = 0.001$ and $a_\sigma = b_\sigma = \mathbf{0.001} \in \mathbb{R}^{Q \times 1}$) are constant [19].

In order to solve the fitting problem in Equation (9), the maximum a posteriori estimate of each unknown variable is established as follows [18]

$$(\hat{S}_E, \hat{L}, \hat{\sigma}^2, \hat{\gamma}^2) = \text{argmax}\,(S_E, L, \sigma^2, \gamma^2 | C) \tag{10}$$

where "$(\cdot | \cdot)$" denotes conditional PDF. Here, the Gibbs sampler [22], a Markov Chain Monte Carlo algorithm, is used to find the above estimate, and then the denoised CSM can be expressed as

$$\hat{C}_A = \hat{L}\hat{S}_E\hat{L}^{\mathrm{H}} \tag{11}$$

2.3. The Output of EFDAS

The output of EFDAS is constructed based on the denoised CSM $\hat{C}_A$. First, the eigenvalue decomposition of the denoised CSM is performed.

$$\hat{C}_A = U\Lambda U^H \tag{12}$$

where $\Lambda = \mathrm{Diag}([\beta_1,\dots,\beta_Q]) \in \mathbb{R}^{Q\times Q}$ is a diagonal matrix, $\beta_1,\dots,\beta_Q$ are the eigenvalues of $\hat{C}_A$ and satisfy $\beta_1 \geq \dots \geq \beta_Q \geq 0$, $U = [u_1,\dots,u_Q] \in \mathbb{C}^{Q\times Q}$ is an unitary matrix with the column u_q being the eigenvector that corresponds to β_q. Subsequently, the exponential function of $\hat{C}_A$ is constructed as follows

$$\hat{C}_A^{1/\xi} = U\mathrm{Diag}\left(\beta_1^{1/\xi},\dots,\beta_Q^{1/\xi}\right)U^H \tag{13}$$

where ξ is an exponent parameter. According to the literature [11], when $\xi = 16$, FDAS has strong sidelobe suppression and high quantization accuracy. In view of this, $\xi = 16$ is also recommended in EFDAS. Finally, the average pressure contribution of the focus point (r_f, Ω_f) [11], the arithmetic average of the microphone auto-spectrum generated by the source located at the focus point (r_f, Ω_f), is taken as the output of EFDAS.

$$W(kr_f,\Omega_f) = \frac{1}{Q}\left(\frac{v^H(kr_f,\Omega_f)\hat{C}_A^{1/\xi}v(kr_f,\Omega_f)}{\|v(kr_f,\Omega_f)\|_2^2}\right)^{\xi} \tag{14}$$

where $v(kr_f,\Omega_f) \in \mathbb{C}^{Q\times 1}$ is the focus column vector of the focus point (r_f, Ω_f). In $v(kr_f,\Omega_f)$, the element $v_q(kr_f,\Omega_f)$ represents the sound pressure generated at the qth microphone by a unit intensity sound source located at the focus point (r_f, Ω_f), and it is defined as

$$v_q(kr_f,\Omega_f) = \sum_{n=0}^{\infty}\sum_{m=-n}^{n} R_n(kr_f,\Omega_f)Y_n^{m*}(\Omega_f)Y_n^{m}(\Omega_q) \tag{15}$$

Equations (2) and (15) are similar, but are used for forward sound field simulation and inverse focus of beamforming, respectively, which both require degree truncation of spherical harmonics [23], i.e., $\infty = 50$ in Equation (2) and $\infty = 17$ in Equation (15). In this case, EFDAS satisfies the high accuracy of sound source identification while maintaining a low computational cost.

3. Simulations

The simulations are carried out with a 36-channel solid spherical microphone array with a radius of 0.0975 m, as shown in Figure 1, which is also used in the subsequent experiment. The simulation steps are as follows: (1) Establish the array model and set the frequencies, intensities, and locations of the incoherent monopole sound sources to be identified. (2) Calculate the theoretical sound pressure p_A emitted by the sound sources, add the noise n according to the SNR setting, and then obtain the total measured pressure $p^{\star}$ and the corresponding CSM C. (3) Reconstruct the measured CSM C with DRec, RPCA, and PFA to acquire the denoised CSM $\hat{C}_A$. (4) Arrange the focus points on a spherical surface 1 m away from the array center with the spacing $\Delta\theta = 5°$ and $\Delta\varphi = 5°$, respectively, thus there creating a total of 37×72 focus points. (5) Calculate the EFDAS output of each focus point according to Equation (14), and thus perform the acoustic imaging.

3.1. Acoustic Imaging Colormaps

Acoustic imaging colormaps can intuitively reflect the localization and quantification of sound sources. In this subsection, acoustic imagings of FDAS and EFDAS are compared to demonstrate the performance enhancement of EFDAS. The maximum outputs of the mainlobes for true sources are labeled in the acoustic imaging maps, which are scaled to dB by referring to 2×10^{-5} Pa. Five sound sources with the coordinates (r, θ, φ) of (1 m, 90°, 180°), (1 m, 130°, 110°), (1 m, 50°, 110°), (1 m, 50°, 250°), and (1 m, 130°, 250°) are simulated, and their theoretical average sound pressure contributions are 94 dB, 91 dB, 88 dB, 85 dB, and 79 dB, respectively. Meanwhile, the display dynamic range of each map is set to 30 dB.

As shown in Figure 2, the ideal acoustic imaging colormaps of FDAS without noise interference are presented, with the columns from left to right corresponding to 1000 Hz, 3000 Hz, and 5000 Hz, in that order. It is clear that all sound sources are accurately located and quantified, which also proves the strong identification capability of FDAS for weak sources in the absence of noise interference.

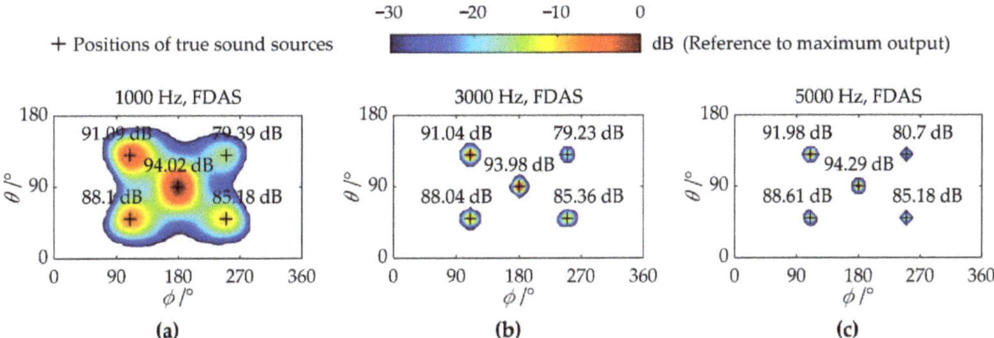

Figure 2. Colormaps of FDAS without noise interference: (**a**) 1000 Hz; (**b**) 3000 Hz; (**c**) 5000 Hz.

Figure 3 shows the imaging colormaps of FDAS at SNR = 0 dB and discloses the influence of the number of snapshots. In Figure 3a–c, 10 snapshots are considered. Although there is no sidelobe contamination, the quantification errors of the sound sources are large, and multiple weak sources get lost. As shown in Figure 3d–f, the quantification errors markedly reduce when the number of snapshots increases to 100, but the imaging results suffer from severe sidelobe contamination. For the identification of weak sources, at 3000 Hz and 5000 Hz, since the amplitude of the weakest source is comparable to the sidelobes, the identification results are less reliable, and at 1000 Hz, there is no indication of the weakest source. The further increase in the number of snapshots fails to improve the imaging results, as illustrated in Figure 3g–i. The reasons why the number of snapshots has such an impact on the results of FDAS are that: (1) The measured CSM, which is the input of FDAS, requires a sufficient number of snapshots for averaging to correctly reflect the actual coherence between sound sources. (2) The insufficient number of snapshots causes part of the noise to remain on the off-diagonal elements of CSM. The residual noise power is equivalent to the sound sources and works with the true sources; hence, the actual output of FDAS deviates from the ideal output. As the number of snapshots increases, both the coherence between sound sources and the noise residual on the off-diagonal of CSM gradually decrease, allowing FDAS to locate and quantify sound sources with higher accuracy.

Figure 4 shows the imaging colormaps of three EFDAS methods at SNR = 0 dB. The number of snapshots is set to 1000 (same with Figure 3g–i). In Figure 4, the columns from left to right are 1000 Hz, 3000 Hz, and 5000 Hz, and the rows from top to bottom correspond to EFDAS-DRec, EFDAS-RPCA, and EFDAS-PFA, respectively. Compared with Figure 3g–i, the sidelobes are much lower and the mainlobes are narrower in Figure 4a–c, which is attributed to the noise reduction with DRec. Moreover, all sources are identified at 3000 Hz and 5000 Hz, while at 1000 Hz, the weakest source gets lost and the mainlobes of strong sources are also seriously fused. Figure 4d–i shows that with the noise removal by RPCA and PFA, the sidelobes are eliminated in the display dynamic range, and the width of mainlobes is significantly decreased, allowing for clearer imaging and higher spatial resolution. Both EFDAS-RPCA and EFDAS-PFA can accurately locate all sources, and EFDAS-RPCA outperforms EFDAS-PFA in quantification accuracy for the weakest source (15 dB less than the strongest source).

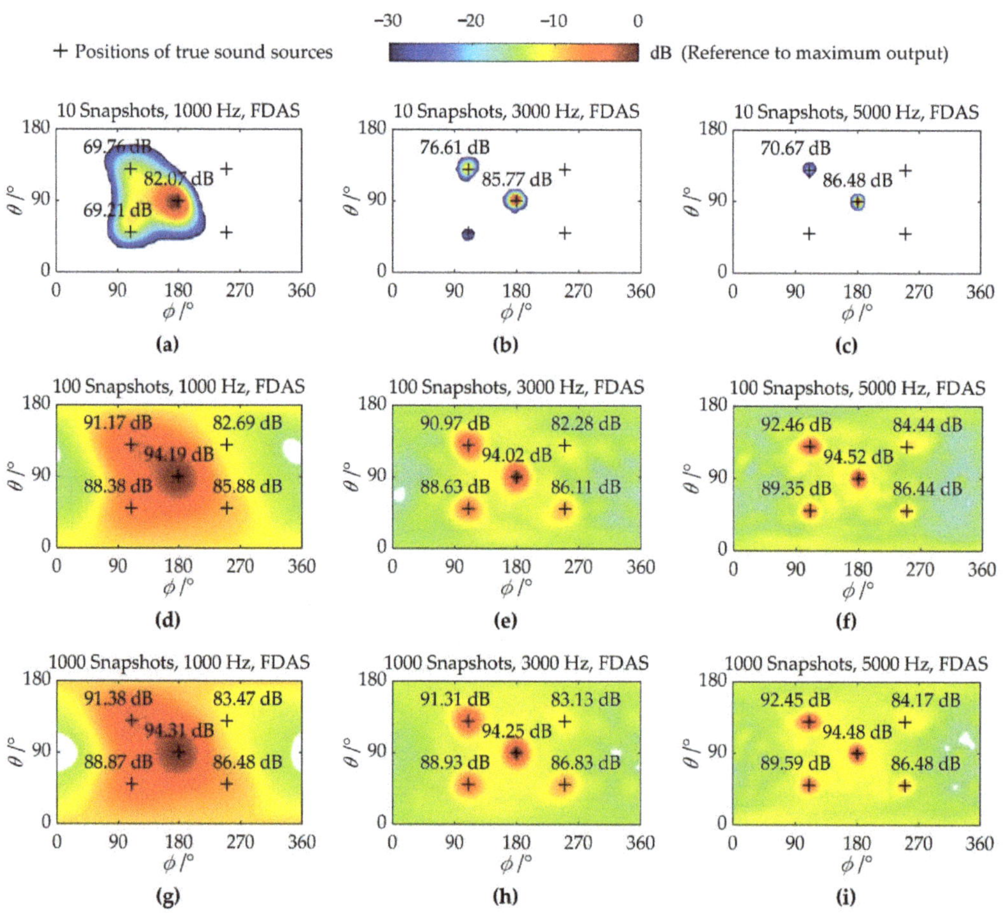

Figure 3. Colormaps of FDAS with SNR = 0 dB and different snapshots: (**a–c**) Maps of FDAS with 10 snapshots at 1000 Hz, 3000 Hz, and 5000 Hz, respectively; (**d–f**) Maps of FDAS with 100 snapshots at 1000 Hz, 3000 Hz, and 5000 Hz, respectively; (**g–i**) Maps of FDAS with 1000 snapshots at 1000 Hz, 3000 Hz, and 5000 Hz, respectively.

The acoustic imaging performance of EFDAS is summarized as follows: (1) EFDAS-RPCA and EFDAS-PFA can eliminate sidelobes in the display dynamic range, while EFDA-DRec still suffers from sidelobe contamination, despite a noticeable reduction in sidelobe amplitudes. (2) EFDAS-RPCA and EFDAS-PFA obviously improve spatial resolution by narrowing the mainlobes, while EFDAS-DRec only slightly enhances spatial resolution. (3) The three EFDAS methods enjoy a strong identification ability for weak sources, as they can accurately locate and quantify the sources that are 9 dB lower than the strongest source at SNR = 0 dB. Even for the sources that are 15 dB lower than the strongest source, the EFDAS methods are still able to locate them accurately, in spite of large quantification errors.

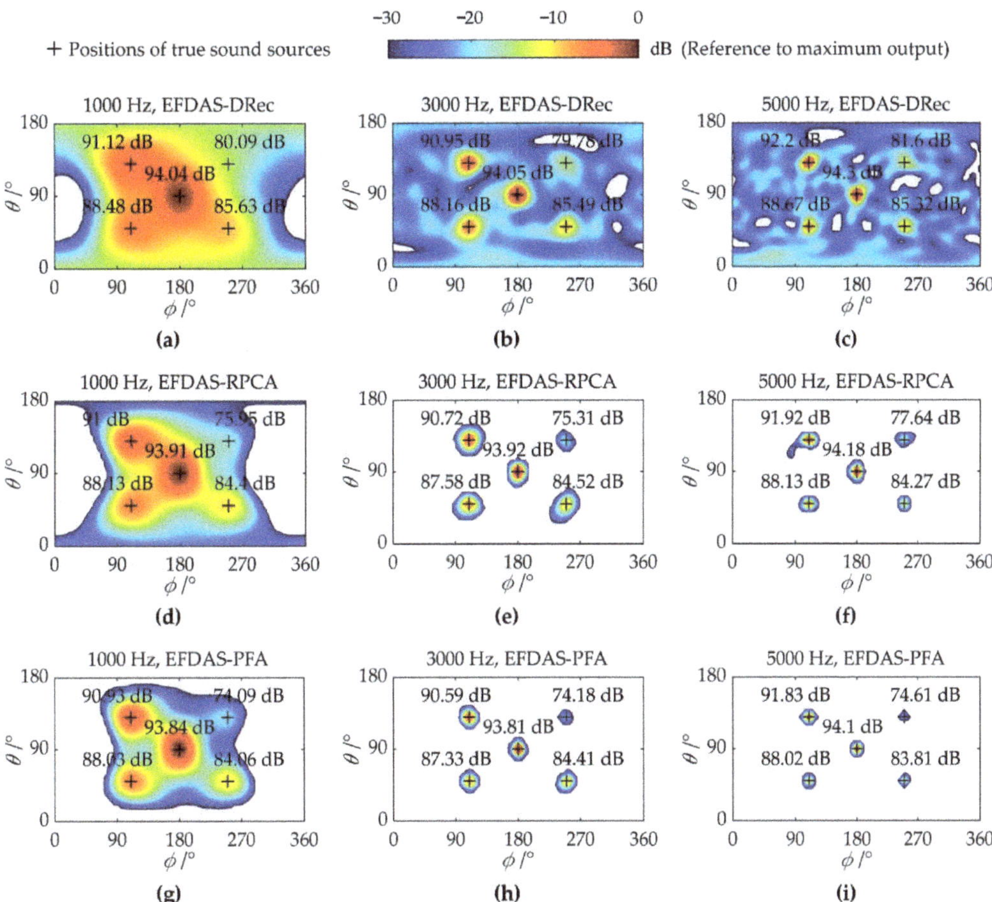

Figure 4. Colormaps of EFDAS with SNR = 0 dB and 1000 snapshots: (**a–c**) Maps of EFDAS-DRec at 1000 Hz, 3000 Hz, and 5000 Hz, respectively; (**d–f**) Maps of EFDAS-RPCA at 1000 Hz, 3000 Hz, and 5000 Hz, respectively; (**g–i**) Maps of EFDAS-PFA at 1000 Hz, 3000 Hz, and 5000 Hz, respectively.

3.2. Performance Analysis

Furthermore, the effects of SNR and number of snapshots on CSM diagonal reconstruction error and the acoustic imaging accuracy are explored. The smaller diagonal reconstruction error of the signal CSM indicates the better noise reduction, and the higher acoustic imaging accuracy indicates the more accurate localization and quantification for sound sources. In this subsection, five sound sources of unequal intensity at 3000 Hz are adopted, which is consistent with those in Figure 2b.

3.2.1. Diagonal Reconstruction Error of Signal CSM

The relative error of diagonal elements between the denoised CSM $\hat{C}_A$ and the noise-free CSM C_A is defined as the diagonal reconstruction error δ_{diag}.

$$\delta_{\mathrm{diag}} = 10 \log_{10} \frac{\|\mathrm{diag}(C_A) - \mathrm{diag}(\hat{C}_A)\|_2}{\|\mathrm{diag}(C_A)\|_2} \tag{16}$$

where "diag($\cdot$)" denotes the extraction of the diagonal elements of a matrix to construct a new vector.

The diagonal reconstruction errors of DRec, RPCA, and PFA are analyzed under different SNRs (from -10 to 10 dB with 2 dB steps) and different numbers of snapshots (100×2^u, $u = 0, 1, \ldots, 7$), respectively. Figure 5a shows the curves of δ_{diag} vs. SNR when 1000 snapshots are considered, and Figure 5b shows the curves of δ_{diag} vs. the number of snapshots at SNR = 0 dB. The curves for other combinations of SNR and the number of snapshots are similar to those in Figure 5. As can be seen in Figure 5, the diagonal reconstruction errors of DRec, RPCA, and PFA increase as SNR drops and decrease as the number of snapshots rises, indicating that more snapshots are required to maintain small diagonal reconstruction errors under lower SNRs. Moreover, the diagonal reconstruction errors of RPCA and PFA are comparable and much smaller than those of DRec, which is in general agreement with the findings in the literature [19].

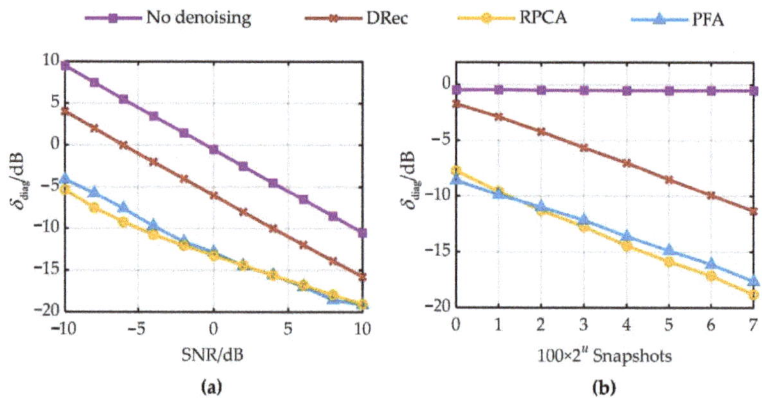

Figure 5. Diagonal reconstruction errors under different SNRs and different numbers of snapshots: (a) different SNRs with 1000 snapshots; (b) different numbers of snapshots with SNR = 0 dB.

3.2.2. Acoustic Imaging Accuracy

The similarity between two vectors can be evaluated by the cosine of their angle. Therefore, define the similarity coefficient ℓ to describe the similarity between the vectorized actual and theoretical beamforming output.

$$\ell = |\langle \boldsymbol{b}_{\text{F}} \cdot \bar{\boldsymbol{b}}_{\text{F}} \rangle| / (\|\boldsymbol{b}_{\text{F}}\|_2 \|\bar{\boldsymbol{b}}_{\text{F}}\|_2), \quad 0 \leq \ell \leq 1 \tag{17}$$

where $\boldsymbol{b}_{\text{F}} = [b_{\text{F}}]$ represents the actual beamforming output vector, $\bar{\boldsymbol{b}}_{\text{F}} = [\bar{b}_{\text{F}}]$ represents the theoretical beamforming output vector with theoretical values at the true source positions and 0 at other positions, and the symbol "$\langle \cdot \rangle$" denotes inner product operator. Define the difference coefficient as $\xi = 1 - \ell$ ($0 \leq \xi \leq 1$). The closer the difference coefficient ξ is to 0, the more similar the beamforming output vector is to the theoretical output vector. When sources are accurately localized, the smaller difference coefficient reflects the fewer sidelobes and the narrower mainlobes, i.e., the better imaging clarity. Since the difference coefficient ξ does not reflect quantification error, we further define the average quantification error η as

$$\eta = \frac{1}{I} \sum_{i=1}^{I} \left| 10 \log_{10} |b_{\text{F}i}| - 10 \log_{10} \left| \bar{b}_{\text{F}i} \right| \right|, \quad \eta \geq 0 \tag{18}$$

where $b_{\text{F}i}$ and $\bar{b}_{\text{F}i}$ are the actual and the theoretical output in the direction of the ith sound source, respectively. When the difference coefficient ξ and the average quantification error η are close to zero, the actual beamforming output is close to the theoretical beamforming output. In that case, the actual acoustic imaging is close to the noise-free acoustic imaging, indicating the high acoustic imaging accuracy, i.e., great imaging clarity and accurate quantification.

Figure 6 shows the histograms of ζ and η vs. SNR and the number of snapshots, processed by FDAS, EFDAS-DRec, EFDAS-RPCA, and EFDAS-PFA, respectively. In general, as the number of snapshots and SNR increase, the difference coefficients and the average quantification errors of each EFDAS method gradually decrease, and the sound source identification performance gradually improves. Compared with FDAS, the difference coefficients of all EFDAS methods decrease. EFDAS-PFA has the smallest difference coefficient, followed by EFDAS-RPCA, and finally EFDAS-DRec, indicating that EFDAS-PFA has the strongest capability to suppress sidelobe contamination and reduce the width of mainlobe contamination, and thus enjoys the best imaging clarity. Moreover, EFDAS-DRec has the highest quantification accuracy and maintains a small quantification error even at SNR = −10 dB. EFDAS-RPCA and EFDAS-PFA show large quantification errors at the small number of snapshots and low SNRs. For the same SNR, EFDAS-PFA requires more snapshots to achieve quantification accuracy comparable to EFDAS-RPCA.

3.3. Improvements of Acoustic Imaging Performance

Acoustic imaging accuracy describes the imaging quality in terms of both the difference coefficient and the average quantification error. Generally, the smaller the difference coefficient and the average quantification error, the better the imaging quality. However, the difference coefficient cannot separately describe the sidelobe suppression and the mainlobe shrinkage, and the average quantification error in Figure 6 cannot separately describe the quantification accuracy of strong and weak sources. Therefore, the maximum sidelobe level (MSL) is adopted to describe the amplitude of the sidelobes, and the number of grids in the mainlobe (NGM) within the 6 dB display dynamic range is defined to describe the size of mainlobe. additionally, the quantification and localization of weak sources are analyzed in this subsection.

3.3.1. Sidelobe Suppression and Mainlobe Reduction

MSL and NGM are simulated with a monopole source at different frequencies (from 1000 to 6000 Hz, with 500 Hz steps). A total of 20 Monte Carlo simulations are performed at each frequency, and in each simulation, the monopole source with an average sound pressure contribution of 94 dB is randomly arranged in a focus point. SNR = 0 dB and the number of snapshots of 1000 are set. Figure 7a shows the curves of MSL vs. frequency, which indicate that the MSL of EFDAS-PFA is the smallest at all frequencies, followed by EFDAS-RPCA, then EFDAS-DRec, and finally FDAS. Compared to FDAS, the average MSL of all frequencies is reduced by 6.4 dB for EFDAS-DRec, 21.6 dB for EFDAS-RPCA, and 53.1 dB for EFDAS-PFA. Figure 7b shows the curves of NGM vs. frequency, which indicate that the NGM of EFDAS-PFA is the smallest at all frequencies, followed by EFDAS-RPCA, then EFDAS-DRec, and finally FDAS. The shrinkage of the mainlobe for the three EFDAS methods is particularly noticeable at low frequencies. Compared to FDAS, the average NGM of all frequencies is reduced by 43.5% for EFDAS-DRec, 69.0% for EFDAS-RPCA, and 80.0% for EFDAS-PFA. Therefore, all three EFDAS methods can effectively suppress sidelobes and shrink the mainlobe, thus improving the imaging clarity. Among the three enhancement methods, EFDAS-PFA enjoys the best imaging clarity.

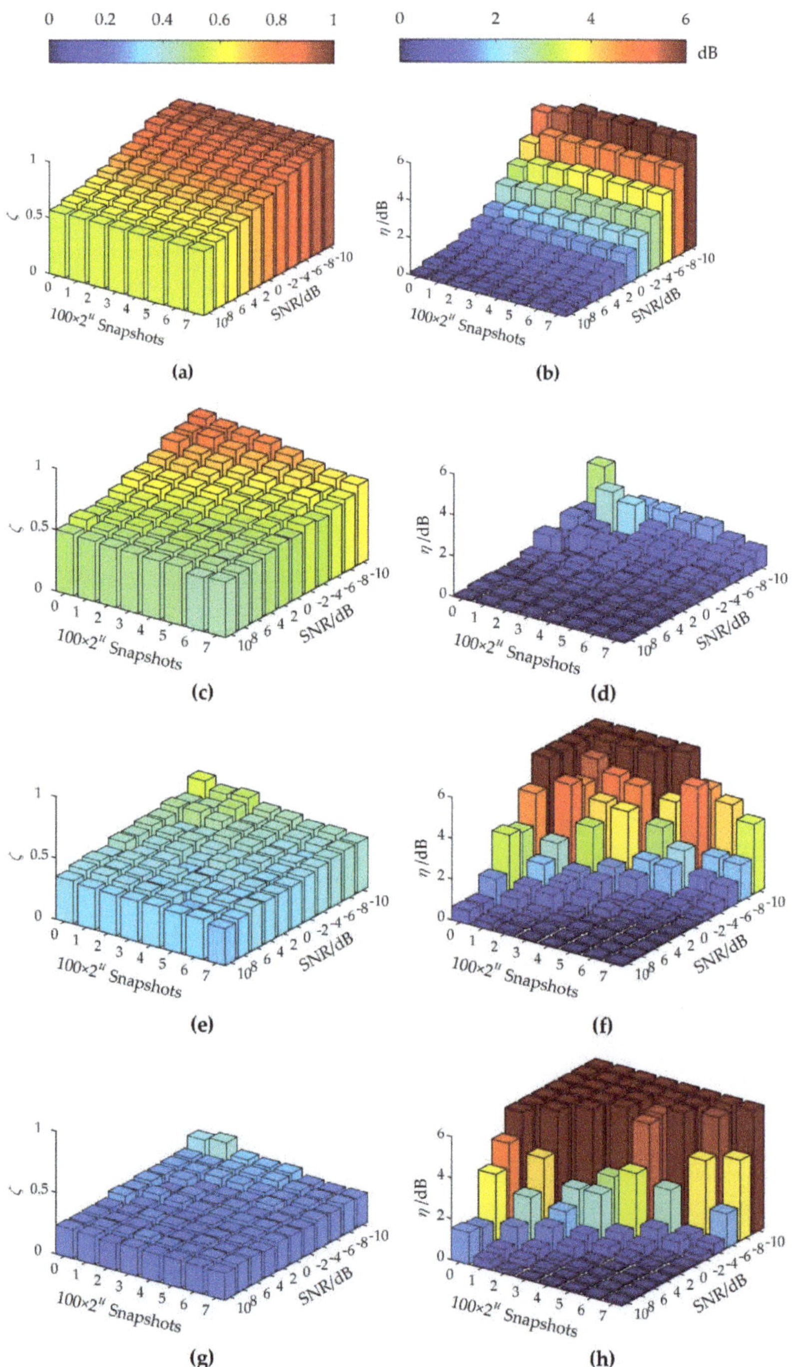

Figure 6. Acoustic imaging errors: (**a,c,e,g**) The difference coefficients of FDAS, EFDAS-DRec, EFDAS-RPCA, and EFDAS-PFA, respectively; (**b,d,f,h**) The average quantification errors of FDAS, EFDAS-DRec, EFDAS-RPCA, and EFDAS-PFA, respectively.

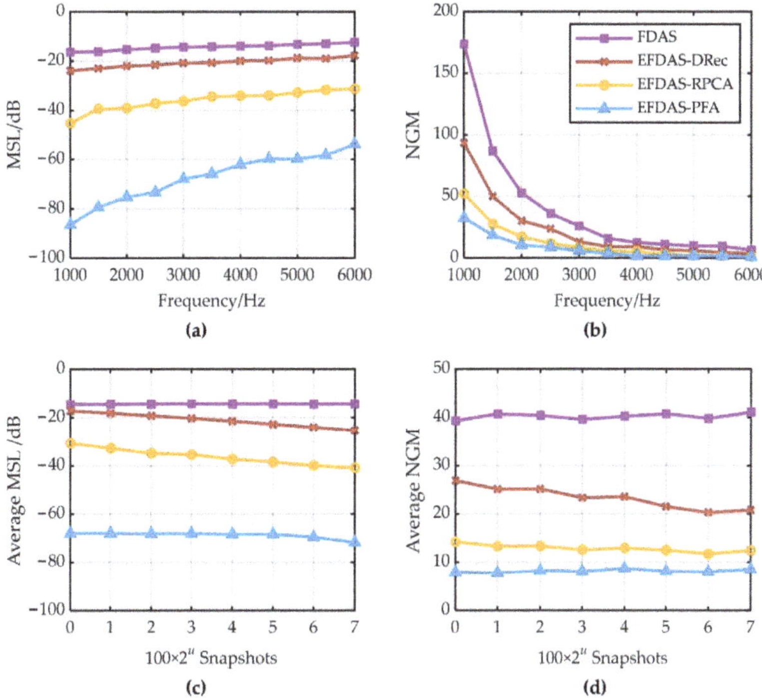

Figure 7. MSL and NGM of FDAS and EFDAS at SNR = 0 dB: (**a**,**b**) MSL and NGM under different frequencies; (**c**,**d**) Average MSL and NGM under different numbers of snapshots.

The effects of the number of snapshots (100×2^{u}, $u = 0, 1, \ldots ,7$) on MSL and NGM averaged over multiple frequencies (from 1000 to 6000 Hz, with 500 Hz steps) are further analyzed. Figure 7c,d shows the curves of the average MSL and the average NGM vs. the number of snapshots. It can be seen that the average MSL and average NGM of EFDAS-DRec and EFDAS-RPCA decrease with the increase in the number of snapshots, while those of FDAS and EFDAS-PFA are basically stable. The simulation results show that when the number of snapshots continues to increase from 100, it still improves the imaging clarity of EFDAS-DRec and EFDAS-RPCA, but has little effect on that of FDAS and EFDAS-PFA.

3.3.2. Quantification and Localization of Weak Sources

The excellent ability to identify weak sources is one of the advantages of FDAS; however, it is affected by noise interference. Although EFDAS is able to suppress noise by reconstructing the CSM, it leads to an underestimation of the intensity of the weak sources, which can be seen in Figure 4. In this subsection, the quantification and localization of weak sources are analyzed.

In Figure 8, five sound sources of varying intensity are simulated at SNR = 0 dB. The five sound sources are divided into two groups, one with the average sound pressure contribution of 94 dB, 91 dB, 88 dB, and 85 dB (strong sources), and the other with that of 79 dB (weak source). As shown in Figure 8, the average quantification errors η of the above two groups of sound sources vary with the number of snapshots. Figure 8 shows that: (1) The quantification errors of FDAS and EFDAS for the weak source are much larger than those for strong sources. (2) The average quantification errors of FDAS and EFDAS-DRec is less affected by the number of snapshots, and EFDAS-DRec consistently maintains a high quantification accuracy. (3) The quantification errors of EFDAS-RPCA and EFDAS-PFA decreases as the number of snapshots increases, reaching a quantification accuracy comparable to that of EFDAS-Rec, when the number of snapshots is 800~1600. Combining the diagonal reconstruction errors in Figure 5 and the average quantification errors in Figure 8, it is clear that EFDAS-RPCA and EFDAS-PFA are prone to decrease the quantification accuracy of weak sources, while significantly removing the noise.

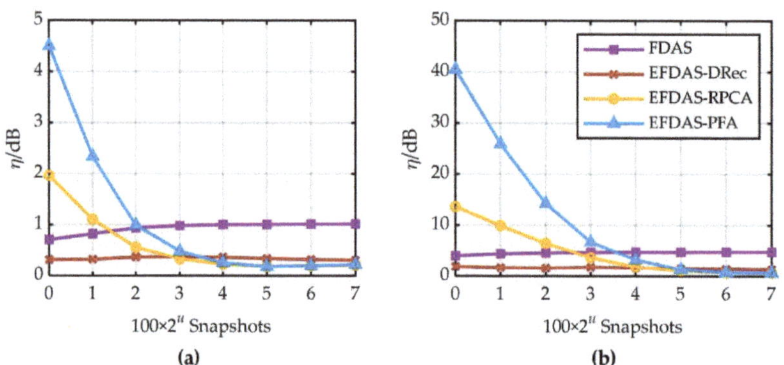

Figure 8. Average quantification error η under different numbers of snapshots: (**a**) Relatively strong sound sources (<-10 dB); (**b**) Weak source (-15 dB).

Although EFDAS-RPCA and EFDAS-PFA are prone to large quantification errors of weak sources, they are still able to locate weak sources effectively, due to the strong sidelobe suppression. Moreover, the quantification errors can be reduced by increasing the number of snapshots. To verify the excellent weak source localization capability of EFDAS, five sound sources with the average sound pressure contribution of 94 dB, 82 dB, 79 dB, 76 dB, and 73 dB are simulated. The coordinates of the sound sources are the same as with those in Figure 4. The imaging colormaps of the above sound sources at SNR = 0 dB are shown in Figure 9, where the number of snapshots is 1000 and the dynamic display range is 60 dB. As shown in Figure 9, all three EFDAS methods have enhanced the localization capability of weak sources for FDAS. Among the three EFDAS methods, EFDAS-PFA have the greatest weak source localization capability, followed by EFDAS-RPCA, and finally, EFDAS-DRec.

EFDAS-DRec, EFDAS-RPCA, and EFDAS-PFA all improve the sound source identification performance of FDAS under low SNRs, and the comparison among the performances of the three methods are summarized in Table 2. EFDAS-DRec has the weakest anti-noise interference capability and still cannot suppress sidelobes within the 30 dB dynamic range, but it has the advantages of high quantification accuracy, no manual setting parameters, and high calculation efficiency. EFDAS-RPCA offers strong noise suppression, clear imaging, accurate localization for weak sources, small quantification errors with sufficient snapshots, and particularly high computational efficiency. However, it involves selecting an appropriate regularization parameter λ, which has been discussed in Reference 14. EFDAS-PFA has slightly lower quantification accuracy than EFDAS-PRCA, but enjoys better imaging clarity and stronger weak source localization ability. Additionally, the computational efficiency of EFDAS-PFA improves with the decrease in the number of equivalent sources κ (as explained in Equation (8)), which is always larger than the number of true sources. In practice, a reasonable κ can be set according to the number of main sound sources of interest.

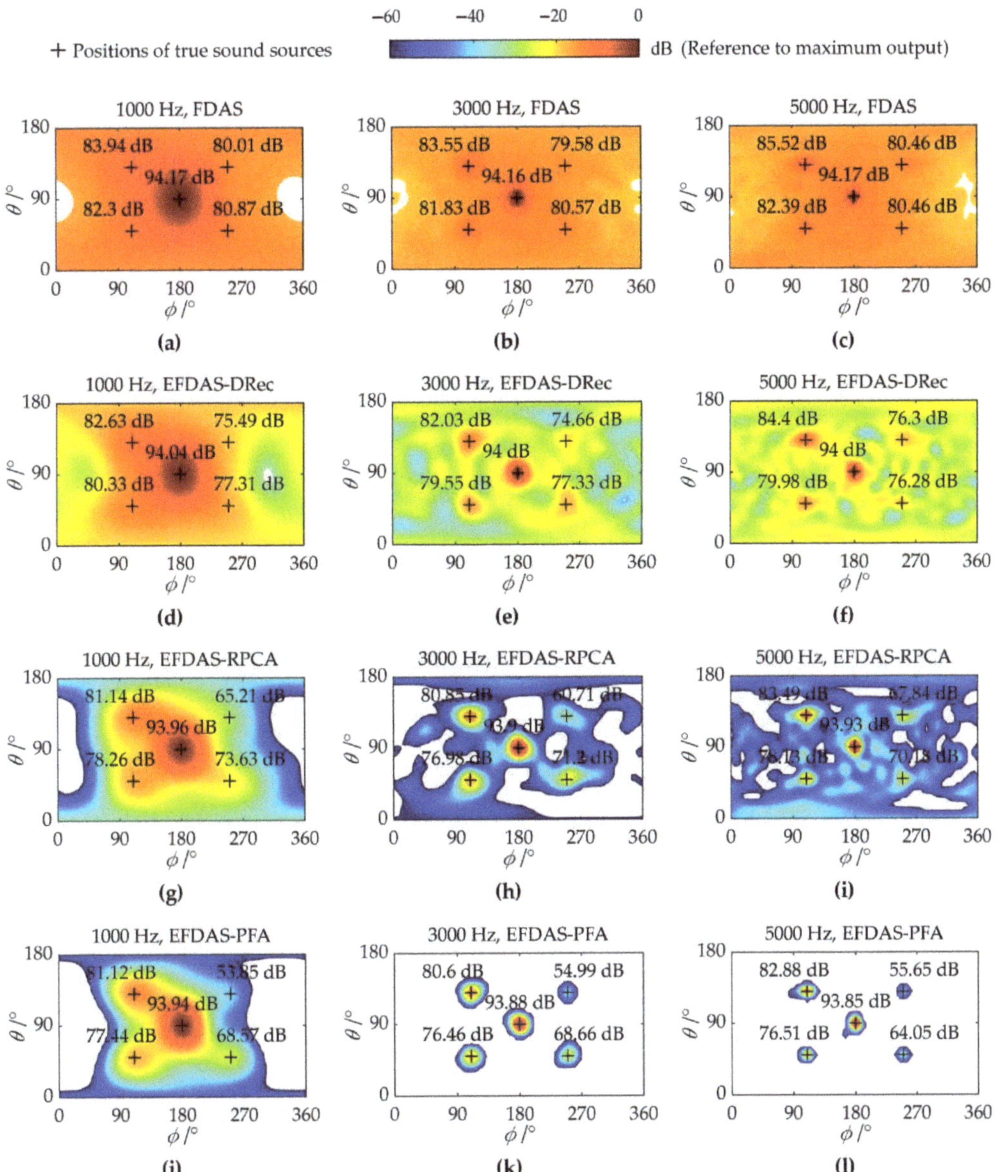

Figure 9. Colormaps of FDAS and EFDAS with SNR = 0 dB and 1000 snapshots: (**a–c**) Maps of FDAS at 1000 Hz, 3000 Hz, and 5000 Hz, respectively; (**d–f**) Maps of EFDAS-DRec at 1000 Hz, 3000 Hz, and 5000 Hz, respectively; (**g–i**) Maps of EFDAS-RPCA at 1000 Hz, 3000 Hz, and 5000 Hz, respectively; (**j–l**) Maps of EFDAS-PFA at 1000 Hz, 3000 Hz, and 5000 Hz, respectively.

Table 2. The comparison among the performances of three EFDAS methods.

Performances	EFDAS-DRec	EFDAS-RPCA	EFDAS-PFA
Parameter setting	none	$\lambda = Q^{-0.5}$	κ
δ_{diag} (@SNR = 0 dB and 1000 Snapshots)	−6.0 dB	−13.3 dB	−12.9 dB
Computational efficiency	0.8 s	0.1 s	1.7 s @κ = 10, 53.6 s @κ = 35
Sidelobe suppression	★ *	★★	★★★
Mainlobe reduction	★	★★	★★★
Quantification accuracy	★★★	★★	★
Weak sources localization	★	★★	★★★

* ★★★ represents the best, ★★ represents the second best, and ★ represents the worst.

4. Experiment

As shown in Figure 10, the validation experiment is conducted to examine the effectiveness of the proposed methods and the correctness of the simulation conclusions. A 36-channel solid spherical array from Brüel & Kjær with a radius of 0.0975 m is used to collect the sound pressure signals, where the coordinates of the array microphones are the same as those used in the simulations.

Figure 10. Experimental layout.

Meanwhile, five loudspeaker sources with a radius of 0.025 m are arranged at different positions, and all are 1 m from the array center. During the experiment, each loudspeaker is independently excited by different steady-state Gaussian white noise, and the acoustic signals received by the array microphones are synchronously sampled by the PULSE Type 3660C data acquisition system at a sampling frequency of 16,384 Hz. The noise with SNR = 0 dB is artificially added to the measured time-domain sound pressure signal to simulate strong noise interference. In post-processing, we perform FFT on time-domain sound pressure signals to obtain the measured CSM averaged over multiple snapshots and apply different beamforming methods for acoustic imaging. A total of 500 snapshots participate in the FFT, with an overlap rate of 66.7%. Each snapshot is weighted with a Hanning window and has a length of 0.125 s, which corresponds to the frequency resolution of 8 Hz.

Figure 11 shows the experimental results. Each column from left to right corresponds to 1000 Hz, 3000 Hz, and 5000 Hz, and each row from top to bottom corresponds to FDAS, EFDAS-DRec, EFDAS-RPCA, and EFDAS-PFA. As shown in Figure 11a–c, the imaging results of FDAS suffer from severe sidelobe contamination, resulting in poor imaging clarity. Since the noise interference is even stronger than the weak sources, some of the sources are totally covered by sidelobes. Compared with FDAS, the imaging results of EFDAS-DRec in Figure 11d–f exhibit a noticeable decrease in the sidelobe magnitudes, resulting in improved imaging clarity at each frequency. Figure 11g–i illustrates the imaging results of EFDAS-RPCA, which reveal that the sidelobe contamination is largely eliminated in the dynamic range, the mainlobe width is significantly reduced, and all sources are accurately identified, except for the weakest source at 1000 Hz. In the imaging results of EFDAS-PFA, as shown in Figure 11j–m, the sidelobe contamination is completely eliminated, and the mainlobe width is further reduced. The above experimental findings are consistent with the simulation conclusions.

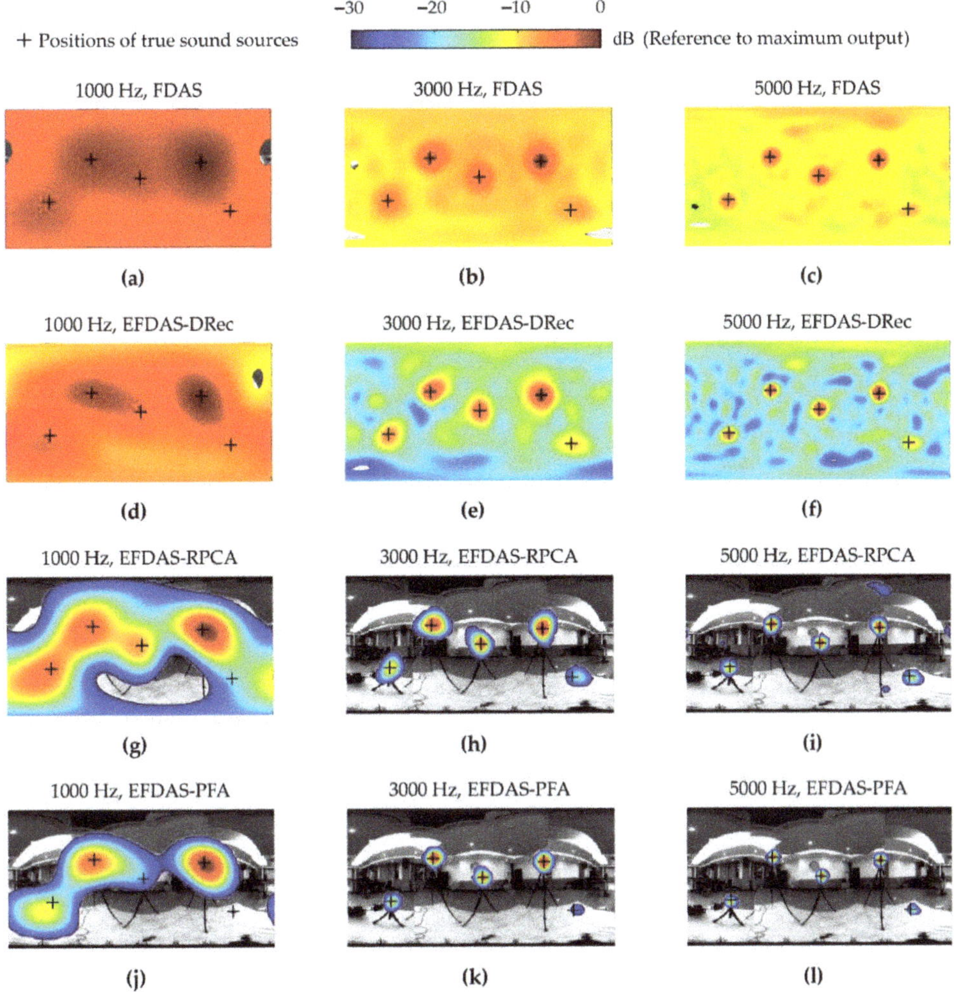

Figure 11. Acoustic imaging colormaps of the experiment: (**a–c**) Maps of FDAS at 1000 Hz, 3000 Hz, and 5000, respectively; (**d–f**) Maps of EFDAS-DRec at 1000 Hz, 3000 Hz, and 5000, respectively; (**g–i**) Maps of EFDAS-RPCA at 1000 Hz, 3000 Hz, and 5000, respectively; (**j–l**) Maps of EFDAS-RPCA at 1000 Hz, 3000 Hz, and 5000 Hz, respectively.

5. Conclusions

To improve the sound source identification performance in low-SNR cabin environments, this paper introduces CSM reconstruction methods such as DRec, RPCA, and PFA, which are widely studied in planar arrays, into spherical arrays-based FDAS. Correspondingly, three enhancement methods, namely EFDAS-DRec, EFDAS-RPCA, and EFDAS-PFA, are established. The main conclusions obtained through simulations and experiments are as follows: (1) EFDAS can significantly improve the sound source identification performance of FDAS under low SNRs, that is, effectively suppress sidelobe contamination, shrink mainlobe contamination, and maintain the strong localization capability for weak sources. (2) Compared with FDAS at SNR = 0 dB and the number of snapshots = 1000, the average MSLs of EFDAS-DRec, EFDAS-RPCA, and EFDAS-PFA are reduced by 6.4 dB, 21.6 dB, and 53.1 dB, respectively, and the mainlobes of sound sources are shrunk by 43.5%, 69.0%, and 80.0%, respectively. (3) All three EFDAS methods can improve the quantification accuracy of FDAS when the number of snapshots is sufficiently large. EFDAS-DRec enjoys the best quantification accuracy and is less affected by the number of snapshots, while the quantification errors of EFDAS-RPCA and EFDAS-PFA are quite large with few snapshots, even worse than FDAS. (4) Among the three EFDAS methods, EFDAS-PFA has the strongest ability to localize weak sources, followed by EFDAS-RPCA, and finally, EFDAS-DRec.

Author Contributions: Conceptualization, Y.Z. and Z.C.; methodology, Y.Z.; software, Y.Z.; validation, Y.Z.; writing—original draft preparation, Y.Z.; writing—review and editing, Y.Z. and Z.C.; supervision, Y.Z., Z.C. and L.L. All authors have read and agreed to the published version of the manuscript.

Funding: This research was funded by the Science and Technology Project of China Southern Power Grid under Grant 036100KK52200034(GDKJXM20201968) and the National Natural Science Foundation of China under Grant 11774040.

Data Availability Statement: The data presented in this study are available on request from the corresponding author.

Conflicts of Interest: The authors declare no conflict of interest.

References

1. Gil-Lopez, T.; Medina-Molina, M.; Verdu-Vazquez, A.; Martel-Rodriguez, B. Acoustic and economic analysis of the use of palm tree pruning waste in noise barriers to mitigate the environmental impact of motorways. *Sci. Total Environ.* **2017**, *584*, 1066–1076. [CrossRef] [PubMed]
2. Amran, M.; Murali, G.; Khalid, N.H.A.; Fediuk, R.; Ozbakkaloglu, T.; Lee, Y.H.; Haruna, S.; Lee, Y.Y. Slag uses in making an ecofriendly and sustainable concrete: A review. *Constr. Build. Mater.* **2021**, *272*, 121942. [CrossRef]
3. Chiariotti, P.; Martarelli, M.; Castellini, P. Acoustic beamforming for noise source localization—Reviews, methodology and applications. *Mech. Syst. Signal Proc.* **2019**, *120*, 422–448. [CrossRef]
4. Sun, H.; Yan, S.; Svensson, U.P. Robust minimum sidelobe beamforming for spherical microphone arrays. *IEEE Trans. Audio Speech Lang. Process.* **2011**, *19*, 1045–1051. [CrossRef]
5. Wang, D.; Guo, J.; Xiao, X.; Sheng, X. CSA-based acoustic beamforming for the contribution analysis of air-borne tyre noise. *Mech. Syst. Signal Proc.* **2022**, *166*, 108409. [CrossRef]
6. Haddad, K.; Hald, J. 3D localization of acoustic sources with a spherical array. In Proceedings of the 7th European Conference on Noise Control 2008, EURONOISE 2008, Pairs, France, 29 June–4 July 2008; pp. 1585–1590.
7. Hald, J. Spherical beamforming with enhanced dynamic range. *SAE Int. J. Passeng. Cars Mech. Syst.* **2013**, *6*, 1334–1341. [CrossRef]
8. Chu, Z.; Yang, Y.; He, Y. Deconvolution for three-dimensional acoustic source identification based on spherical harmonics beamforming. *J. Sound Vib.* **2015**, *344*, 484–502. [CrossRef]
9. Yang, Y.; Chu, Z.; Shen, L.; Ping, G.; Xu, Z. Fast fourier-based deconvolution for three-dimensional acoustic source identification with solid spherical arrays. *Mech. Syst. Signal Proc.* **2018**, *107*, 183–201. [CrossRef]
10. Dougherty, R.P. Functional beamforming. In Proceedings of the 5th Berlin Beamforming Conference, Berlin, Germany, 19–20 February 2014; BeBeC-2014-S01.
11. Yang, Y.; Chu, Z.; Shen, L.; Xu, Z. Functional delay and sum beamforming for three-dimensional acoustic source identification with solid spherical arrays. *J. Sound Vib.* **2016**, *373*, 340–359. [CrossRef]
12. Chu, Z.; Zhao, S.; Yang, Y.; Yang, Y. Deconvolution using CLEAN-SC for acoustic source identification with spherical microphone arrays. *J. Sound Vib.* **2019**, *440*, 161–173. [CrossRef]
13. Hald, J.; Kuroda, H.; Makihara, T.; Ishii, Y. Mapping of contributions from car-exterior aerodynamic sources to an in-cabin reference signal using Clean-SC. In Proceedings of the 45th International Congress and Exposition on Noise Control Engineering: Towards a Quieter Future, INTER-NOISE 2016, Hamburg, Germany, 21–24 August 2016; pp. 3732–3743.

14. Dinsenmeyer, A.; Leclère, Q.; Antoni, J.; Julliard, E. Comparison of microphone array denoising techniques and application to flight test measurements. In Proceedings of the 25th AIAA/CEAS Aeroacoustics Conference, Delft, The Netherlands, 20–23 May 2019; pp. 2744–2753.
15. Hald, J. Denoising of cross-spectral matrices using canonical coherence. *J. Acoust. Soc. Am.* **2019**, *146*, 399–408. [CrossRef] [PubMed]
16. Dougherty, R.P. Cross spectral matrix diagonal optimization. In Proceedings of the 6th Berlin Beamforming Conference, Berlin, Germany, 29 February–1 March 2016; BeBeC-2016-S02.
17. Hald, J. Removal of incoherent noise from an averaged cross-spectral matrix. *J. Acoust. Soc. Am.* **2017**, *142*, 846–854. [CrossRef] [PubMed]
18. Dinsenmeyer, A.; Antoni, J.; Leclère, Q.; Pereira, A. On the denoising of cross-spectral matrices for (aero) acoustic applications. In Proceedings of the 7th Berlin Beamforming Conference, Berlin, Germany, 5–6 March 2018; BeBeC-2018-S02.
19. Dinsenmeyer, A.; Antoni, J.; Leclère, Q.; Pereira, A. A probabilistic approach for cross-spectral matrix denoising: Bechmarking with some recent methods. *J. Acoust. Soc. Am.* **2020**, *147*, 3108–3123. [CrossRef] [PubMed]
20. Jin, Z.; Han, Q.; Zhang, K.; Zhang, Y. An intelligent fault diagnosis method of rolling bearings based on Welch power spectrum transformation with radial basis function neural network. *J. Vib. Control* **2020**, *26*, 629–642. [CrossRef]
21. Wright, J.; Peng, Y.; Ma, Y. Robust principal component analysis: Exact recovery of corrupted low-rank matrices by convex optimization. In Proceedings of the 23rd Annual Conference on Neural Information Processing Systems 2009, NIPS 2009, Vancouver, BC, Canada, 7–10 December 2009; pp. 2080–2088.
22. Antoni, J.; Vanwynsberghe, C.; Magueresse, T.L.; Bouley, S.; Gilquin, L. Mapping uncertainties involved in sound source reconstruction with a cross-spectral-matrix-based Gibbs sampler. *J. Acoust. Soc. Am.* **2019**, *146*, 4947–4961. [CrossRef] [PubMed]
23. Chu, Z.; Yang, Y.; Yang, Y. A new insight and improvement on deconvolution beamforming in spherical harmonics domain. *Appl. Acoust.* **2021**, *177*, 107900. [CrossRef]

Article

Driver Cardiovascular Disease Detection Using Seismocardiogram

Gediminas Uskovas [1], Algimantas Valinevicius [1], Mindaugas Zilys [1], Dangirutis Navikas [1], Michal Frivaldsky [2], Michal Prauzek [3], Jaromir Konecny [3] and Darius Andriukaitis [1,*]

[1] Department of Electronics Engineering, Kaunas University of Technology, Studentu St. 50–438, LT-51368 Kaunas, Lithuania; gediminas.uskovas@ktu.edu (G.U.); algimantas.valinevicius@ktu.lt (A.V.); mindaugas.zilys@ktu.lt (M.Z.); dangirutis.navikas@ktu.lt (D.N.)

[2] Department of Mechatronics and Electronics, Faculty of Electrical Engineering and Information Technologies, University of Zilina, 010 26 Zilina, Slovakia; michal.frivaldsky@fel.uniza.sk

[3] Department of Cybernetics and Biomedical Engineering, VSB—Technical University of Ostrava, 708 00 Ostrava, Czech Republic; michal.prauzek@vsb.cz (M.P.); jaromir.konecny@vsb.cz (J.K.)

* Correspondence: darius.andriukaitis@ktu.lt; Tel.: +370-37-300-519

Abstract: This article deals with the treatment and application of cardiac biosignals, an excited accelerometer, and a gyroscope in the prevention of accidents on the road. Previously conducted studies say that the seismocardiogram is a measure of cardiac microvibration signals that allows for detecting rhythms, heart valve opening and closing disorders, and monitoring of patients' breathing. This article refers to the seismocardiogram hypothesis that the measurements of a seismocardiogram could be used to identify drivers' heart problems before they reach a critical condition and safely stop the vehicle by informing the relevant departments in a nonclinical manner. The proposed system works without an electrocardiogram, which helps to detect heart rhythms more easily. The estimation of the heart rate (HR) is calculated through automatically detected aortic valve opening (AO) peaks. The system is composed of two micro-electromechanical systems (MEMSs) to evaluate physiological parameters and eliminate the effects of external interference on the entire system. The few digital filtering methods are discussed and benchmarked to increase seismocardiogram efficiency. As a result, the fourth adaptive filter obtains the estimated HR = 65 beats per min (bmp) in a still noisy signal (SNR = −11.32 dB). In contrast with the low processing benefit (3.39 dB), 27 AO peaks were detected with a 917.56-ms peak interval mean over 1.11 s, and the calculated root mean square error (RMSE) was 0.1942 m/s^2 when the adaptive filter order is 50 and the adaptation step is equal to 0.933.

Keywords: arrhythmia; driving restrictions; adaptive digital filter; noninvasive method; heart rate

Citation: Uskovas, G.; Valinevicius, A.; Zilys, M.; Navikas, D.; Frivaldsky, M.; Prauzek, M.; Konecny, J.; Andriukaitis, D. Driver Cardiovascular Disease Detection Using Seismocardiogram. *Electronics* **2022**, *11*, 484. https://doi.org/10.3390/electronics11030484

Academic Editor: Jose Eugenio Naranjo

Received: 15 January 2022
Accepted: 4 February 2022
Published: 7 February 2022

Publisher's Note: MDPI stays neutral with regard to jurisdictional claims in published maps and institutional affiliations.

1. Introduction

Quality of life can be expressed by many factors, some of which are health and the opportunity to participate in favorite activities. Carrying out one's favorite work may be restricted or completely suspended due to health impairments [1]. For staff of various professions, one of the main monitoring objects is related to cardiovascular diseases, resulting in mortalities in Europe of about 4 million [2] and up to 46% of all deaths in the United States [3]. Cardiovascular diseases are a leading cause of death in the world [4], and 523.3 million people had cardiovascular disease (CVD) in 2019 [5,6]. The future prognoses of researchers are plaintive because the COVID-19 pandemic and fast aging will allegedly make the CVD numbers worse in the coming years [5,6].

Cardiovascular disease is hard to predict and can render people unfit to perform some daily or professional activities or cause them to suddenly lose control for short time periods and harm others [1,7]. This transient loss of consciousness event is called syncope and can result from cardiac conditions, mostly consisting of arrhythmic events, bradyarrhythmia, or tachyarrhythmias [8,9]. For this reason, the American Heart Association and Heart

Rhythm Society developed recommendations to prevent a person from putting others at risk of harm and at the same time have the opportunity to work according to a society and its culture [1,7,10,11].

Athletes, pilots, drivers, and physicians are among the most prominent professions whose health statuses are regulated by strict legal provisions or directives. Arrhythmia is identified as one of the main problems of heart activity whose duration cannot exceed the prescribed time limits for electrocardiogram (ECG) short and long QT syndromes (SQTS and LQTS, respectively), namely SQRT $\leq$ 320 ms and LQTS $\geq$ 470 ms for males and LQTS $\geq$ 480 ms for females [12–14]. Athletes check their arrhythmic levels before competition and in training during exercise, beyond which they cannot be allowed to participate in competitions [12,15]. The international aviation agency has regulations stating that pilots must check that their arrhythmic levels do not reach certain limits which result in sinus bradycardia (<40 beats per min (bmp)), sinus tachycardia (>100 bpm at rest), or sino-atrial block (>3 s during the day or >4 s at night). Otherwise, they must reschedule flights and contact their doctors. According to the New Standards for Driving and Cardiovascular Diseases [16] and European Commission Directive 2016/1106 [17], the threshold QT interval is LQTS > 500 ms. The European Society of Cardiology (ESC) and the Canadian Cardiovascular Society (CCS) suggest driving restrictions in case of recurrent or unexplained syncope or substantial cardiovascular comorbidities, unless a definitive treatment can be ensured and controlled [18]. Moreover, the Japanese Circulation Society (JCS) describes driving restrictions for private and commercial patients with reflex syncope, which recommend restricting private driving until their symptoms are controlled and for commercial driving unless an effective treatment has been established [19]. The restrictions of activity unite these three risky professions and point out the importance of using a wide range of preventive actions before a sudden event happens. Furthermore, the population over the age of 65 continues to increase globally [20], which implies an increasing average age of active people with a higher risk of cardiovascular disease [14,21] in workplaces, driving vehicles, or traveling by plane. The acceptable risk of harm (RH) estimated by the CCS created a risk estimation formula [22]. Because aging continues, the solutions require finding and using additional diagnostic facilities that prevent sudden cardio events and protect the public from injury [23].

This article deals with the proposal to introduce a preventive diagnosis of heart disease in drivers to avoid critical health conditions that accompany road incidents. Studies of long-distance drivers' working conditions show that a driver's health is exposed to various stressors, both physical and psychological [24,25]. As a result, many working hours are lost, and there is the threat of an accident on the road due to long working hours, weak metabolism, social isolation, low control, and work shifts [14,26]. It enforces the investigations of Canadian, German, Finnish, Swiss, Japanese [14,19], Spanish [27], and Czech Republic roads [28]. Car manufacturers have a wide range of improvements that improve drivers' working conditions and facilitate vehicle management, ergonomic solutions, and the integration of various automatic control, navigation, and alert systems [9,29,30]. Most of these automotive electronic systems are related to car control and diagnostics, but a very small part of the driver and machine interfaces is oriented toward real-time driver health diagnostics, except in sports cars [31–35]. For example, [28] points out the increase in living standards and suggests using various driver response systems which can measure unobtrusively various life functions and health conditions of drivers in order to monitor their alertness.

The most popular measurement of cardiac work is the recording of an electrocardiogram, which has advanced signal processing methods [36,37], but the daily activities of patients are constrained during measurement. Alternatively, the progress of technology allows the detecting of weak cardio signals with a better diagnostic capacity without the requirement of attaching any sensor directly to the patient's skin by using a microelectromechanical system (MEMS) [38,39]. In that case, the recording of a seismocardiogram (SCG) is possible by non-invasive measurement of low-frequency vibrations from 0 to 50 Hz of cardio cycle mechanics along the timings of corresponding cardiac events [40]. This article analyzes using the seismocardiogram method for warning a driver about possible cardiovascular disease during driving. Additionally, this article is one of the non-multiple types of research dealing with a moving patient [41] without simultaneously measuring the ECG.

The novelty of this article reflects a not yet examined location of the SCG sensor when the MEMS is integrated into the driver's safety belt for measuring the cardio mechanical vibrations. Hence, the SCG signal is affected not only by the already known respiratory and body movement artifacts but also by the movement of the car [28] and the acoustic components [42]. The aim is to obtain a useful SCG signal sufficient for heart rate calculation. Therefore, this signifies that it is necessary to select an appropriate adaptive filter algorithm and optimal settings to rapidly find an AO peak during driving. Consequently, the benchmark of the fourth adaptive digital filters has been examined with different settings.

The rest of the article is organized as follows. Section 2 discusses the related works and use methods, describes the system, and introduces SCG signal processing with adaptive filters. Section 3 presents the experimental data and analysis. Section 4 concludes the work of this paper and briefly introduces future works.

2. Materials and Methods

2.1. Related Works

Numerous researchers working on various driver response systems encounter autonomous car system requirements and advance them for safer driving. Technological progress over the past few generations let researchers set up unattached and attached sensor systems inside vehicles [28]. Researchers use galvanic, capacitive, mechanical, optical, or electromagnetic types of contact in different locations for monitoring the vital signs of a driver [43]. Based on this, a few monitoring methods and systems like video motion, a capacitive electrocardiogram, electroencephalogram, balistocardiogram, seismocardiogram, thermography, photoplethysmography, magnetic induction, and radar-based methods are possible [28,43].

An increase in the speed of signal processing allows one to return to the mechanocardiography previously proposed and analyzed by scientists, also known as a balistocardiogram, seismocardiogram, or gyrocardiogram. Unlike the most popular clinical cardiac monitoring methods such as electrocardiogram, ultrasound cardiogram, phonocardiogram, or photoplethysmography, the measurement of mechanocardiogram is oriented toward the recording and analysis of mechanical heart transmitters.

The micromechanical system (sensor), which is micro in five types—metal, semiconductor, ceramics, polymer, and composite—is used for the measurement of these twins. In this case, the greatest attention shall be paid to the characteristics of the accelerometers and their type, of which there are five: piezo resistance, receptacle, tunnel, optical, and piezoelectric [44].

A mechanocardiogram, a balistocardiogram, seismocardiogram, or gyrocardiogram may be recorded at the same time as several parameters of a heart condition, as the transmitter of the heart muscle consists of the atrial, lung, mitral, and triple valves' opening and bruising, the movement of a heart muscle caused by a sinus node, and the propagation of the heart sound. The surveyor may provide a lot of information, thus increasing the sensitivity and resilience of an MEMS to interference, reduced dimensions (0.08 g, 5 mm × 5 mm × 1.6 mm [32] and 3 mm × 3 mm × 1 mm [45]), and lowering energy costs, enabling measurements with various types of smartphones (e.g., portable phones and clocks) or specialized devices. Specialized devices have been developed according to the future anchorage site, such as a chest, a chair, or a bed [46]. The ability to see heart valve tremors in MEMS sensors has led scientists to investigate the possibility of diagnosing early-stage myocardial infarction, heart failure, atrial fibrillation, and other diseases related to cardiac valves.

Previously, the greatest attention was focused on the measurement of the balloon cardiogram when the waves of the surfaces were generated by the rise of blood pressure slopes and during the fall. Attempts have been made to monitor the work of a driver's heart by creating a special occupant evaluating the vibration of the breath at 0.13–0.73 Hz [47] and the vibration of the body due to ballistic forces [45]. Following a further examination of the seismocardiogram, the recording of mechanical fluctuations was caused by the movement of the heart and circulation through the main channels of the upper body. The mathematical model described clearly reflects the mechanical nature of the seismocardiogram signal and the calculation method that evaluates the forces operating in the vascular walls [48]. The balistocardiogram and seismocardiogram signals are not identifiable due to their mechanical twisting nature, although they can be measured with the same accelerometers. In both cases, it is necessary to properly select the location of the suitable sensor, such as the seat suitable for the balloon cardio signal and the limbs when the most suitable place to measure the seismocardiogram signal is to the left of the crude side rather than the chest. The measurement of the seismocardiogram signal shall include all three axes of the acceleration direction to obtain the desired accuracy [49].

The experimental studies described in relevant scientific articles show that the most extensive mechanocardiogram signals are recorded and analysed together with an electrocardiogram to better detect the beginning of systolic time intervals and to combine the observed deviations of the electrocardiogram and the seismocardiogram [50,51]. Such systems are not flexible and are not convenient for home use or during activity because the electrodes attached to the body must be used, which causes discomfort for patients. Therefore, this variant of cardiac disorders is more suitable for clinical use when there is a limited number of patients involved or sitting. One of the greatest advantages of such complementary systems is that it is possible to evaluate the dependency of the heart on the body. Lying on the left side is the riskiest heart disease, as the heart muscle is the most unloaded due to the mechanical pressure of the warrior in the heart [52,53].

Investigators have increasingly tried to perform non-invasive seismocardiogram or gyrocardiogram measurements with a smart phone without measuring the electrocardiogram. To be properly diagnosed, it is necessary to know the result of the systolic time intervals (from 149 ms to 1091 ms when the heartbeat drops from 220 to 30 bpm [54]) and to further analyze the signal obtained by comparing the theoretical classification requirements with the practical ones. One of the first articles on automatic identification of seismocardiogram signals appeared in 2016, with an emphasis on the statistical calculation of systolic time intervals [55]. In accordance with the method proposed in this article in 2017, Finnish scientists have applied the automatic recognition of seizures to the experimental test for atrial fibrillation [45].

The seismocardiogram signal is classified as a nonstationary signal that can lead to sudden abnormalities that indicate possible signs of disease. The analysis and processing of such biosignals is problematic, requiring a range of statistical methods, including fast Fourier transformation, wavelength theory, machine learning, decision tree, and analysis of the related and unrelated attributes [56,57]. In summary, scientists seek to automatically detect the attribute in the biosignal so that large and small data do not accumulate. The choice of methods used for analysis depends heavily on the characteristics of the signal that characterize the frequency, time intervals, number of attributes, operation principle, and purpose of interpretation. The purpose of the interpretation is to understand the real-time alerts, diagnostics, prophylactic monitoring, and planning [56]. The analysis performed is faster and more reliable in advance of a prior signal classification based on the frequency of intervals, frequency, form of attributes, and the distribution of the signal in the separate segments [58,59].

As we see in the literature analysis, measurement of the mechanocardiogram signals is not a new topic and has garnered interest from investigators since recording of the reactionary forces of Gordon's invented organism in 1877, known as a balistocardiogram, but the tests have been reduced due to the simplicity of the application of electrocardiograms and the more accurate diagnosis of cardiac diseases [60]. Thanks to technological progress, micromechanical systems have become smaller and more accurate, so researchers are again encouraged to deal with the development of mechanocardiogram signals and the extension of applications, such as the use of seismocardiogram signals for the nonclinical prevention of cardiac diseases, allowing the first signs of heart problems to be observed [61]. It is a very real topic to solve health problems during human daily activity and not in hospitals by using wearable technologies with the ability to alert users to the presence of health diseases. Moreover, this investigation is a small step in big research which requires creativity, know-how, and a wide range of knowledge from different study fields.

2.2. The System Discription

The idea of the system described in this article is based on a provision that the measurement of a driver's heart rate is carried out in a non-invasive manner without causing a sense of discomfort, avoiding ethical and private data protection aspects. Additionally, the electronic system must evaluate and predict signal changes in the noisy environment with many uncertainties, such as vibrations and body movements. The main requirement of this system is to prevent the driver from carrying out his direct function to manage the car safely.

The designed system was for the seismocardiogram measurement via an accelerometer, which was integrated in safety belt of the driver. There is a different measuring principle for measuring mechanical displacement of the chest load to that previously described in [3,7], where an accelerometer was integrated in the driver seat for measuring vibration of the back of the chest. The MEMS sensor had very good contact with the body, which was closer to the heart and provided a bigger signal amplitude of the forward chest vibration. The driver could not see the integrated MEMS sensor in the safety belt, and he or she did not have to enter any personal data in order to be in line with the requirements.

A more detailed comparison of the vital sign monitoring methods is summarized in Table 1, and the present research method is included too. The advantages and disadvantages are discussed in Table 1 based on the work in [28,43].

In order to reduce the system's interference and ensure the reliability of the diagnostics, the use of two accelerometers is described. Figure 1 shows the idea for where the accelerometers would be located.

The hardware of this system consists of two MPU9250 accelerometers, an ESP32 WROOM microcontroller, and an SD card reader.

The main accelerometer is integrated in the safety belt and is located at the chest during driving. The location of the second accelerometer is the seat of the driver. This location was chosen with respect to the need to measure the movement of the car and the movement of the driver's body. A second accelerometer was attached to the driver's seat because the chair is directly related to the driver and the weak vibrations of the car and the potholes on the road, which are measured for the acceleration amplitudes.

When measuring the seismocardiogram with an accelerometer, it is important to calculate the resultant of all accelerations of the accelerometer coordinates, because in this case, the most accurate reading of the seismocardiogram was obtained, with the evaluation being as a close to human accelerometer as possible [62]. This way, it was not necessary to accurately record the position of the accelerometer and perform precise corrections of the position in each case. Then, the driver wanted to change the angle of the seat back while driving. Therefore, the drive felt free and could concentrate attention only on the road. Moreover, the system did not require any additional adjustments before or after driving.

Table 1. Advantages and disadvantages of vital sign monitoring systems used in vehicles.

Vital Sign Monitoring Method	Sensor Location	Advantages	Disadvantages
Seismocardiogram	Safety belt (this research)	Allows measuring cardiac and respiratory activity unobtrusively. The signal of the three-axis accelerometer characterizes specific events of the heart's activity. The SCG signal is measured from the front of the chest. The additional reference signal is not required. The safety belt obliges using the vital sign monitoring system automatically.	Requires solving noise issues.
Seismocardiogram	Worn sensor	Allows measuring cardiac and respiratory activity unobtrusively. The signal of the three-axis accelerometer characterizes specific events of the heart's activity. Does not require a reference signal.	Driver required to wear sensor on the body. Requires solving noise issues.
Seismocardiogram	Back of the car seat	Allows measuring cardiac and respiratory activity unobtrusively. The signal of the three-axis accelerometer characterizes specific events of the heart's activity [28,43].	The seat attenuates the SCG signal. Requires a reference signal, which increases signal processing duration.
Balistocardiogram	Car seat	Allows measuring cardiac and respiratory activity unobtrusively. Records the cardio, mechanic, and lung vibrations and the momentum of the blood pulse traveling down to the aorta [28].	The noise of car motor vibrations may make measurement difficult [28]. Requires a reference signal.
Capacitive electrocardiogram	Steering wheel Car seat Back of car seat	Records electrical activity of the heart muscle. Still the most valuable physiological signal. No galvanic contact with the body. Electrically insulated and remains stable in long-term applications [43].	Both hands have to touch two different parts of the wheel. Requires an infinitely high ohmic resistance [43].
Video monitoring	Camera-based	Allows measuring cardiac activity unobtrusively. No contact required to monitor a driver or passengers. Monitoring of respiratory and temperature can happen in complete darkness. Driver drowsiness and attention detection. Driver stress and pain detection by analyzing facial expressions [28,43].	Requires free line of sight. Absence of privacy. Sufficient light cannot be guaranteed for operating in the far infrared spectrum. Shadows from other cars and trees can rapidly change the signal [28,43].
Radar system transmitter—receiver system Doppler radar	Front radar Back of car seat	Allows measuring cardiac and respiratory activity unobtrusively. No contact required. Uses high-frequency electromagnetic waves that are emitted and reflected by the chest's surface [28,43].	The heart-related motions are very small and hard to detect [43].
Electroencephalogram	Special helmet	Allows measuring concentration, reaction time, and cognitive state, as well as drowsiness of drivers [43].	The measurement system is complex.

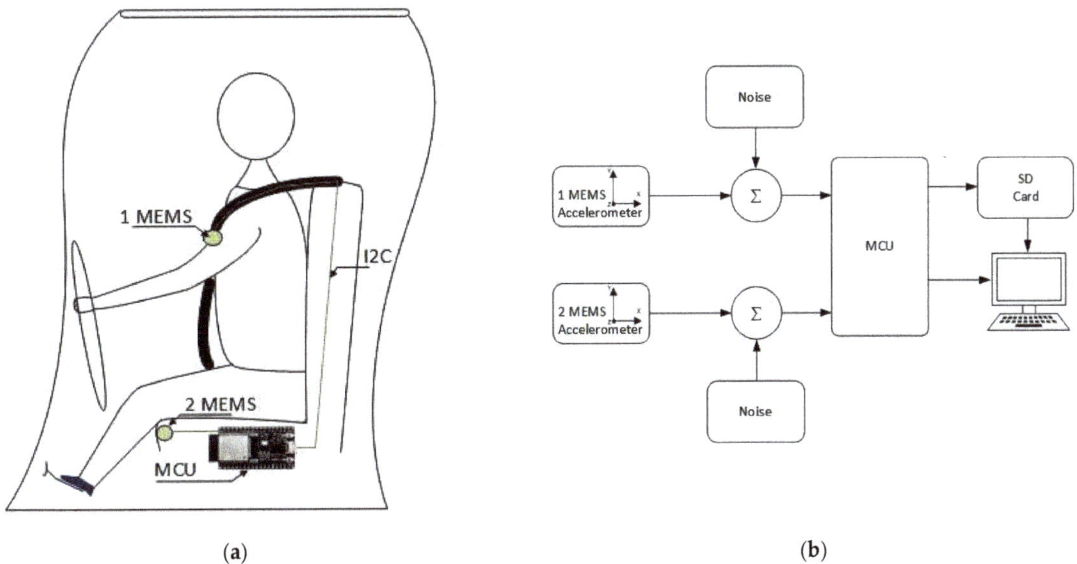

(a)   (b)

Figure 1. Driver seismocardiogram measurement system with two accelerometers. (**a**) Driver seismocardiogram measurement system inside car. (**b**) Structure of measurement data collection and processing system.

$$s(t) = \sqrt{s_x(t)^2 + s_y(t)^2 + s_z(t)^2} \tag{1}$$

The total acceleration (Equation (1)) required was calculated on the basis Figure 2 that a better SCG signal could be obtained instead of only the z-axis, as in other previous studies [54,63]. In addition to preparing this signal for processing, normalization of the total acceleration was performed. As a result, the power of the big vibrations reduced and impacted the input signal processing.

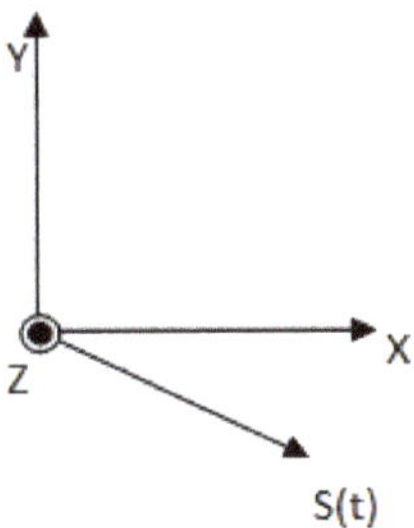

Figure 2. Resultant of three accelerometer coordinates.

This did not disturb the driver during the trip, and this system avoided personal identity regulations. The data were collected on an SD card if needed.

2.3. The Signal Processing

Adaptive filtering has many advantages and is useful for biomedical applications. Unlike other filters, an adaptive filter can self-adjust the filter coefficients to a rapidly and unpredictably changed signal. The principle of operation of the adaptive filter is defined by Equation (2) [64,65]:

$$e(n) = d(n) - y(n) \tag{2}$$

The operation of this filter is based on the tendency of the filter output signal $y(n)$ to correspond as closely as possible to the affected signal $d(n)$ through the feedback response of the error signal $e(n)$ to the coefficients H(z) of the filter transfer function so that $e(n)$ is zero [64,65]:

$$y(n) = \sum_{k=1}^{L} b_n(n)x(n - k) \tag{3}$$

The structure shows in Figure 3 that, based on the fact that the SCG signal is measured in the frequency band of 1–20 Hz, it feeds the signal of the frequency band of interest to the adaptive filter to make the adaptive filter much more efficient and effective [66,67]. For this, a finite impulse response filter (FIR) or an infinite impulse response filter (IIR) can be used. As a result, one accelerometer can be used for measurement. Apart from the fundamental adaptive filter configuration, the frequency band is 5–45 Hz for the adaptive filter with a delay (Figure 4). Additionally, the performance for detecting AO peaks was analyzed with the FIR and IRR filters:

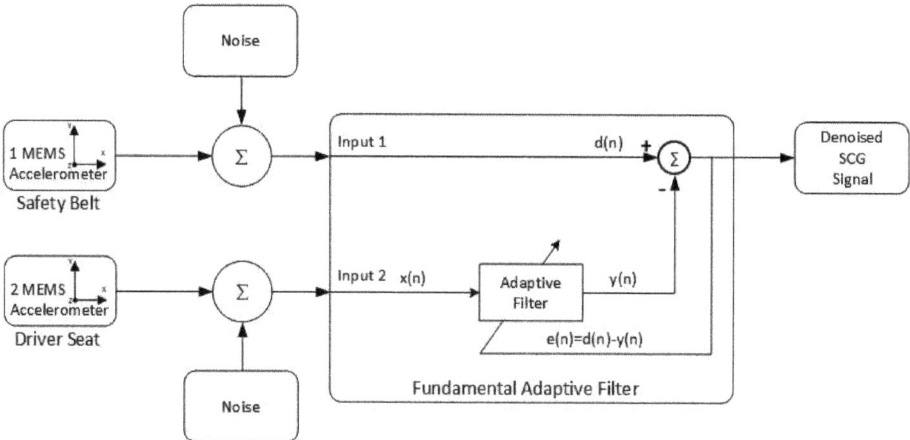

Figure 3. Fundamental adaptive filter configuration.

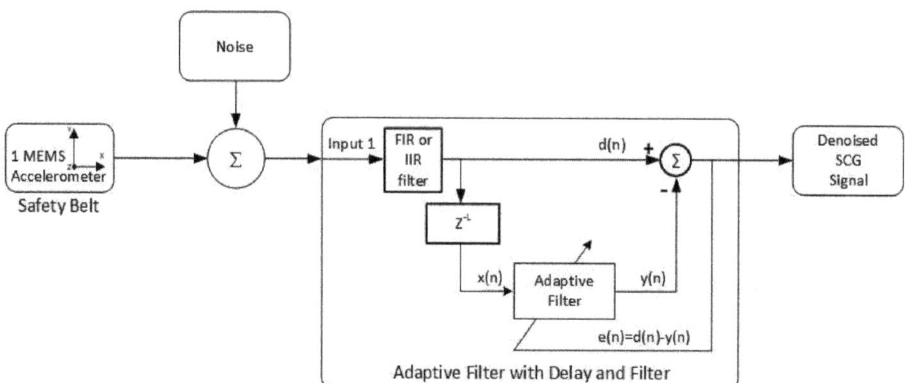

Figure 4. Adaptive filter configuration.

The adaptive filter modifies the filter coefficients $b_n(k)$ according signal property and they can be calculated with Equation (4) [64]. The estimation accuracy of coefficients $b_n(k)$

depends to rule how changes the convergence parameter Δ, which lie in the range described in Equation (5) [64].

$$b_n(k) = b_{n-1}(k) + \Delta e(n)x(n-k) \tag{4}$$

$$0 < \Delta < \frac{1}{10LP_x} \tag{5}$$

where: L is length of the FIR filter and P_x power of the signal in the input.

The power is calculated using Equation (6) [64]:

$$P_x \approx \frac{1}{N-1} \sum_{n=1}^{N} x^2(n) \tag{6}$$

3. Results and Discussion

The proposed system collects data without additional measurements, except for seismocardiogram detection. The main aim of this system is to monitor the heart rate with one system and, at the same time, avoid disturbing human activity during the driving process.

The duration of the interval between two peaks has a few requirements. The first requirement sets a 410-ms minimal time interval between two peaks. This requirement helps to avoid incorrectly detecting mistakes in the opening moments of the aorta. The second limitation relates to the signal's minimal level, which means that the peak has to be above this minimal level, which is equal to the total signal power. Each adaptive filter has its own minimal level, depending on the signal processing results.

4. Experimental Results

The data were collected to perform a deeper analysis of the proposed system's performance. Figure 5 shows the signal processing results, where the top diagram indicates the seismocardiogram signal in the input. The second diagram shows the adaptive filter learning processes that calculate the adaptive error. As a result, the third diagram indicates the filtered signal and looks like the driver's seismocardiogram. The fourth diagram shows that the third adaptive filter could recognize patterns of heart beats which represent the aorta opening in the seismocardiogram.

The calculated signal-to-noise ratio of the third adaptive filter (SNR = -8.0627 dB) was negative, indicating huge noise in the environment. Thus, the fourth adaptive filter was analyzed, which first filtrated the SCG signal with the IIR filter (filter order of five) and afterward with an adaptive filter with a delay (order of 5) (Figure 4). In that case, Figure 6 presents three signal processing diagrams of the fourth adaptive filter, showing the adaptation error, signal in the fourth adaptive filter output, and efficiency of the adaptive filter.

Figure 6 shows that the denoised signal y(t) in the output of the fourth adaptive filter was clearer and could mark places with similar form (black ellipse), which repeated but had different amplitudes. As a result, this partly conformed to publications of other authors and enforced the importance of continuing research by creating and developing new methods.

Figure 7 shows how the variation of the signal-to-nose ratios of the adaptive filters depended on the filter order. An interesting fact is that the signal-to-noise ratio was negative. This means that the adaptive filter did not perform as well as the filtration. Therefore, the filtrated signal had noise. This situation is not typical compared with other signals, such as with electrocardiogram measurement.

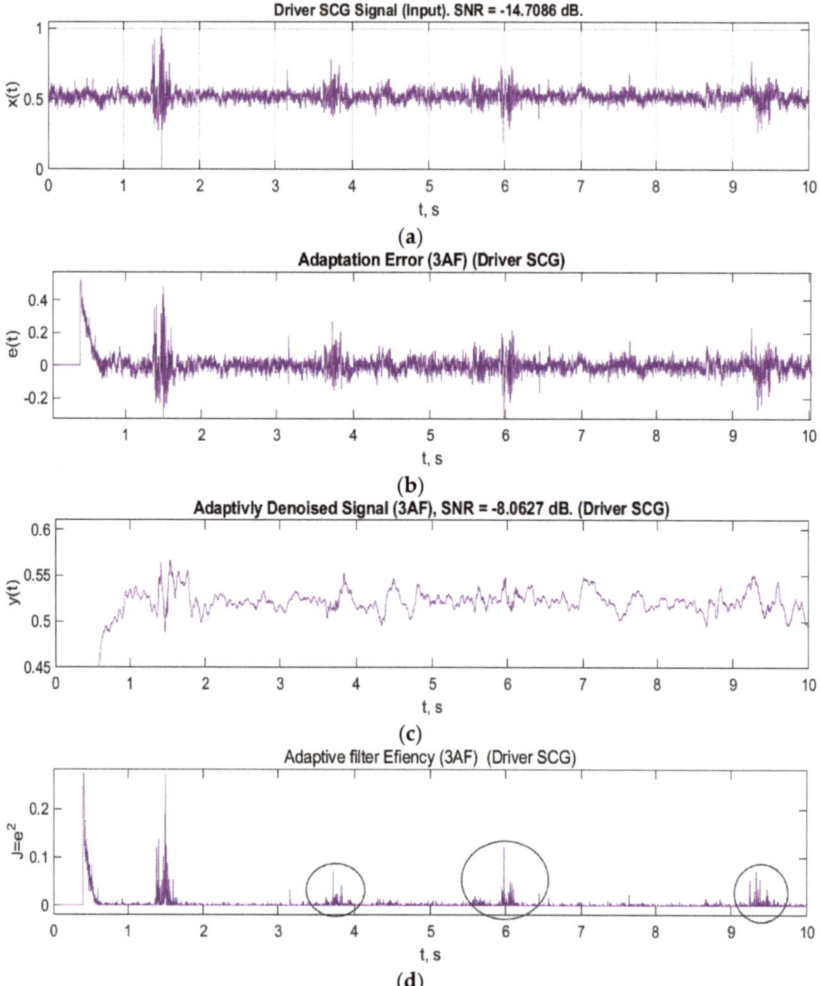

Figure 5. Processing of the driver's seismocardiogram signal in the third adaptive filter with sample delay order of 5 and the FIR filter in the input. (**a**) SCG signal on input. (**b**) Adaptation error of the third adaptive filter. (**c**) Adaptively filtrated signal. (**d**) The efficiency of the adaptative filter.

On the other hand, the received signal managed to find the peak markings and heart rate calculations, while the signal had a negative value for the SNR in all adaptive filters from the first one to the fourth −7.61 dB, −7.22 dB, −8.06 dB, and −11.32 dB, respectively. The processing benefit of the fourth filter was the smallest (3.39 dB), and the others were about 7 dB.

Figure 8 shows the driver's seismocardiogram with detected peaks and a calculated heart rate of 65 beats per minute.

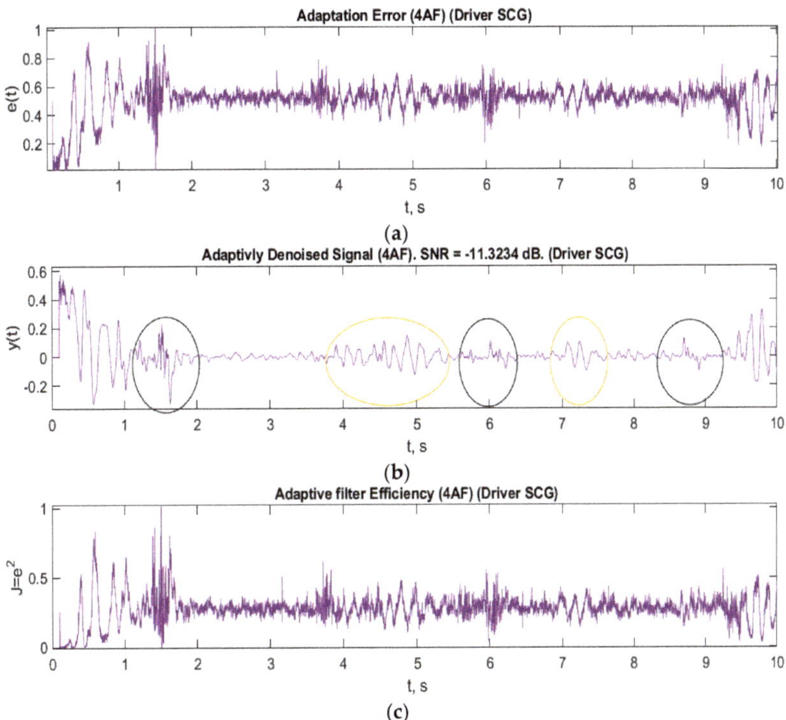

Figure 6. The fourth adaptive filter. (**a**) Adaptation error of the fourth adaptive filter. (**b**) The fourth adaptively filtrated signal. (**c**) Adaptation efficiency.

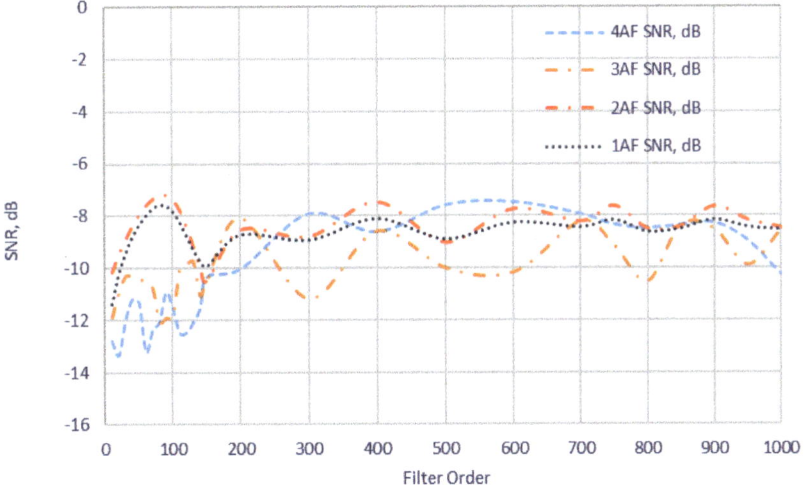

Figure 7. Signal-to-noise rate's dependence on the adaptive filter order.

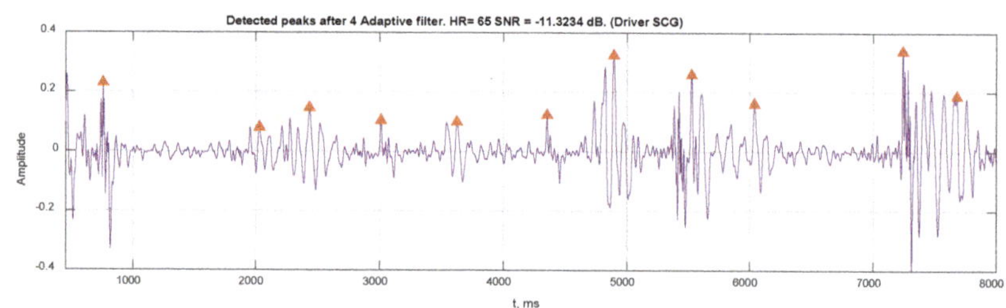

Figure 8. Driver seismocardiogram after the fourth adaptive filter and detected peaks.

Additionally, calculations have been performed to better understand the seismocardiogram signal and the behaviour of four different adaptive filter configurations. The settings of the adaptive filters and signal processing data are tabulated in Table 2.

Table 2. Adaptive filter processing data.

	Adaptive Filter 1	Adaptive Filter 2	Adaptive Filter 3	Adaptive Filter 4
Filter order	90	90	200	50
mu AF step	1.0133×10^{-3}	1.0133×10^{-3}	5.0855×10^{-3}	9.3302×10^{-1}
Heart rate (beats/min)	109	107	111	65
RMS (m/s^2)	0.5212	0.5208	0.5198	0.0684
SNR (dB)	-7.61	-7.22	-8.06	-11.32
RMSE (m/s^2)	0.0472	0.0370	0.0248	0.1942
Detected peaks number	53	52	53	27
Peak interval mean (ms)	549.07	558.21	536.15	917.56
Peak Interval STD	117.82	127.11	117.22	635.12
Processing time (s)	57.69	1.09	1.09	1.11
Processing benefit (dB)	7.10	7.49	6.65	3.39

The data in the table show that increasing the minimum duration limit between pulses decreased the heart rate and coincided with the value of the standard deviation between the detected peaks, which means greater reliability for the measurement. In all cases, the root mean square error (RMSE) of the third adaptive filter was 0.0248 (m/s^2), and it was the smallest value compared with the other filters, indicating the stability of the signal peak level of the received signal at this adaptive filter output. The RMSE of the fourth adaptive filter (0.1942 (m/s^2) indicates smaller stability of the received signal peaks' levels. Hence, the seismocardiogram signal was not stable.

Figure 9 shows the benchmark of all the adaptive filter autocorrelations and presents that the signals after processing did not change enough from the input signal in the first three adaptive filters. Alternatively, the signal in the fourth was evidently changed from 5 to 20 time lags.

Additionally, the data show that the fourth filter detected 27 fewer peaks than the other 53 peaks. The standard deviation of 635.12 for the peak intervals shows that in these cases, the system did not find any peak.

Measurement of the processing duration was performed, which shows that the typical adaptive filter without delay spent the longest time in processing (57.69 s). As a result, this adaptive filter is not recommended for signal processing of seismocardiograms.

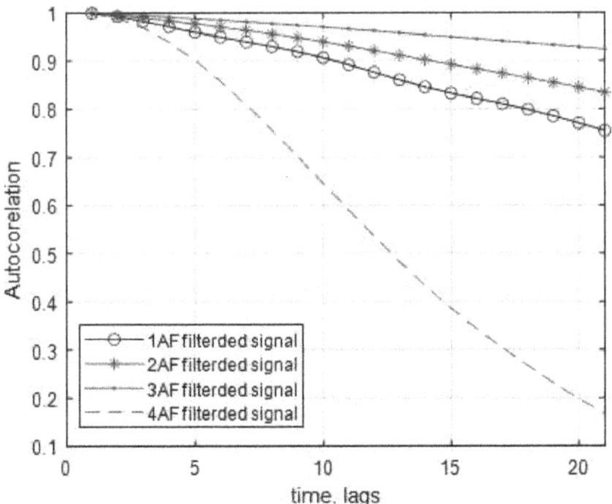

Figure 9. Autocorrelations of the filtered seismocardiogram signals.

5. Conclusions

A system monitoring driver cardiovascular disease has been described. Several advantages and disadvantages of the SCG were described and compared with alternative investigations. The results of the experiments show that the developed system collected sufficient data and could interpret them later using an adaptive filter. As a result, the fourth adaptive filter obtained an estimated HR = 65 beats per min (bpm) in a still-noisy signal (SNR = −11.32 dB). In contrast with the low processing benefit (3.39 dB), 27 AO peaks were detected with a 917.56-ms peak interval mean over 1.11 s, and the calculated root mean square error (RMSE) was 0.1942 m/s^2 when the adaptive filter order was 50 and the adaptation step was equal to 0.933.

Signal processing was performed using four adaptive filter algorithms and compared. The results show that the filtered signal was noisy, and more advanced adaptive filter algorithms and a sensor system with few accelerometers are needed for better signal acquisition.

The proposed system and methodology can be useful and integrated in car safety systems. For better performance, more research and improvements to the system are needed. Furthermore, future works will relate to improving the processing methods by including machine learning with an additional mathematical model of the seismocardiogram.

Author Contributions: Conceptualization, G.U., A.V. and D.N.; methodology, D.A., M.Z. and M.P.; software, G.U. and J.K.; validation, D.A., A.V., M.Z., M.P. and M.F.; visualization, G.U., D.N. and A.V.; investigation, D.N., G.U., D.A., A.V. and M.Z.; resources, M.Z., M.F., J.K. and M.P.; data curation, D.N., M.Z. and G.U.; writing—original draft preparation, G.U., A.V., D.A. and M.P.; writing—review and editing, D.N., D.A., J.K. and M.F.; supervision, A.V.; funding acquisition, M.Z. and D.A. All authors have read and agreed to the published version of the manuscript.

Funding: This research received no external funding.

Conflicts of Interest: The authors declare no conflict of interest.

References

1. Epstein, A.E.; Miles, W.M.; Benditt, D.G.; Camm, A.J.; Darling, E.J.; Friedman, P.L.; Garson, A.; Harvey, J.C.; Kidwell, G.A.; Klein, G.J.; et al. Personal and public safety issues related to arrhythmias that may affect consciousness: Implications for regulation and physician recommendations. A medical/scientific statement from the American Heart Association and the North American Society of Pacing. *Circulation* **1996**, *94*, 1147–1166. [CrossRef] [PubMed]
2. Pencina, M.J.; Navar, A.M.; Wojdyla, D.; Sanchez, R.J.; Khan, I.; Elassal, J.; D'agostino, R.B.; Peterson, E.D.; Sniderman, A.D. Quantifying Importance of Major Risk Factors for Coronary Heart Disease. *Circulation* **2019**, *139*, 1603–1611. [CrossRef] [PubMed]

3. Virani, S.S.; Alonso, A.; Benjamin, E.J.; Bittencourt, M.S.; Callaway, C.W.; Carson, A.P.; Chamberlain, A.M.; Chang, A.R.; Cheng, S.; Delling, F.N. Heart disease and stroke statistics—2020 update: A report from the American Heart Association. *Circulation* **2020**, *141*, E139–E596. [CrossRef] [PubMed]
4. Cardiovascular Diseases (CVDs). Available online: https://www.who.int/news-room/fact-sheets/detail/cardiovascular-diseases-(cvds) (accessed on 28 December 2021).
5. Virani, S.S.; Alonso, A.; Aparicio, H.J.; Benjamin, E.J.; Bittencourt, M.S.; Callaway, C.W.; Carson, A.P.; Chamberlain, A.M.; Cheng, S.; Delling, F.N. Heart Disease and Stroke Statistics–2021 Update A Report from the American Heart Association. *Circulation* **2021**, *143*, E254–E743. [CrossRef]
6. Ruthmann, N. AHA: Heart Disease Remains Leading Cause of Death Worldwide; Trends Discouraging. Available online: https://www.healio.com/news/cardiology/20210127/aha-heart-disease-remains-leading-cause-of-death-worldwide-trends-discouraging (accessed on 28 December 2021).
7. Epstein, A.E.; Baessler, C.A.; Curtis, A.B.; Estes, N.A.A.M.; Gersh, B.J.; Grubb, B.; Mitchell, L.B. Addendum to "Personal and Public Safety Issues Related to Arrhythmias That May Affect Consciousness: Implications for Regulation and Physician Recommendations: A Medical/Scientific Statement From the American Heart Association and the North American Society of Pacing and Electrophysiology". *Circulation* **2007**, *115*, 1170–1176.
8. Brignole, M.; Moya, A.; De Lange, F.J.; Deharo, J.C.; Elliott, P.M.; Fanciulli, A.; Fedorowski, A.; Furlan, R.; Kenny, R.A.; Martín, A.; et al. 2018 ESC Guidelines for the diagnosis and management of syncope. *Eur. Heart J.* **2018**, *39*, 1883–1948. [CrossRef]
9. Margulescu, A.D.; Anderson, M.H. A review of driving restrictions in patients at risk of syncope and cardiac arrhythmias associated with sudden incapacity: Differing global approaches to regulation and risk. *Arrhythmia Electrophysiol. Rev.* **2019**, *8*, 90–98. [CrossRef]
10. Sonmezocak, T.; Kurt, S. Detection of EMG Signals by Neural Networks Using Autoregression and Wavelet Entropy for Bruxism Diagnosis. *Elektron. Elektrotechnika* **2021**, *27*, 11–21. [CrossRef]
11. Nguyen, T.-N.; Nguyen, T.-H. Deep Learning Framework with ECG Feature-Based Kernels for Heart Disease Classification. *Electron. Electr. Eng.* **2021**, *27*, 48–59. [CrossRef]
12. Drezner, J.A.; Ackerman, M.J.; Cannon, B.C.; Corrado, D.; Heidbuchel, H.; Prutkin, J.M.; Salerno, J.C.; Anderson, J.; Ashley, E.; Asplund, C.A.; et al. Abnormal electrocardiographic findings in athletes: Recognising changes suggestive of primary electrical disease. *Br. J. Sports Med.* **2013**, *47*, 153–167. [CrossRef]
13. Guettler, N.; Bron, D.; Manen, O.; Gray, G.; Syburra, T.; Rienks, R.; D'Arcy, J.; Davenport, E.D.; Nicol, E.D. Management of cardiac conduction abnormalities and arrhythmia in aircrew. *Heart* **2019**, *105*, S38–S49. [CrossRef]
14. Klein, H.H.; Sechtem, U.; Trappe, H.-J. Fitness to Drive in Cardiovascular Disease. *Dtsch. Arztebl. Int.* **2017**, *114*, 692–701. [CrossRef]
15. Al-Khatib, S.M.; Stevenson, W.G.; Ackerman, M.J.; Bryant, W.J.; Callans, D.J.; Curtis, A.B.; Deal, B.J.; Dickfeld, T.; Field, M.E.; Fonarow, G.C.; et al. 2017 AHA/ACC/HRS Guideline for Management of Patients with Ventricular Arrhythmias and the Prevention of Sudden Cardiac Death: A Report of the American College of Cardiology/American Heart Association Task Force on Clinical Practice Guidelines and the Heart. *J. Am. Coll. Cardiol.* **2018**, *72*, e91–e220. [CrossRef]
16. Vijgen, J.; Albrecht, M.; Kumar, A.; Steen, T.; Tant, M.; Lerecouvreux, M.; Jung, W.; Rugina, M.; Falk, V.; Moerz, R.; et al. New Standards for Driving and Cardiovascular Diseases. *Eur. Work. Gr. Driv. Cardiovasc. Dis.* **2013**, *1*, 59.
17. EUR-Lex–32016L1106–EN–EUR-Lex. Available online: https://eur-lex.europa.eu/eli/dir/2016/1106/oj/deu (accessed on 3 January 2022).
18. Numé, A.K.; Gislason, G.; Christiansen, C.B.; Zahir, D.; Hlatky, M.A.; Torp-Pedersen, C.; Ruwald, M.H. Syncope and motor vehicle crash risk: A Danish nationwide study. *JAMA Intern. Med.* **2016**, *176*, 503–510. [CrossRef]
19. Sumiyoshi, M. Driving restrictions for patients with reflex syncope. *J. Arrhythmia* **2017**, *33*, 590–593. [CrossRef]
20. Jaul, E.; Barron, J. Age-Related Diseases and Clinical and Public Health Implications for the 85 Years Old and Over Population. *Front. Public Health* **2017**, *5*, 335. [CrossRef]
21. Kassebaum, N.J.; Arora, M.; Barber, R.M.; Brown, J.; Carter, A.; Casey, D.C.; Charlson, F.J.; Coates, M.M.; Coggeshall, M.; Cornaby, L.; et al. Global, regional, and national disability-adjusted life-years (DALYs) for 315 diseases and injuries and healthy life expectancy (HALE), 1990–2015: A systematic analysis for the Global Burden of Disease Study 2015. *Lancet* **2016**, *388*, 1603–1658. [CrossRef]
22. Simpson, C.; Ross, D.; Dorian, P. CCS Consensus Conference 2003: Assessment of the cardiac patient for fitness to drive and fly–Executive summary. *Can. J. Cardiol.* **2004**, *20*, 1313–1323.
23. Pockevicius, V.; Markevicius, V.; Cepenas, M.; Andriukaitis, D.; Navikas, D. Blood Glucose Level Estimation Using Interdigital Electrodes. *Elektron. Elektrotechnika* **2013**, *19*, 71–74. [CrossRef]
24. Jain, P.K.; Tiwari, A.K. Heart monitoring systems–A review. *Comput. Biol. Med.* **2014**, *54*, 1–13. [CrossRef]
25. Barbic, F.; Casazza, G.; Zamunér, A.R.; Costantino, G.; Orlandi, M.; Dipaola, F.; Capitanio, C.; Achenza, S.; Sheldon, R.; Furlan, R. Driving and working with syncope. *Auton. Neurosci. Basic Clin.* **2014**, *184*, 46–52. [CrossRef]
26. Chen, C.C.; Shiu, L.J.; Li, Y.L.; Tung, K.Y.; Chan, K.Y.; Yeh, C.J.; Chen, S.C.; Wong, R.H. Shift Work and Arteriosclerosis Risk in Professional Bus Drivers. *Ann. Epidemiol.* **2010**, *20*, 60–66. [CrossRef]
27. García Lledó, A.; Valdés Rodríguez, E.; Ozcoidi Val, M. Heart Disease and Vehicle Driving: Novelties in European and Spanish Law. *Rev. Española Cardiol. (Engl. Ed.)* **2018**, *71*, 892–894. [CrossRef]

28. Sidikova, M.; Martinek, R.; Kawala-Sterniuk, A.; Ladrova, M.; Jaros, R.; Danys, L.; Simonik, P. Vital sign monitoring in car seats based on electrocardiography, ballistocardiography and seismocardiography: A review. *Sensors* **2020**, *20*, 5699. [CrossRef]
29. Surgailis, T.; Valinevicius, A.; Markevicius, V.; Navikas, D.; Andriukaitis, D. Avoiding forward car collision using stereo vision system. *Elektron. Elektrotechnika* **2012**, *18*, 37–40. [CrossRef]
30. Soni, N.; Malekian, R.; Andriukaitis, D.; Navikas, D. Internet of Vehicles based approach for road safety applications using sensor technologies. *Wirel. Pers. Commun.* **2019**, *105*, 1257–1284. [CrossRef]
31. Ieremeiev, O.; Lukin, V.; Okarma, K.; Egiazarian, K. Full-Reference Quality Metric Based on Neural Network to Assess the Visual Quality of Remote Sensing Images. *Remote Sens.* **2020**, *12*, 2349. [CrossRef]
32. Paterova, T.; Prauzek, M. Estimating Harvestable Solar Energy from Atmospheric Pressure Using Deep Learning. *Elektron. Elektrotechnika* **2021**, *27*, 18–25. [CrossRef]
33. Sotner, R.; Domansky, O.; Jerabek, J.; Herencsar, N.; Petrzela, J.; Andriukaitis, D. Integer-and Fractional-Order Integral and Derivative Two-Port Summations: Practical Design Considerations. *Appl. Sci.* **2020**, *10*, 54. [CrossRef]
34. Prauzek, M.; Konecny, J. Optimizing of Q-Learning Day/Night Energy Strategy for Solar Harvesting Environmental Wireless Sensor Networks Nodes. *Elektron. Elektrotechnika* **2021**, *27*, 50–56. [CrossRef]
35. Skovierova, H.; Pavelek, M.; Okajcekova, T.; Palesova, J.; Strnadel, J.; Spanik, P.; Halašová, E.; Frivaldsky, M. The Biocompatibility of Wireless Power Charging System on Human Neural Cells. *Appl. Sci.* **2021**, *11*, 3611. [CrossRef]
36. Saini, S.K.; Gupta, R. Artificial intelligence methods for analysis of electrocardiogram signals for cardiac abnormalities: State-of-the-art and future challenges. *Artif. Intell. Rev.* **2021**, 1–47. [CrossRef]
37. Maršánová, L.; Ronzhina, M.; Smíšek, R.; Vítek, M.; Němcová, A.; Smital, L.; Nováková, M. ECG features and methods for automatic classification of ventricular premature and ischemic heartbeats: A comprehensive experimental study. *Sci. Rep.* **2017**, *7*, 1–11. [CrossRef]
38. Taymanov, R.; Sapozhnikova, K. What makes sensor devices and microsystems 'intelligent' or 'smart'? In *Smart Sensors and MEMS*, 2nd ed.; Elsevier: Amsterdam, The Netherlands, 2018; pp. 1–22, ISBN 9780081020562.
39. Yang, C.; Tavassolian, N. Combined Seismo- and Gyro-Cardiography. *IEEE J. Biomed. Health Inform.* **2018**, *22*, 1466–1475. [CrossRef]
40. Sahoo, P.K.; Thakkar, H.K.; Lin, W.Y.; Chang, P.C.; Lee, M.Y. On the design of an efficient cardiac health monitoring system through combined analysis of ECG and SCG signals. *Sensors* **2018**, *18*, 379. [CrossRef]
41. Javaid, A.Q.; Ashouri, H.; Dorier, A.; Etemadi, M.; Heller, J.A.; Roy, S.; Inan, O.T. Quantifying and reducing motion artifacts in wearable seismocardiogram measurements during walking to assess left ventricular health. *IEEE Trans. Biomed. Eng.* **2017**, *64*, 1277–1286. [CrossRef]
42. Pandia, K.; Ravindran, S.; Kovacs, G.T.A.; Giovangrandi, L.; Cole, R. Chest-accelerometry for hemodynamic trending during valsalvarecovery. In Proceedings of the 2010 3rd International Symposium on Applied Sciences in Biomedical and Communication Technologies, Rome, Italy, 7–10 November 2010; pp. 1–5.
43. Leonhardt, S.; Leicht, L.; Teichmann, D. Unobtrusive vital sign monitoring in automotive environments—A review. *Sensors* **2018**, *18*, 3080. [CrossRef]
44. Yang, C.; Tavassolian, N. Motion Noise Cansellation in Seismocardiographic Monitoring of Moving Subjects. In *IEEE Biomedical Circuits and Systems Conferences (BioCAS)*; IEEE: Piscataway, NJ, USA, 2015.
45. Hurnanen, T.; Lehtonen, E.; Jafari Tadi, M.; Kuusela, T.; Kiviniemi, T.; Saraste, A.; Vasankari, T.; Airaksinen, J.; Koivisto, T.; Tadi, M.J.; et al. Automated Detection of Atrial Fibrillation Based on Time-Frequency Analysis of Seismocardiograms. *IEEE J. Biomed. Health Inform.* **2017**, *21*, 1233–1241. [CrossRef]
46. Mohammed, Z.; Elfadel, I.; Abe, M.; Rasras, M. Monolithic multi degree of freedom (MDoF) capacitive MEMS accelerometers. *Micromachines* **2018**, *9*, 602. [CrossRef]
47. Cooper, S.; Cant, R.; Sparkes, L. Respiratory rate records: The repeated rate? *J. Clin. Nurs.* **2014**, *23*, 1236–1238. [CrossRef]
48. Mizuno, N.; Washino, K. A model based filtering technique for driver's heart rate monitoring using seat-embedded vibration sensors. In *ISCCSP 2014—2014 The 6th International Symposium on Communications, Control, and Signal Processing Proceedings*; IEEE: Piscataway, NJ, USA, 2014. [CrossRef]
49. Holcik, J.; Moudr, J. Mathematical model of seismocardiogram. *World Congr. Med. Phys. Biomed.* **2007**, *14*, 3415–3418.
50. Casanella, R.; Inan, O.T.; Migeotte, P.; Park, K.; Member, S.; Etemadi, M.; Member, S.; Tavakolian, K.; Casanella, R.; Zanetti, J.; et al. Ballistocardiography and Seismocardiography: Ballistocardiography and Seismocardiography: A Review of Recent Advances. *J. Biomed. Health Inform.* **2014**, *19*, 1414–1427. [CrossRef]
51. Leitão, F.; Moreira, E.; Alves, F.; Lourenço, M.; Azevedo, O.; Gaspar, J.; Rocha, L.A. High-Resolution Seismocardiogram Acquisition and Analysis System. *Sensors* **2018**, *18*, 3441. [CrossRef]
52. Okada, S.; Fujiwara, Y.; Yasuda, M.; Ohno, Y.; Makikawa, M. Non-restrictive heart rate monitoring using an acceleration sensor. In Proceedings of the 2006 International Conference of the IEEE Engineering in Medicine and Biology Society, New York, NY, USA, 30 August–3 September 2006.
53. Jafari Tadi, M.; Lehtonen, E.; Saraste, A.; Tuominen, J.; Koskinen, J.; Teräs, M.; Airaksinen, J.; Pänkäälä, M.; Koivisto, T. Gyrocardiography: A New Non-invasive Monitoring Method for the Assessment of Cardiac Mechanics and the Estimation of Hemodynamic Variables. *Sci. Rep.* **2017**, *7*, 6823. [CrossRef]

54. Luu, L.; Dinh, A. Using Moving Average Method to Recognize Systole and Diastole on Seismocardiogram without ECG Signal. In Proceedings of the Annual International Conference of the IEEE Engineering in Medicine and Biology Society, EMBS, Honolulu, HI, USA, 18–21 July 2018; pp. 3796–3799.

55. Di Rienzo, M.; Vaini, E.; Lombardi, P. An algorithm for the beat-to-beat assessment of cardiac mechanics during sleep on Earth and in microgravity from the seismocardiogram. *Sci. Rep.* **2017**, *7*, 1–12. [CrossRef]

56. Shafiq, G.; Tatinati, S.; Ang, W.T.; Veluvolu, K.C. Automatic Identification of Systolic Time Intervals in Seismocardiogram. *Sci. Rep.* **2016**, *6*, 37524. [CrossRef]

57. Georgoulas, G.; Chudacek, V.; Rieger, J.; Stylios, C.; Lhotska, L. Methods and Tools for Prosessing Biosignals: A Survey Paper. In Proceedings of the the the 3rd European Medical & Biological Engineering Conference, Prague, Czech Republic, 20–25 November 2005; Volume 11, pp. 20–25.

58. Fong, S.; Hang, Y.; Mohammed, S.; Fiaidhi, J. Stream-based Biomedical Classification Algorithms for Analyzing Biosignals. *J. Inf. Process. Syst.* **2012**, *7*, 717–732. [CrossRef]

59. Fong, S.; Lan, K.; Sun, P.; Mohammed, S.; Fiaidhi, J. A Time-Series Pre-Processing Methodology for Biosignal Classification using Statistical Feature Extraction. In Proceedings of the 10th IASTED International Conference on Biomedical Engineering (Biomed'13), Innsbruck, Austria, 13–15 February 2013; pp. 207–214.

60. Sircar, P. *Mathematical Aspects of Signal Processing*; Cambridge University Press: Cambridge, UK, 2016; ISBN 9781107175174.

61. Di Rienzo, M.; Meriggi, P.; Rizzo, F.; Vaini, E.; Faini, A.; Merati, G.; Parati, G.; Castiglioni, P. A wearable system for the seismocardiogram assessment in daily life conditions. In Proceedings of the 2011 Annual International Conference of the IEEE Engineering in Medicine and Biology Society, Boston, MA, USA, 30 August–3 September 2011; pp. 4263–4266.

62. Kaisti, M.; Tadi, M.J.; Lahdenoja, O.; Hurnanen, T.; Saraste, A.; Pankaala, M.; Koivisto, T. Stand-Alone Heartbeat Detection in Multidimensional Mechanocardiograms. *IEEE Sens. J.* **2019**, *19*, 234–242. [CrossRef]

63. Jafari Tadi, M.; Lehtonen, E.; Hurnanen, T.; Koskinen, J.; Eriksson, J.; Pänkäälä, M.; Teräs, M.; Koivisto, T. A real-time approach for heart rate monitoring using a Hilbert transform in seismocardiograms. *Physiol. Meas.* **2016**, *37*, 1885–1909. [CrossRef]

64. Semmlow, J.L.; Griffel, B. *Biosignal and Medical Image Processing MATLAB-Based Application*, 3rd ed.; Taylor & Francis Group: New York, NY, USA, 2014; ISBN 9781466567368.

65. Poularikas, A.D. *Discrete Random Signal Processing and Filtering Primer with MATLAB*; Taylor & Francis Group: New York, NY, USA, 2008; ISBN 9781420089332.

66. Sørensen, K.; Schmidt, S.E.; Jensen, A.S.; Søgaard, P.; Struijk, J.J. Definition of Fiducial Points in the Normal Seismocardiogram. *Sci. Rep.* **2018**, *8*, 15455. [CrossRef] [PubMed]

67. Mora, N.; Cocconcelli, F.; Matrella, G.; Ciampolini, P. Detection and Analysis of Heartbeats in Seismocardiogram Signals. *Sensors* **2020**, *20*, 1670. [CrossRef] [PubMed]

Article

Rescheduling of Distributed Manufacturing System with Machine Breakdowns

Xiaohui Zhang [1], Yuyan Han [2], Grzegorz Królczyk [3], Marek Rydel [4], Rafal Stanislawski [4] and Zhixiong Li [3,*]

1   School of Electrical and Control Engineering, Xuzhou University of Technology, Xuzhou 221018, China; xh_zhang@xzit.edu.cn
2   School of Computer Science, Liaocheng University, Liaocheng 252000, China; hanyuyan@lcu-cs.com
3   Department of Manufacturing Engineering and Automation Products, Opole University of Technology, 45-758 Opole, Poland; g.krolczyk@po.opole.pl
4   Department of Electrical, Control and Computer Engineering, Opole University of Technology, 45-758 Opole, Poland; m.rydel@po.opole.pl (M.R.); r.stanislawski@po.edu.pl (R.S.)
*   Correspondence: z.li@po.edu.pl

Abstract: This study attempts to explore the dynamic scheduling problem from the perspective of operational research optimization. The goal is to propose a rescheduling framework for solving distributed manufacturing systems that consider random machine breakdowns as the production disruption. We establish a mathematical model that can better describe the scheduling of the distributed blocking flowshop. To realize the dynamic scheduling, we adopt an "event-driven" policy and propose a two-stage "predictive-reactive" method consisting of two steps: initial solution pre-generation and rescheduling. In the first stage, a global initial schedule is generated and considers only the deterministic problem, i.e., optimizing the maximum completion time of static distributed blocking flowshop scheduling problems. In the second stage, that is, after the breakdown occurs, the rescheduling mechanism is triggered to seek a new schedule so that both maximum completion time and the stability measure of the system can be optimized. At the breakdown node, the operations of each job are classified and a hybrid rescheduling strategy consisting of "right-shift repair + local reorder" is performed. For local reorder, we designed a discrete memetic algorithm, which embeds the differential evolution concept in its search framework. To test the effectiveness of DMA, comparisons with mainstream algorithms are conducted on instances with different scales. The statistical results show that the ARPDs obtained from DMA are improved by 88%.

Keywords: distributed manufacturing; rescheduling; memetic algorithm

Citation: Zhang, X.; Han, Y.; Królczyk, G.; Rydel, M.; Stanislawski, R.; Li, Z. Rescheduling of Distributed Manufacturing System with Machine Breakdowns. *Electronics* **2022**, *11*, 249. https://doi.org/10.3390/ electronics11020249

Academic Editors: Darius Andriukaitis and Peter Brida

Received: 5 December 2021
Accepted: 10 January 2022
Published: 13 January 2022

Publisher's Note: MDPI stays neutral with regard to jurisdictional claims in published maps and institutional affiliations.

1. Introduction

With the advancement of economic globalization and the intensification of mergers between enterprises, the emergence of large-scale or concurrent production makes the pattern of distributed manufacturing necessary [1,2]. Distributed manufacturing decentralizes tasks into factories or workshops from different geographical locations. This pattern can help the manufacturers raise productivity, reduce cost, control risks, and adjust marketing policies more flexibly [3]. As an important part of distributed manufacturing, scheduling directly affects the efficiency and competitiveness of enterprises. Generally speaking, to solve such problems, a problem-specific model with production constraints should be first established to describe the scheduling problem considered. Then, optimization methods (e.g., mathematical programming, intelligent optimization, etc.) of operational research are developed to search for an optimal solution. For systems with large-scale and high complexity, mathematical programming such as integer programming, branch and bound, dynamic programming, or cut plane can rarely find an optimal solution (ranking) in the target space due to enumeration concept, but the efficiency decreases with the increment of the number of jobs/tasks to be scheduled.

At present, most studies use intelligent optimization algorithms to approximate the optimal solution for scheduling problems. The intelligent optimization algorithm, also called the evolutionary optimization algorithm, or metaheuristic, reveals the design principle of optimization algorithm through the understanding of relevant behavior, function, rules, and action mechanism in biological, physical, chemical, social, artistic, and other systems or fields. It refines the corresponding feature model under the guidance of the characteristics of specific problems and designs an intelligent iterative search process. That is, these kinds of algorithms do not rely on the characteristics of problems, but obtain near-optimal solutions through continuous iterations of global and local search. When an intelligent algorithm is applied for scheduling problems, it can express the schedule as a permutation model in the form of coding, and further compress the solution space into a very flat space, so that a large number of different permutations (schedules) correspond to the same target. Hence, the permutation model-based algorithm can search more different schedules in the target space range in tens of milliseconds to tens of seconds, so as to obtain a solution better than the traditional mathematical programming method.

The object of this study is related to the distributed blocking flowshop scheduling problem (DBFSP) [4]. Figure 1 illustrates DBFSP, which considers f parallel factories that contain the same machine configurations and technological processes [5]. The jobs can be assigned to any factories and each job follows the same blocking manufacturing procedure [6]. Although the machines configured in each distributed factory are the same, the processing time of each operation of each job is assumed to be different, thereby the processing tasks assigned to each distributed factory and their completion time are also different. The idea of solving DBFSP is to reasonably allocate the jobs to the factory through optimization algorithms, and then sequence the jobs in each distributed factory, to optimize the manufacturing objectives of the whole work order. Currently, researchers have made great efforts on solving DBFSP in a static environment, the existing researches mainly focused on the construction of mathematical models and the design of optimization algorithms. Zhang et al. [7] have established two different mathematical models using forward and reverse recursion approaches. A hybrid discrete differential evolution (DDE) algorithm was proposed to minimize the maximum completion time (makespan). Zhang et al. [8] constructed the mixed-integer model for DBFSP and developed a discrete fruit fly algorithm (DFOA) with a speed-up mechanism to minimize the global makespan. Additionally, Shao et al. [9] proposed a hybrid enhanced discrete fruit fly optimization algorithm (HEDFOA) to optimize the makespan. A new assignment rule and an insertion-based improvement procedure were developed to initialize the common central location of different fruit fly swarms. Li et al. [10] investigated a special case of DBFSP, in which a transport robot was embedded in each factory. The loading and unloading times are considered and different for all of the jobs conducted by the robot. An improved iterated greedy (IIG) algorithm was proposed to improve productivity. Moreover, Zhao et al. [11] proposed an ensemble discrete differential evolution (EDE) algorithm, in which three initialization heuristics that consider the front delay, blocking time, and idle time were designed. The mutation, crossover, and selection operators are redesigned to assist the EDE algorithm to execute in the discrete domain.

The above researches on DBFSP have formed a certain system, but they assumed that no explicit disruptions occur during the manufacturing process. In fact, a series of uncertainties often happened during the manufacturing process [12]. These uncertainties, which are sudden and uncontrollable, can change the state of the system strongly and affect the scheduling activities continuously [13]. As a result, the original static schedules are no longer suitable for real-time scheduling. To eliminate the impact of sudden uncertainties, rescheduling operations are generally performed in response to disruptions [14,15]. Rescheduling refers to the procedure of modifying the existing schedule to obtain a new feasible one after uncertain events occur [16]. One of the most important rescheduling strategies for traditional flowshop is "predictive-reactive" scheduling [17]. "predictive-reactive" scheduling defines a two-stage "event-driven" scheduling operation: the first stage generates an initial schedule that provides a baseline reference for other manufactural activities such as procurement and distribution of raw materials [18]. Influenced by the disruptions, the second stage explicitly quantifies the disruptions, constructs the management model with the disruption information gathered by the cyber-physical smart manufacturing technology [19–21], adjusts the initial schedule, and makes an effective trade-off between the initial optimization objective and the disturbance objective [22]. Since little literature is on the rescheduling of DBFSP, we review only the rescheduling strategies and algorithms developed for traditional and single flowshop. To realize the rescheduling, a suitable strategy should be determined in advance according to the scenario. Framinan et al. [23] discussed the problem of high system tension caused by continuous rescheduling of multi-stage flow production. A rescheduling strategy was described by estimating the availability of the machines after disruptions and a reordering algorithm based on the critical path was proposed. Katragjini et al. [24] analyzed eight types of uncertainties and designed rescheduling strategies through the classification of job status, which considers the completed, in processed and unprocessed operations. Iris et al. [25] designed a recoverable strategy taking the uncertainty of crane arrival to the ship and the fluctuation of loading and unloading speeds into account. The rescheduling strategy used a proactive baseline with reactive costs as the objective. Ma et al. [26] took the overmatch time (difference between real manufacturing time and the estimated time of the initial schedule) as one of the objectives in the rescheduling model to handle production emergencies in parallel flowshops. Li et al. [27] discussed both machine breakdown and processing change interruptions for a hybrid flowshop. The authors have proposed a hybrid fruit fly optimization algorithm (HFOA) with processing-delay, cast-break erasing, and right-shift strategy to minimize different rescheduling objectives in a steelmaking-foundry system. Li et al. [28] also considered five types of interruption events in the flowshop, namely machine breakdown, new jobs arrival, jobs cancellation, job processing change, and job release time change. A rescheduling strategy based on job status was designed for each event. A discrete teaching and learning optimization (DTLO) algorithm was proposed to optimize the makespan and stability. Valledor et al. [29] applied the Pareto optimum to solve the multi-objective flowshop rescheduling problem with makespan, total weighted tardiness, and steadiness as objectives. Three classes of disruptions (appearance of new jobs, machine faults, and changes in operational times) were discussed and an event management model was constructed. A restarted iterated Pareto greedy (RIPG) metaheuristic is used to find the optimal Pareto front.

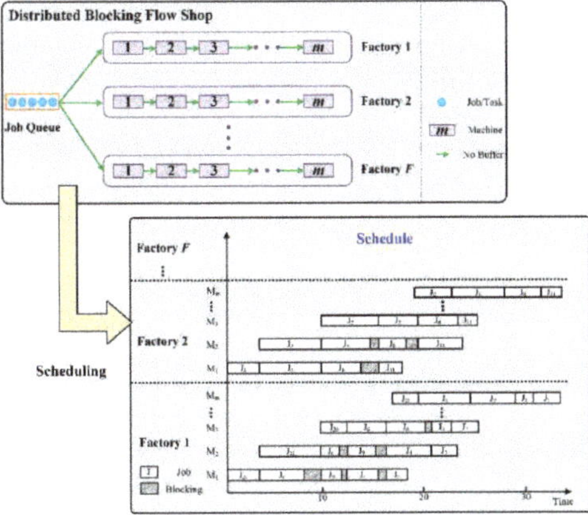

Figure 1. An example of DBFSP with the Gantt chart.

From the above review, it can be concluded that current researches focused mostly on the rescheduling of a single flowshop with various constraints. Little literature has considered rescheduling from the distributed manufacturing perspective. Though the Industry 4.0 wireless networks [30,31] have quickly developed in recent years, they are involved more in distributed information interconnection rather than decision making in scheduling fields. Likewise, the big data-driven technology [32,33] may provide real-time decisions or schedule rules for small-scale manufacturing, but has not formed a sound system. Moreover, big data technology relies strongly on a large amount of historical data, it is difficult to apply to new products due to the highly discrete, stochastic, and distributed properties of scheduling problems. Therefore, with the in-depth application of distributed manufacturing, distributed rescheduling strategies and approaches need to be formulated prudently so that effective references could be provided for modern decision-makers.

On the other hand, the objects of job shop scheduling are usually individual jobs, products, or other resources in the manufacturing process. Such resources have typical discrete characteristics, which need to be marked and expressed through special information carriers, and then obtain new combinations (ranking) by constantly updating the information carriers. These optimization characteristics are similar to the optimization process of the intelligent algorithm based on the permutation model. Therefore, the intelligent algorithm based on the evolution concept is more suitable for solving scheduling problems.

According to the above analysis and good applicability of the intelligent algorithm, we use an intelligent algorithm to reschedule the distributed blocking flowshop scheduling problem in a dynamic environment (DDBFSP). In the last decade, the application of intelligent algorithms for solving scheduling problems has been extensively investigated. Memetic algorithm (MA), also called the Lamarckian evolutionary algorithm, is attracting increasing concern. The concept of "meme" refers to contagious information patterns proposed by Dawkins in 1976 [34]. "Memes" are similar to genes in GA, but there are differences: Memetic evolution is characterized by Lamarckism, while genetic evolution is characterized by Darwinism. Meanwhile, neural system-based memetic information is more malleable than genetic information, so memes are more likely to change and spread more quickly than genes. In evolutionary computing, MA can combine various global and local strategies to construct different search frameworks, which possess the characteristics of GA but with stronger merit-seeking ability. MA was widely applied in many engineering problems, such as vehicle path planning [35], home care routing [36], bin packing problem [37], broadcast resource allocation [38], and production scheduling optimization [39–41]. Until now, MA has not been applied to solve DBFSP in a dynamic environment(DDBFSP), it will be of significance to extend MA as a solver for DDBFSP.

In summary, this paper aims to optimize DDBFSP with both makespan and stability measures as the objectives. The machine breakdown is defined as the disruption and assumed to happen stochastically in any distributed factories. To handle such dynamic events, a problem-specific disruption management model is constructed. A rescheduling framework that includes a job status-oriented classification strategy and a reordering algorithm-discrete Memetic algorithm (DMA) is proposed. For DMA, the differential evolution (DE) operators have been embedded to execute the neighborhood search. A simulated annealing (SA)-based reference local search framework is designed to help the algorithm escape from local optimums. Finally, the effectiveness of DMA is validated through comparative experiments. It is expected that the effect after rescheduling is to highly maintain the level of optimization of the original manufacturing objective (makespan) while ensuring the stability of the newly generated schedules.

The remainder of the paper is organized as follows. Section 2 states DDBFSP and constructs the mathematical model and objective function for DDBFSP. Section 3 designs the corresponding rescheduling framework. Section 4 elaborates the details of the DMA reordering algorithm. Section 5 verifies the performance of DMA and analyzes the results; Section 6 summarizes the research content of this paper.

2. Method

In this section, the mathematical models for DBFSP with optimization objectives in both static and dynamic environments are proposed. The classifications of job status after breakdown events are also introduced.

2.1. Statement of DBFSP in Static Environment

As can be seen in Figure 1, DBFSP not only needs to consider the correlation between processing task characteristics and blocking constraints but also needs to consider the coupling of global scheduling and local scheduling of each distributed factory, the solving process is more complex. As illustrated in Figure 1, a set of jobs $J = \{J_j|1,2,\ldots,n\}$ will be assigned to a set of factories $F = \{F_k|k=1,2,\ldots,f\}$, each of which contains a set of machines $M = \{M_i|1,2,\ldots,m\}$. The blocking constraint determines that no buffers are allowed between two adjacent machines. Therefore, the job can only be released to the next operation when the subsequent machine is free; otherwise, the job must be blocked on the current machine. We assume the processing time for each job is stochastic and different. After a job is assigned to a processing plant, it is not allowed to move to other factories.

Assume $n_k(n_k \leq n)$ jobs are assigned to factory k, and the job sequence in this factory is denoted as π_k, where $\pi_k(l)$ represents the l-th job in π_k. The operation $O_{\pi_k(l),i}$ has a processing time $P_{\pi_k(l),i}$. Assume $S_{\pi_k(l),0}$ is the start time of $\pi_k(l)$ on the first machine of factory k, and $d_{\pi_k(l),i}$ is defined as the departure time of operation $O_{\pi_k(l),i}$ on machine i. The recursive formulas of DBFSP can be derived as follows:

$$S_{\pi_k(1),0} = 0 \tag{1}$$

$$D_{\pi_k(1),i} = D_{\pi_k(1),i-1} + P_{\pi_k(1),i}, \quad i = 2,3,\ldots,m \tag{2}$$

$$S_{\pi_k(l),0} = D_{\pi_k(l-1),1}, \quad i = 2,\ldots,n_k \tag{3}$$

$$D_{\pi_k(l),i} = \max\left\{D_{\pi_k(l),i-1} + P_{\pi_k(l),i}, \ D_{\pi_k(l-1),i+1}\right\}, \quad l = 2,3,\ldots,n_k \ \ i = 1,2,\ldots,m-1 \tag{4}$$

$$D_{\pi_k(l),m} = D_{\pi_k(l),m-1} + P_{\pi_k(l),m}, \quad l = 2,3,\ldots,n_k \tag{5}$$

In the above equations, Equations (1) and (2) calculate the start and departure time of the first job π_k from machine 1 to the last machine m in factory k. Equations (3) and (4) calculate the start and departure time of job $\pi_k(l)$ from machine 1 to machine $m - 1$. Equation (5) gives the departure time of $\pi_k(l)$ on machine m. If we take makespan $C(\pi_k)$ as the optimization objective, $C(\pi_k)$ of factory k can be expressed as:

$$C(\pi_k) = D_{\pi_k(n_k),m} \tag{6}$$

As a result, the global makespan for DBFSP is defined through the comparison between $C(\pi_k)$ of all distributed factories:

$$C_{\max}(\Pi) = \max_{k=1}^{f}(C(\pi_k)) \tag{7}$$

The detailed recursive process and example refer to our previous group work [8] for solving DBFSP.

2.2. Statement of DDBFSP

When characterizing the machine breakdown event of DBFSP in a dynamic and stochastic environment, the following questions should be marked: (1) Which factory has happened the breakdown event and when (the probability of breakdown)? (2) Which machine in that factory breaks (the distributivity of the breakdown)? (3) When will the machine resume?

In fact, machine breakdowns are difficult to simulate since the probabilistic model of breakdown occurrence could hardly cover the real manufacturing situation. Moreover, the recovery time is mostly predicted based on a priori knowledge, which cannot guarantee accuracy. With consideration of randomness and distribution, this paper triggers machine breakdowns in a randomly selected distributed factory at time t. The breakdown time is defined to follow a discrete uniform distribution function which is expressed as follows:

$$E(B_{k,i}) = rand()\%P(T_i) + p_{\pi_k(l)} \times i, \quad i = 1, 2, 3 \ldots, m, \; k = 1, 2, \ldots, f, \; l = 1, 2, \ldots, n_k \quad (8)$$

where $rand()$ represents the random function between $[0, \omega]$ and ω denotes the maximum constant of the system; "%" is the remainder operator; $P(T_i)$ represents the total processing time of all jobs on machine i. Equation (8) defines the time range of the breakdown in the factory k during the processing time of all jobs, i.e., $[P_{\pi_k(l),i}, C_{\pi_k(l),i}]$.

To maintain the distribution of breakdowns and the convenience of experimental statistics, this study assumes that each distributed factory occurs β times random breakdowns during manufacturing. Additionally, to maintain the randomness of breakdown occurrence, each machine has the same probability to break down. To ensure a stochastic dynamic environment, the duration of breakdowns are generated randomly and uniformly following the interval $[0, \omega]$ and are determined immediately after the event. Moreover, other constraints for the breakdown event in DDBFSP are defined as follows:

(1)   All machines exist three statuses during manufacturing: idle, processing and blocked, a breakdown event occurs in the processing period.
(2)   The system triggers one machine breakdown each time, and the process stops immediately when the breakdown occurs.
(3)   After the machine is recovered, the affected process can continue with the remaining processing without re-processing.

2.3. Optimization Objectives of DDBFSP

In DBFSP, it is generally necessary to consider the production efficiency-related objective, e.g., makespan. While in a dynamic environment, from the decision point of view, stability becomes one importantly practical metric for manufacturing systems. If rescheduling optimization is performed only considering the production efficiency-related indicators, it may generate new schedules that deviate significantly from the initial plan, which in turn affects other planning activities, such as material management and manpower planning. Therefore, in the rescheduling phase, in addition to makespan, the stability of the new schedules should be considered. In this study, the initial schedule of one distributed factory before a breakdown occurs is denoted by B (Baseline), and the schedule after rescheduling is denoted by $B*$. The goal of rescheduling is to optimize both the makespan and the stability of the distributed factory at each breakdown node. The first objective (makespan) of the DDBFSP is expressed as follows:

$$f_1 = C_{\max}(B*) \quad (9)$$

The second objective (stability) of the DDBFSP is derived as follows.

$$f_2 = \min\{\sum_{i=1}^{m} \sum_{l=1}^{n_k} Z_{\pi_k(l),i}\}, \quad k = 1, 2, \ldots, f, \; n_k = 1, 2, \ldots, n \quad (10)$$

where the decision variable $Z_{\pi_k(l),i}$ indicates whether the relative position of the job B and $B*$ has changed. $Z_{\pi_k(l),i} = 1$ represents that the position of job $\pi_k(l)$ on machine i in factory k has changed in the new schedule $B*$ and vice versa $Z_{\pi_k(l),i} = 0$.

To simplify the optimization process and avoid redundant calculations, a weighting mechanism is applied to combine both objective functions:

$$f(B*) = w_1 * f_1 + w_2 * f_2 \quad (11)$$

In Equation (11), w_1 and w_2 represent the weight coefficients of f_1 and f_2, respectively. Since f_1 and f_2 have different distribution ranges of dimensions, to avoid the results being dominated by the data with larger or smaller distribution ranges, the normalization method proposed in [24] is applied so that the value range of each objective falls in the interval. The normalization function is defined as follows:

$$f(B*) = w_1 * N(f_1) + w_2 * N(f_2) \tag{12}$$

where:

$$N(f_1) = \frac{f_1(B*) - low(f_1)}{up(f_1) - low(f_1)} \tag{13}$$

$$N(f_2) = \frac{f_2(B*) - low(f_2)}{up(f_2) - low(f_2)} \tag{14}$$

In Equations (13) and (14), $up(.)$ and $low(.)$ represent the upper and lower bounds of f_1 and f_2 for the two extreme rankings of the jobs at the breakdown node, respectively. The specific calculation procedure refers to [24].

2.4. Statement of Job Status after Breakdown Event

After a machine breaks down, the jobs are categorized to construct the event management model:

$$C^*_{\pi_k(l),i} - S^*_{\pi_k(l),i} - P^*_{\pi_k(l),i} + (1 - y_{\pi_k(l),i,1})\omega \geq 0 \tag{15}$$

$$C^*_{\pi_k(l),i} - S^*_{\pi_k(l),i} - P^*_{\pi_k(l),i} - (1 - y_{\pi_k(l),i,1})\omega \leq 0 \tag{16}$$

$$C^*_{\pi_k(l),i} - S^*_{\pi_k(l),i} - P^*_{\pi_k(l),i} - B_{e,i} + B_{s,i} + (1 - y_{\pi_k(l),i,2})\omega \geq 0 \tag{17}$$

$$C^*_{\pi_k(l),i} - S^*_{\pi_k(l),i} - P^*_{\pi_k(l),i} - B_{e,i} + B_{s,i} - (1 - y_{\pi_k(l),i,2})\omega \leq 0 \tag{18}$$

$$C^*_{\pi_k(l),i} - \max\left\{S^*_{\pi_k(l),i}, B_{e,i}\right\} - P^*_{\pi_k(l),i} + (1 - y_{\pi_k(l),i,3})\omega \geq 0 \tag{19}$$

$$C^*_{\pi_k(l),i} - \max\left\{S^*_{\pi_k(l),i}, B_{e,i}\right\} - P^*_{\pi_k(l),i} - (1 - y_{\pi_k(l),i,3})\omega \leq 0 \tag{20}$$

$$\sum_{g=1}^{3} Y_{\pi_k(l),i,g} = 1, \quad i = \{1,2,\ldots,m\}, \ k = \{1,2,\ldots,f\}, \ g = \{1,2,3\} \tag{21}$$

$$Y_{\pi_k(l),i,g} = \{0,1\}, \quad i = \{1,2,\ldots,m\}, \ k = \{1,2,\ldots,f\}, \ g = \{1,2,3\} \tag{22}$$

In the above equations, $C^*_{\pi_k(l),i}$ denotes the completion time of job $\pi_k(l)$ on the machine i in $B*$. $S^*_{\pi_k(l),i}$ and $P^*_{\pi_k(l),i}$ represent the corresponding start time and processing time of $C^*_{\pi_k(l),i}$, respectively. $B_{e,i}$ and $B_{s,i}$ denote the occurrence time and completion time of the breakdown on the machine i. Equation (15) to Equation (20) defines three statuses in which an operation of a job is in when a breakdown occurs: Equations (15) and (16) determine that the operation was completed before the breakdown occurs; Equations (17) and (18) determine that the operation is being processed when the breakdown occurs; Equations (19) and (20) determine that the operation was originally scheduled to start after the machine is recovered. Equation (21) represents that the job can only be in a state in case a breakdown occurs. Equation (22) is a binary decision variable set for three cases: (1) $Y_{\pi_k(l),i,1} = 1$ means the operation is completed before the breakdown occurs; (2) $Y_{\pi_k(l),i,2} = 1$ denotes the operation overlaps with the machine breakdown node; (3) $Y_{\pi_k(l),i,3} = 1$ means the operation begins after the machine is resumed.

For a better understanding, an example is presented in Figure 2 to illustrate the classification of the operation status at the moment that a machine breaks down. In case 1 of Figure 2, machine 2 of factory k breaks down at time 55, at which time the operations $O_{\pi_k(1),1}$, $O_{\pi_k(1),2}$ and $O_{\pi_k(2),1}$ have already completed processing. Their start and completion times are not affected and do not need to be adjusted in the rescheduling phase. In case 2 of Figure 2, machine 1 breaks down at node 55 and operation $O_{\pi_k(3),1}$ is being processed at this time. The breakdown divides $O_{\pi_k(3),1}$ into two parts: the finished and the remaining part, the remaining part is completed after the machine is recovered. Hence, the start time $O_{\pi_k(3),1}$ remains unchanged in the rescheduling phase, but its finish time is affected by both the breakdown time and the recovery time. In case 3 of Figure 2, machine 1 breaks down at node 38 which is before the startup of job J_3 and J_4. Therefore, the start and finish times of J_3 and J_4 are affected by both the breakdown time and the recovery time.

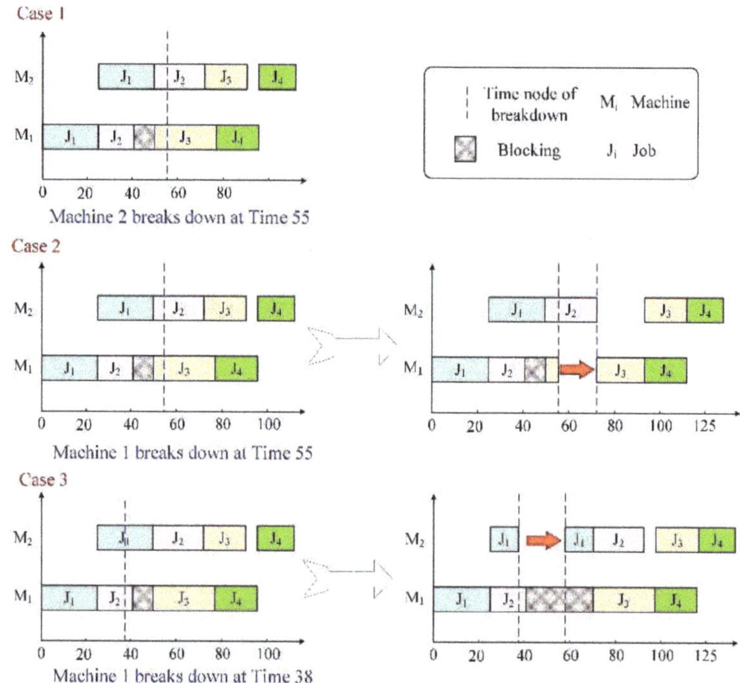

Figure 2. Classification of job status at machine breakdowns.

3. Rescheduling Framework for DDBFSP

Since the breakdowns occur during the manufacturing procedure, on which each job has already been fixed in a certain factory for processing, a change of jobs between different factories is unrealistic. Hence, the individual factory is defined as the decision subject in response to the events.

3.1. Rescheduling Strategy

We envision a stochastic and dynamic distributed scheduling environment. A two-stage "predictive-reactive" method is proposed for DDBFSP: initial solution pre-generation and rescheduling. In the first stage, the initial schedule for DDBFSP is generated by considering a static environment without machine breakdowns. After the breakdown occurs, the initial schedule may no longer be optimal. Therefore, in the second stage, the "event-driven" rescheduling is triggered to evaluate the breakdown, and a new schedule is provided in response to the events. Since the new schedule will be executed until the next breakdown occurs, the rescheduling strategy proposed in this study should have a dual objective: on one hand, to adapt and minimize the impact of the event; on the other hand, to generate a schedule that gives a good tradeoff between scheduling quality and stability when the event is resumed.

3.2. Rescheduling Method

According to the classification of operation status in Section 2.3, we propose a hybrid rescheduling method: "right-shift repair + local reorder". At the breakdown node, no adjustment is made to the completed operations; for the directly affected operations, the right-shift strategy is adopted for a local repair; for the jobs that have not yet started their processing, the reordering algorithm is performed to seek a better partial schedule. The proposed "right-shift repair + local reorder" method is described as follows:

First, determine the jobs that are to be rescheduled. According to the constraint limitation of the flowshop manufacturing, once the processing sequence of the jobs on the first machine is determined, their sequence on other machines must be the same. Therefore, we mark the jobs $\pi_k(l)$ at the breakdown node by using the first machine as the reference. Suppose that there is a breakdown event at time $B_{s,i}$ when job $\pi_k(l)$ is processed on machine i, then the jobs in this factory of which the first operation has been started are counted in the set N_c; relatively, the jobs of which the first operation has not been started are counted in the set of unprocessed jobs N_n.

Subsequently, the jobs N_c are further divided by taking $\pi_k(l)$ as the midpoint: the jobs sequencing before $\pi_k(l)$ are included in the set $N_{c,c} = \{\pi_k(1),\dots,\pi_k(l-1)\}$; the remaining jobs N_c are included in the set $N_{c,n}$. The rescheduling system maintains the same order and time points for the operations of jobs $N_{c,c}$. For the operations of jobs in $N_{c,n}$, the unaffected operations remain unchanged; the affected and other unprocessed operations are adjusted using the right-shift repair method [42], which shifts the start time to right by certain matching time units. The right-shift repair is essentially a FIFO-based heuristic.

At the breakdown node, all jobs N_n have not been started processing. Their initial order may be no longer optimal after the recovery. Thus, we propose an improved algorithm to reorder the jobs. Eventually, the new schedule is merged into the global schedule and is executed as the baseline until the next breakdown occurs.

To describe the proposed "right-shift repair + local reorder" method more clearly, Figure 3 shows a comparative example with different rescheduling methods at the time of breakdown in one distributed factory. As shown in Figure 3a, the initial schedule of the factory has a desired makespan of 36. During manufacturing, Machine 2 breaks down at time $B_{s,i} = 8$ and is assumed to be recovered at $B_{e,i} = 11$. At this point, the jobs that have already been processed on machine 1 are J_2 and J_1 (marked in gray), and these two jobs are counted in the set N_c. In contrast, jobs J_3, J_4 and J_5 are counted in N_n. The framework adjusts the affected operations in N_c based on the breakdown information and reorders the jobs in N_n. As seen in Figure 3b, the right-shift method is implemented on partial operations of the jobs (marked in yellow), the process order remains the same as $J_2 \to J_1 \to J_5 \to J_3 \to J_4$ and the makespan is delayed to 41. In Figure 3c while using the "right-shift repair + local reorder" method, job J_3, J_4 and J_5 (marked in green) is reordered using the reordering algorithm. The process order changes to $J_2 \to J_1 \to J_3 \to J_4 \to J_5$ and the makespan is 37, which absorbs 4 units of the recovery time. This example proves that the proposed rescheduling method is more efficient and flexible than the single rule-based (right-shift repair) heuristics.

3.3. Rescheduling Procedure

We illustrate the proposed optimization procedure for DDBFSP in Figure 4. First, with makespan as the optimization objective, a global schedule for DBFSP in a static environment is generated using DFOA [8]; then, each distributed factory executes manufacturing tasks according to their initial schedules; the breakdowns are triggered following the discrete generation mechanism proposed in Section 2.2. Each time the breakdown happens, rescheduling is implemented by the single distributed factory, with makespan and stability as optimization objectives. According to process status at the breakdown node, the jobs are classified and different methods (right-shift repair or reordering algorithm) are performed. The rescheduling results of different job sets are integrated, and the updated schedule under a single breakdown is provided as the initial schedule before the next event occurs. The above procedure is repeated until the termination condition of the breakdown trigger is met, and the final schedule is output.

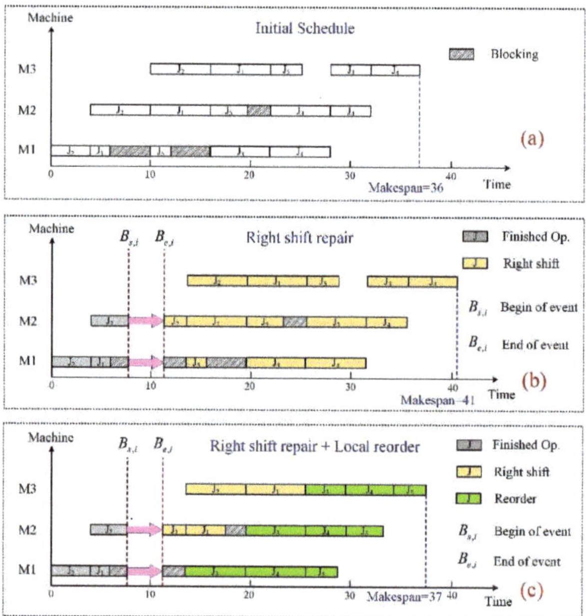

Figure 3. Comparison of different rescheduling strategies.

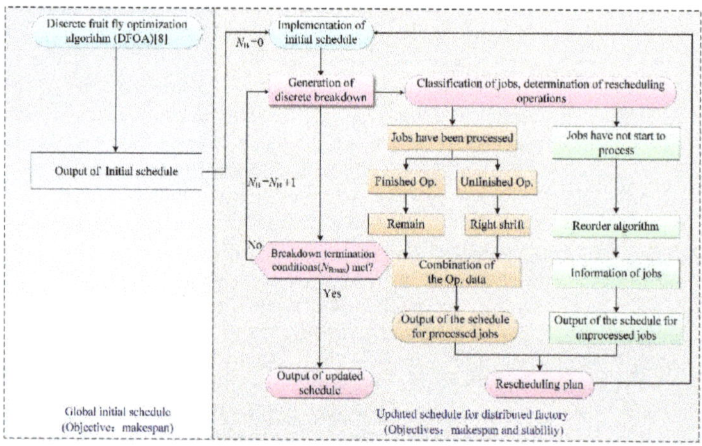

Figure 4. Proposed rescheduling procedure for DDBFSP.

4. Reordering Algorithm-DMA

Since the "right-shift repair" is introduced in [42], this section introduces the proposed reordering algorithm-DMA for the jobs in the set N_n.

4.1. Introduction of Standard MA

MA was initially defined as an improvement of GA, it can combine different global and local strategies to construct various search frameworks, which has stronger flexibility, and merit-seeking ability than GA. The flow diagram of the basic MA is shown in Figure 5. MA starts from initializing the population, operates on memes with evolutionary thought, generates new individuals using generating functions (e.g., crossover and variation operators, etc.), and finally forms new populations using updating functions (e.g., selection operators, etc.).

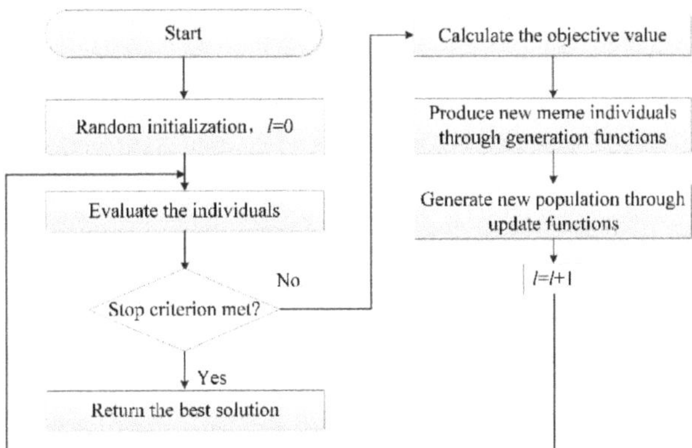

Figure 5. Flow diagram of standard MA.

Although standard MA has a strong optimization-seeking capability, it also encounters problems such as insufficient global exploration capability [43]. On the other hand, DE has proven its powerful search capability since being proposed. Inspired by this, we embed the DE operators into MA and proposed a discrete MA (DMA). DMA mainly contains the following parts: population initialization, DE (mutation and crossover), and local search.

4.2. Population Initialization

In the initialization phase, we use the well-known job sequence-based method [9] to encode. The position of each job in the sequence represents its processing order on corresponding machines. An initialization method WPNEH (Weighted position NEH method, WPNEH) considering the weighted position of the job is proposed under the premise of determining the reordering jobs.

Step 1: Using the PFT_NEH(x) heuristic [9] and the initial schedule to generate two seeds π^P and π^B. The total weighted position of a single job is defined as follows:

$$\phi_j = \chi_1 \times \varphi_j(\pi^P) + \chi_2 \times \varphi_j(\pi^B) \tag{23}$$

In Equation (23), $\varphi_j(\pi^P)$ and $\varphi_j(\pi^B)$ represent the absolute positions of job j in two seeds. χ_1 and χ_2 represent the weight coefficients occupied by two seeds. The values of χ_1 and χ_2 are defined adaptively by the population size Ps and the order l ($l = 1, 2, \ldots, Ps$) of the newly generated individuals in the whole population:

$$\chi_1 = \frac{l}{Ps - 1} \tag{24}$$

$$\chi_2 = 1 - \frac{l}{Ps - 1} \tag{25}$$

Step 2: Arrange all jobs in decreasing order of the total weighted positions ϕ_j to obtain a sequence π^0.

Step 3: Create a new empty sequence π^{emp}. The jobs in π^0 are inserted into each position π^{emp} in turn. The solutions obtained from each insertion are evaluated based on Equations (11) and (12) to determine the optimal order. Continue the operations until all jobs finish insertion.

Step 4: Let $l = l + 1$, and continue steps 1–3 until all Ps individuals are generated.

The initialization procedure of WPNEH is shown as Algorithm 1.

Algorithm 1 WPNEH initialization

Input: job set N_n, population size Ps, initial schedule B of a distributed factory
Output: initial population POP
01:    Define the order of jobs in N_n, generate seed π^B
02:    Apply PFT_NEH(x) method to rearrange the jobs in N_n, generate seed π^P
03:    **While** $l \leq Ps$ **do**
04:        Set the coefficient $\chi_1 = l/Ps - 1$ and $\chi_2 = 1 - l/(Ps - 1)$
05:        Calculate ϕ_j for each job according to Equation (22)
06:        Generate a new sequence π^0 by arranging all jobs in descending order based on their ϕ_j
07:        Execute the NEH insertion procedure, evaluate the solutions obtained, find the
best order
08:        Finish insertions of all jobs, obtain a new sequence π^c, count in POP
09:        $l = l + 1$
10:    **End While**

4.3. DE Operation

DE [44] drives the search directions through the mutual synergistic and competitive behaviors among individuals of the population. Its overall structure is similar to GA, but the evolution principle is quite different. DE generates new individuals through perturbing the difference vectors of two or more individuals. It can obtain more operation space and enhance the global search capability when the individuals are significantly different from each other in the early stage of the search. In DMA, two key operators of DE (mutation and crossover) are embedded to perform the global search.

4.3.1. Mutation

The weighted difference vector of two individuals is first selected. Then, the weighted difference is summed with another individual vector. Hence, three individuals are randomly selected from the population POP, the optimal one is defined as π_a, and the other two are defined as π_b and π_c. The mutation operator can be expressed as:

$$V_a = \pi_a \oplus \kappa \otimes (\pi_b - \pi_c) \tag{26}$$

where κ is the mutation scaling factor used to control the magnitude of the differences. "$\otimes$" represent the weighted differences between π_b and π_c:

$$\pi_a - \pi_b = \Delta \Leftrightarrow \delta(j) = \begin{cases} \pi_b(j), & \text{if } \pi_b(j) \neq \pi_c(j) \\ 0, & \text{otherwise} \end{cases} \quad j = 1, 2, \ldots, n \tag{27}$$

$$\kappa \otimes \Delta = \Phi \Leftrightarrow \varphi(j) = \begin{cases} \delta(j), & \text{if } rand() < \kappa \\ 0, & \text{otherwise} \end{cases} \tag{28}$$

In Equations (27) and (28), $\Delta = [\delta(1), \delta(2), \ldots, \delta(n)]$ and $\Phi = [\varphi(1), \varphi(2), \ldots, \varphi(n)]$ are the two temporary vectors used for the calculation. "$\oplus$" means the mutation individual π_V is obtained through adding Φ with the target individual π_{best}:

$$\pi_V = \pi_{best} \oplus \Phi \tag{29}$$

The generation process of π_V is described as follows.
Step 1: Select π_a, set $j = 1$.
Step 2: If $\varphi(j) = 0$, set $j = j + 1$, go to step 3; otherwise, generate a random number between (0, 1). If $rand() < \kappa$, update π_a by swapping the job $\pi_a(j)$ and $\varphi(j)$; else, insert the job $\pi_a(j)$ into all different positions of $\varphi(j)$, take the optimal solution and update π_a. Let $j = j + 1$.
Step 3: If $j \leq n$, return to step 2; otherwise, return $\pi_V = \pi_a$.
The mutation procedure of DMA is sketched in Algorithm 2.

Algorithm 2 Mutation Operation

Input: population POP, job number n, population size Ps, mutation factor κ, temporary set Δ and Φ
Output: mutation individual π_V
01:   Select 3 individuals (π_a, π_b and π_c) from POP randomly
02:      **For** $j = 1$ to n **do**
03:         Calculate the vector difference $\delta(j)$ between two individuals and save in Δ
04:         Generate $rand()$ between (0,1), calculate the mutation difference $\varphi(j)$ and save in Φ
05:      **End For**
06:      Output Φ
07:      **For** $j = 1$ to n do
08:         **If** $\varphi(j) = 0$
09:            $j = j + 1$
10:         **Else**
11:            generate $rand()$ between (0, 1)
12:              **If** $rand() < \kappa$
13:                 exchange job $\pi_a(j)$ and $\varphi(j)$ in π_a
14:                 **Else** insert $\varphi(j)$ from π_a into all positions of after $\pi_a(j)$, take the optimal solution
15:              **End If**
16:           return the π_a
17:         **End If**
18:      **End For**
19:      Let $\pi_V = \pi_a$, output π_V

4.3.2. Crossover

When solving discrete scheduling problems based on job sequence, the probability factor is mainly used to determine the crossed jobs [38]. In this study, we improved the determination method of the crossed jobs by eliminating the crossover probability factor and proposed a random crossover operator: First, select two jobs randomly from the mutated individual π_V; second, put these two jobs and jobs between them in a temporary set N_{temp}; third, determine the positions in π_d of all jobs from N_{temp}, remove all jobs from N_{temp} and keep their positions. Finally, insert the jobs in π_d with original order to obtain a new sequence π'. The crossover process for π_d is the same. When the crossover operation of the two parents is completed, the new individual obtained is evaluated and the best one is retained. The crossover operation is sketched in Algorithm 3.

Algorithm 3 Crossover Operation

Input: mutation individual π_V, target individual π_d, temporary job set N_{temp}
Output: new individual π_{new}
01: # Crossover operation on π_d
02:   Select two jobs from π_V randomly, put the jobs between them in N_{temp} in turn
03:   Remove the jobs belonging to N_{temp} from π_d, keep the vacant position unchanged
04:   Insert the jobs in N_{temp} into the free positions of π_d to obtain the new solution π'
05: # Crossover operation on π_V
06:   Ensure π_d has the job $j \in N_{temp}$, clear N_{temp}, put j in N_{temp} orderly
07:   Remove jobs belonging to N_{temp} from π_V, keep the vacant position unchanged
08:   Insert the jobs in N_{temp} into the free positions of π_d to obtain the new solution π''
09:   Evaluate new solutions:
10:      **If** $f(\pi'') < f(\pi')$
11:         Let $\pi_{new} = \pi''$, return π_{new}
12:      **Else**
13:         Let $\pi_{new} = \pi'$, return π_{new}
14:      **End If**

To facilitate understanding, **Example 4-1** presents the procedure of DE operation in detail.

Example 4-1

Mutation: Three individuals are randomly selected from the initial population: $\pi_a = [J_6, J_3, J_2, J_4, J_1, J_5]$, $\pi_b = [J_1, J_4, J_6, J_2, J_5, J_3]$ and $\pi_c = [J_3, J_4, J_2, J_1, J_5, J_6]$. With Equation (26) it yields $\Delta = \pi_b - \pi_c = [J_1, 0, J_6, J_2, 0, J_3]$. A set of random numbers $[0.7, 0.6, 0.9, 0.4, 0.1, 0.3]$ are generated according to Equation (27), so that the mutation scaling factor is $\kappa = 0.5$, then obtain $\Phi = [0, 0, 0, J_2, 0, J_3]$. It can be deduced that when $j = 4$ and $j = 6$, there are $\varphi(4) = J_2$ and $\varphi(6) = J_3$. For $j = 4$, a random number 0.2 (<0.5) is generated between the interval (0,1) satisfying the uniform distribution. Then, two jobs $\pi_a(4) = J_4$ and $\varphi(4) = J_2$ in π_a are swapped and a new solution $\pi_a = [J_6, J_3, J_4, J_2, J_1, J_5]$ is obtained; for $j = 6$, a random number 0.7 (>0.5) is also generated, the job $\varphi(6) = J_3$ in π_a is inserted into the latter position of $\pi_a(6) = J_5$ and a new solution $\pi_a = [J_6, J_4, J_2, J_1, J_5, J_3]$ is obtained.

Crossover: The mutation individual $\pi_V = [J_6, J_4, J_2, J_1, J_5, J_3]$ and the target individual $\pi_d = [J_5, J_3, J_2, J_4, J_6, J_1]$ are crossed as parents. First, two jobs J_2 and J_5 are randomly selected from π_V, and set $N_{temp} = [J_2, J_1, J_5]$. For π_d, remove the same jobs from N_{temp}, obtain $\pi_d = [X, J_3, X, J_4, J_6, X]$. The new solution is obtained by reinserting $N_{temp} = [J_2, J_1, J_5]$ into π_d. The derivation of the new solution for π_V is the same as π_d, and the result is $\pi_V = [J_6, J_4, J_5, J_2, J_1, J_3]$.

4.4. Job Block-Based Random Reference Local Search

As mentioned above, the two key operations of DE can improve the individuals concerning the vector difference of the population. Along with the iteration of an algorithm, the difference between individuals becomes smaller, which tends to lead the algorithm to local optimum easily. Therefore, DMA needs to equip with a local search framework to enhance its exploitation capability. For a long time, reference local search (RLS) has proved to be an effective local search algorithm and is often used to enhance the exploitation of metaheuristics [9]. RLS firstly generates a random reference sequence $\pi_r = [\pi_r(1), \pi_r(2), \ldots, \pi_r(z)]$, and uses it as a reference to guide the direction of the local search; subsequently, the jobs $\pi_r(j)$ are sequentially removed and inserted into all remaining positions of π_r to obtain new solutions. The optimal solution is compared with the incumbent solutions of the population, and if it is better, it is replaced with the population. The insertion process is repeated until all the jobs are traversed. Though RLS has a strong local exploitation capability, it still has some problems. On one hand, RLS uses a single job insertion operation, which may destroy the good information of incumbent solutions and cause the loss of other good solutions. On the other hand, the fixed order of the reference sequence and the fixed insertion process of the jobs will result in a fixed path of local search. If $\pi_r(j)$ is constant for a long time, it will cause a large number of repeated searches, which directly affects the search efficiency of the algorithm.

Boejko et al. [45] have pointed out that in job sorting scheduling problems, compound moves (insertion and swap) based on the job block can retain excellent sequence information during the evolution of an algorithm. It thus expands the neighborhood structure and search space, which is better than single job insertion and swap operations. Inspired by this idea, we hybridize RLS and compound moves of job block, and propose a random reference local search based on job block (BRRLS): firstly, generate a reference sequence $\pi_r = [\pi_r(1), \pi_r(2), \ldots, \pi_r(z)]$ randomly, where n_r represents the number of jobs to be rescheduled; secondly, select two jobs J_a and J_b randomly, construct the job block (including J_a and J_b) and take out all jobs in the individual π_{block} that needs local search; then, insert the job block into all possible positions of π, evaluate the generated solutions, and select the optimal one. Repeat the above procedure (each time select two unselected jobs) until all jobs are traversed. The BRRLS process is sketched in Algorithm 4.

Algorithm 4 BRRLS Procedure

Input: job set N_n, individual π, temporary set N_{temp}, temporary set Λ
Output: new individual π_{new}
01:   Randomly sort the jobs in N_n to generate a reference sequence π_r
02:   Randomly select two jobs J_a and J_b from N_n
03:   Determine the block π_{block} between J_a and J_b in π_r (including J_a and J_b), save π_{block} in N_{temp}
04:   Remove all jobs belonging to N_{temp} from π
05:   Insert π_{block} in all positions of π, evaluate and select the optimal solution, save in Λ
06:   Clear N_{temp}, delete J_a and J_b from N_n
07:   Repeat the above operation (Line 03–07) until $len(N_n) \leq 1$
08:   Evaluate the individuals in Λ, return the optimal solution to π_{new}

To further improve the algorithmic performance, a simulated annealing-like (SA) mechanism [46] is introduced as an acceptance criterion for BRRLS, which guides DMA to receive a certain percentage of poor solutions during the search to avoid being trapped in local optimums. The idea of SA is to compare the neighborhood solution π' obtained by BRRLS with the incumbent solution π. If $f(\pi')$ is better than $f(\pi)$, SA replaces π with π', otherwise, the decision of whether to accept π' is based on a reception probability μ: A random number $rand()$ satisfying a uniform distribution is generated between $(0, 1)$; if $rand() < \mu$, π is replaced by the worse neighborhood solution obtained by the search. μ is expressed as follows:

$$\mu = e^{-(f(\pi') - f(\pi))/Temp} \tag{30}$$

where *Temp* represents the temperature constant:

$$Temp = T_0 \frac{\sum\limits_{j=1}^{n_k} \sum\limits_{i=1}^{m} P_{\pi_k(l),i}}{10 \times m \times n}, k \in \{1, 2, \ldots, f\} \tag{31}$$

In Equation (31), T_0 is the temperature adjustment parameter preset by SA. It can be seen from Equation (31) that the closer $f(\pi')$ is to $f(\pi)$, the closer the value μ is to 1 and π' will be accepted with a higher probability. Conversely, if $f(\pi')$ is much worse than $f(\pi)$, the value μ will be close to 0, and the solution π' will be dropped with a higher probability. Hence, the SA-based reception mechanism ensures that the population does not deviate from the current search position, but can additionally absorb a certain percentage of non-quality solutions to avoid the algorithm from falling into local optimums.

4.5. Update Strategy of the Population

To maintain the diversity of individuals, the following strategy is applied to update the population: first, a new individual is generated using the mutation and crossover operators; subsequently, a local search is performed on these individuals, and the individuals are updated. finally, the incumbent solutions are replaced by the newly generated solutions, and the uniqueness of these new solutions is ensured to complete a single iteration of the whole population.

4.6. Flowchart of DMA

According to previous descriptions, Algorithm 5 presents the flowchart of DMA.

The flow diagram of DMA is sketched in Figure 6. In general, DMA contains population initialization, DE operation, and local search. In the initialization phase, the population is generated using the WPNEH method; in the DE phase, the discrete differential mutation and crossover operators are executed to obtain the child individuals; in the local search phase, BRRLS is performed, and the simulated annealing mechanism is adopted as the reception criterion. Finally, the population is updated and the optimal solution is output.

Algorithm 5 Flowchart of DMA

Input: population size Ps, mutation scaling factor κ, reordered job number n
Output: best solution π
01:   **While** termination condition not met **do**
02:      # Initialization (**Section 4.2**)
03:         Use WPNEH method to create new individuals
04:         Construct *POP*, evaluate the individuals
05:      # DE (**Section 4.3**)
06:         Perform DE according to κ, generate $Ps/2$ individuals
07:         Perform random crossover operation on mutated individuals, and evaluate
08:      # BRRLS (**Section 4.4**)
09:         Implement the SA-based BRRLS procedure on the $Ps/2$ individuals
10:         Replace the incumbent solutions using solutions obtained by local search
11:         Update the population (**Section 4.5**)
12:   **End While**

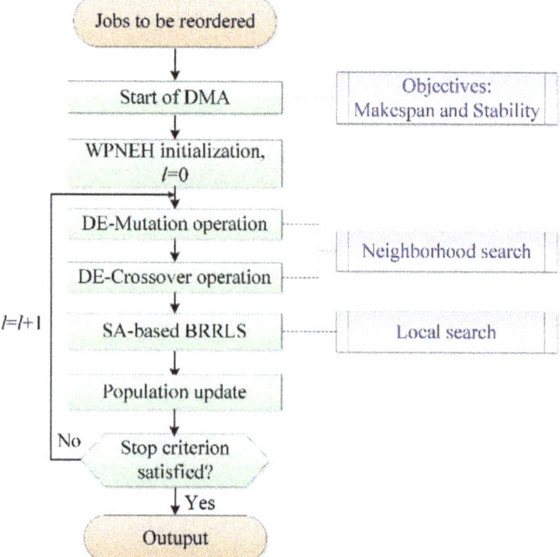

Figure 6. Flow diagram of DMA.

5. Experimental Comparison and Analysis

5.1. Experimental Settings

Since few pieces of literature and public benchmarks have been developed for DDBFSP, we apply DFOA on its benchmark [8] to obtain test instances. These test instances are used as the initial schedules for each distributed factory. To fulfill different experimental requirements, the range of variable intervals of the DPFSP benchmark is set as $n \in \{50, 100, 200\}$, $m \in \{5, 10, 20\}$ and $f \in \{2, 3, 4\}$. There are 27 combinations of different parameters, each containing 10 instances. The termination criterion of the algorithm is set to $T_{\max} = 90 \times n \times m$ milliseconds.

The breakdown events are simulated according to the mechanism introduced in Section 2.2. When an event occurs on machine i, the trigger node is firstly limited to the interval $[P_{\pi_k(l),i}, C_{\pi_k(l),i}]$ to ensure the timeliness of the breakdown. Since DMA performs only on partial jobs which have not started processing, we compress the breakdown interval to $[P_{\pi_k(l),i}, C_{\pi_k(n_k-2),1}]$, i.e., the start time of processing to the completion time of the penultimate job at the first machine, to ensure a feasible execution space for DMA.

The experiments are conducted on a PC with Intel(R) Core(TM) i7-8700 CPU and 16G RAM configuration, and the involved programs are compiled by Python. To balance the objective functions (makespan and stability), both weight coefficients w_1 and w_2 are set to 0.5. The algorithm is repeated 10 times for each breakdown in each factory, and the experiments use the average relative percent deviation (ARPD) as the metric to evaluate the mean quality of the obtained solutions. Since DMA is implemented for each distributed factory, the sum of ARPDs of all factories is firstly calculated, and the mean value is defined as the ARPD value for a single case of DDBFSP. The experiments will be conducted in the following three perspectives:

(1)   Key parameter calibration;
(2)   Effectiveness of the proposed optimization strategy;
(3)   Comparison with other intelligent algorithms.

5.2. Parameter Calibration

DMA contains three key parameters: population size Ps, mutation scaling factor κ, and temperature adjustment parameter T_0. The design of the experiment (DOE) [9] method is used for parameter calibrations, and in total 36 sets of initial schedules with factory number $f = 3$ were generated. The number of random breakdowns for each distributed factory is set to 2. The key parameters and ARPD values are defined as the control factors and response variables, respectively. The candidate values of the above parameters are set as: $Ps = \{30, 50, 80, 100\}$, $\kappa = \{0.4, 0.7, 0.9\}$ and $T_0 = \{0.1, 0.2, 0.3, 0.4\}$, which generate 48 configuration combinations. We use the "Analysis of Variance" (ANOVA) method to analyze the statistical results, as shown in Table 1.

Table 1. ANOVA results of DMA parameter combinations.

Source	Sum of Squares	Degree of Freedom	Mean Square	*F*-Ratio	*p*-Value
Ps	38.4	3	12.8	134.4	**0.007**
κ	144.3	2	72.2	382.4	**0.000**
T_0	6.8	3	2.3	17.6	**0.012**
$Ps \times \kappa$	17.29	6	2.9	8.8	0.354
$Ps \times T_0$	4.8	9	0.5	0.55	0.492
$\kappa \times T_0$	0.5	6	0.1	2.46	0.087

As can be seen in Table 1, the *p*-values of Ps, κ and T_0 are all less than 0.05 confidence level, which means all the parameters have important impacts on the performance of DMA. Among them, the corresponding *F*-ratio value of κ is the largest, which indicates that κ has the greatest impact on DMA. Moreover, Table 1 shows that the *p*-values of the interactions between every two parameters are greater than 0.05, which means the parameter interactions do not have a significant effect on DMA, and the parameter can be selected directly from the main effects plot in Figure 7.

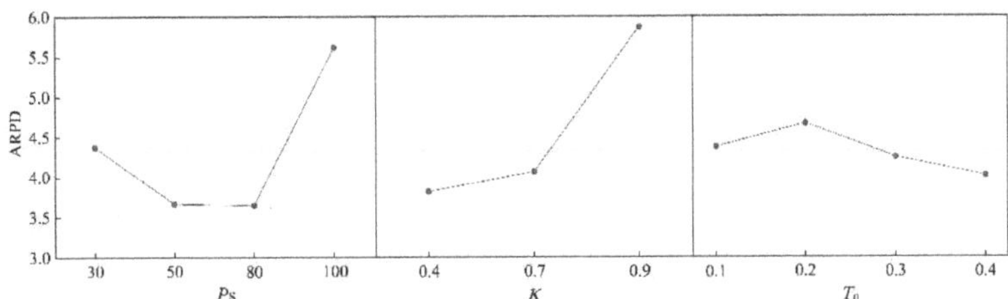

Figure 7. Main effects plot of the parameters for DMA.

From Figure 7, it can be observed that the performance of DMA decreases with the increment of κ, and DMA obtains the best results when $\kappa = 0.4$. A relatively overlarge mutation scaling factor increases the randomness of search and leads to degradation of the mutation. The effect of Ps ranks second, the performance of DMA first increases as the number of Ps increases, it starts to decrease after reaching the optimum. This indicates that increasing Ps appropriately will enhance the diversity of the population. However, an overlarge Ps consumes too much running time of a single iteration, which in turn compresses the number of iterations and leads to a decrease in the probability of obtaining the optimal solution. The effect of T_0 on the algorithmic performance ranked the 3rd, and the main effect plot shows that the performance fluctuates with the growth of T_0, DMA obtained the best results when $T_0 = 0.4$. Based on the above analysis, the parameter combinations of DMA are set as: $Ps = 80$, $\kappa = 0.4$, $T_0 = 0.4$.

5.3. Effectiveness of the Proposed Algorithmic Component

DMA contains three important components: WPNEH initialization, DE operators, and BRRLS. To verify their effectiveness, we mask one corresponding part of DMA each time, and generate three types of variant algorithms from DMA: (1) DMA_RI: random initialization, which is used to verify the effectiveness of WPNEH initialization; (2) DMA_ND: without neighborhood search, which is used to verify the effectiveness of DE operators; (3) DMA_NL: without BRRLS, which is used to verify the effectiveness of BRRLS. The parameter settings and instances used in Section 5.2 were adopted. As the randomness of breakdown has a large impact on the results, to ensure the fairness of the comparison, only one complete set of breakdowns is simulated, and all variant algorithms are tested under this scenario. Table 2 shows the comparison results.

Table 2. Comparison results between DMA and its variant algorithms.

Instance	ARPD			
$n \times m \times f$	DMA_RI	DMA_ND	DMA-NL	DMA
$50 \times 5 \times 3$	0.56	1.24	0.92	**0.54**
$50 \times 10 \times 3$	0.61	1.55	0.98	**0.36**
$50 \times 20 \times 3$	1.17	1.73	1.12	**0.48**
$100 \times 5 \times 3$	1.33	2.26	2.60	**0.18**
$100 \times 10 \times 3$	2.05	2.44	2.38	**0.11**
$100 \times 20 \times 3$	2.84	3.23	3.15	**0.05**
$200 \times 5 \times 3$	3.97	5.58	4.19	**0.00**
$200 \times 10 \times 3$	4.63	6.67	5.49	**0.00**
$200 \times 20 \times 3$	4.17	6.04	5.32	**0.00**

From Table 2, we can observe that the performance of DMA outperforms the other variants in all scenarios. In specific analysis, DMA outperforms DMA_RI, representing that the WPNEH initialization strategy can provide a better search starting point for DMA; DMA outperforms DMA_ND, indicating that the DE operators can improve the performance of DMA effectively. The results of DMA_NL are inferior to those of DMA, which means that the proposed BRRLS and SA-based reception criterion have an important influence on the optimization. BRRLS has retained the "greedy search" idea from RLS but improved the selection of jobs in the reference sequence. It ensures the inconsistency of local search step length through random selection of job blocks strategy, which makes the solutions obtained by local search more diverse and helps DMA to jump out of the local optimum. On the other hand, it can be observed from Table 2 that the differences between the compared algorithms are not significant when the scale of the instance is relatively small (e.g., $n = 50$). If the processing tasks (number of jobs) assigned to a distributed factory are small, the corresponding reorder execution space is also smaller, and the comparison algorithms are more likely to obtain optimal or suboptimal solutions in a given time. With the gradual increase in the problem size, the differences between algorithms start to manifest. The performance of each variant algorithm decreases more on the large-scale instances, while DMA remains relatively more stable.

To verify whether the differences were significant, we conducted a significance test from a statistical point of view. Each algorithm and ARPD value were chosen as the control variable and the response variable. Figure 8 demonstrates the mean plot of the 95% confidence interval. As can be seen, the ARPD values of each algorithm are ranked from top to bottom as DMA_ND, DMA_RI, DMA_NL, and DMA. The confidence intervals of each two algorithms do not overlap, which indicates that DMA is significantly better than its variants. The experimental results reveal that the proposed optimization strategies for each search phase jointly ensure the performance of DMA.

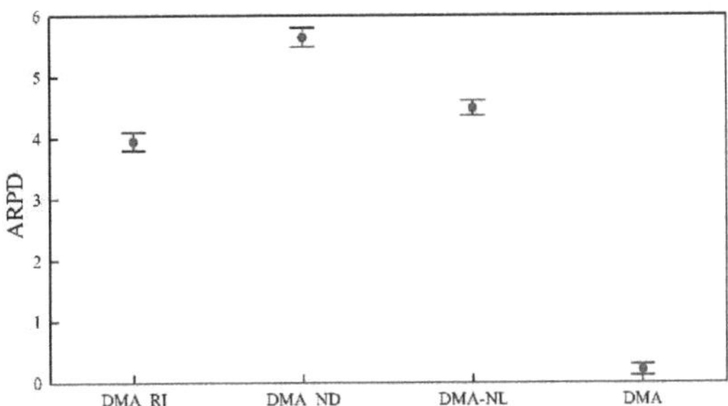

Figure 8. Mean plots with 95% confidence intervals for DMA and its variant algorithms.

5.4. Comparison with Other Intelligent Algorithms

In this section, DMA is compared with the algorithms for solving traditional flow shop rescheduling problems. The compared algorithms are (1) Iterative Local Insertion Search (ILS) [25]; (2) Iterative Greedy Algorithm (IG) [25]; (3) Improved Migratory Bird Algorithm (IMBO) [47]; (4) Discrete Teaching and Learning Optimization (DTLO) Algorithm [29]. We follow strictly the literature to compile all the comparison algorithms, which share the same data structure, objective function, and termination criterion. To ensure comparison fairness, we apply the same rescheduling process, i.e., the comparison algorithms share the same initial schedules, the same time node distribution of breakdown events, and the same rescheduling strategy. The algorithms are only used to reorder unprocessed jobs as a way to test their optimization efficiencies. The calculation of ARPD is consistent with previous sections. The parameter combinations of the comparison algorithm were pre-adjusted and the specific settings are shown in Table 3. According to [25], as a single local search algorithm, ILS stops immediately after obtaining a local optimum, so no special parameters or termination conditions need to be set for ILS.

Table 3. Parameter determination of the compared algorithms.

Algorithm	Parameters
ILS	--
IG	Job numbers of destruction 4
IMBO	Population size 50; Neighborhood set size 7; Migratory birds pass on the number of solutions 3; Number of lead bird iterations 20; Number of population iteration 100; weighted factor 0.8
DTLO	Population size 50; learning factor of teacher 1

Table 4 shows the statistical results of different instances under single breakdown ($\beta = 1$), the optimal values were bolded. It can be seen that DMA outperforms other algorithms in all instances of distributed scenarios. DTLO, IG, IMBO, and ILS obtain results in 2nd, 3rd, 4th and 5th place. It indicates that DMA is more suitable as a reordering algorithm. Specifically, ILS is an algorithm containing only a local insertion search framework, which lacks key structures such as population initialization and neighborhood search. It is difficult for ILS to balance exploration and exploitation, and therefore has the worst performance. As metaheuristics, DTLO, IG, and IMBO are more competitive in terms of performance on small-scale instances. For example, when the number of the distributed factory is set as $f = 4$, and the total number of jobs is $n = 50$, the results of other algorithms do not differ from DMA. The main reason is that when the size of the initial schedule is small, fewer processing tasks are assigned to each distributed factory. The search space for the solution becomes smaller, so the efficiency of each algorithm in finding the best solutions in a given time becomes high, thus reducing the variability of the comparison results. The same phenomenon is reflected in the scenarios of $\beta = 2$ and $\beta = 3$, as shown in Tables 5 and 6. DMA also performs the best among all comparison algorithms. As the number of breakdowns increases, the performances of the compared algorithms are not affected too much by small-scale instances. The performances on large-scale instances ($n = 200$) showed different degrees of degradation. The errors of single rescheduling accumulate with the increment of breakdowns, which causes a deterioration effect, and thus decreases the algorithmic performances relatively. In comparison, the statistical results of DMA under different breakdown scenarios are less different, which also reveals that DMA is more stable and robust on both small- and large-scale instances.

Table 4. Comparison results of different algorithms by $\beta = 1$.

Initial Instance	ARPD				
$n \times m \times f$	ILS	IG	IMBO	DTLO	DMA
$50 \times 5 \times 2$	2.40	1.46	1.43	1.06	**1.02**
$50 \times 5 \times 3$	2.18	0.55	0.85	0.51	**0.39**
$50 \times 5 \times 4$	0.00	0.00	0.00	0.00	**0.00**
$50 \times 10 \times 2$	2.93	1.77	1.48	1.13	**1.08**
$50 \times 10 \times 3$	2.46	0.72	0.83	0.66	**0.28**
$50 \times 10 \times 4$	0.14	0.00	0.00	0.00	**0.00**
$50 \times 20 \times 2$	3.17	1.04	1.47	1.19	**1.05**
$50 \times 20 \times 3$	2.89	0.85	1.04	0.69	**0.34**
$50 \times 20 \times 4$	1.74	0.31	0.49	0.00	**0.00**
$100 \times 5 \times 2$	4.23	2.53	2.28	2.24	**0.53**
$100 \times 5 \times 3$	3.44	1.56	1.71	1.47	**0.17**
$100 \times 5 \times 4$	2.56	1.42	1.38	1.25	**0.08**
$100 \times 10 \times 2$	4.49	2.06	2.33	2.09	**0.58**
$100 \times 10 \times 3$	3.23	1.60	1.63	1.71	**0.22**
$100 \times 10 \times 4$	2.91	1.53	1.24	1.39	**0.13**
$100 \times 20 \times 2$	5.44	2.55	2.70	2.31	**0.32**
$100 \times 20 \times 3$	3.69	1.97	1.80	1.88	**0.04**
$100 \times 20 \times 4$	3.28	1.11	1.31	1.78	**0.37**
$200 \times 5 \times 2$	5.78	2.88	3.53	2.90	**0.00**
$200 \times 5 \times 3$	4.95	2.19	2.61	2.13	**0.12**
$200 \times 5 \times 4$	4.16	2.28	2.39	2.11	**0.08**
$200 \times 10 \times 2$	5.57	3.54	3.72	3.02	**0.06**
$200 \times 10 \times 3$	5.11	2.05	2.47	2.00	**0.00**
$200 \times 10 \times 4$	4.39	2.29	2.45	2.03	**0.00**
$200 \times 20 \times 2$	6.06	2.98	3.84	3.23	**0.00**
$200 \times 20 \times 3$	6.08	2.57	2.79	1.97	**0.00**
$200 \times 20 \times 4$	4.47	2.56	2.94	2.19	**0.00**
Ave.	3.62	1.71	1.87	1.59	**0.25**

To further verify the superiority of DMA, the differences between the compared algorithms are observed through statistical tests. ANOVA is used to describe the mean plot with a 95% confidence interval of the results obtained by the algorithms for different f, as shown in Figure 9.

It can be observed that the ARPD of DMA falls below that of other algorithms, and none of the confidence intervals overlap. It indicates again that the optimization performance of DMA is significantly better than its comparators. Moreover, it can be seen from Figure 9 that the performance of each algorithm gradually becomes better as f increases.

This is mainly because the reordering algorithm does not focus on the assignment of jobs to plants in the initial schedule, rather on the reordering within the distributed factories. An increase in f leads to a decrease in assigned jobs and a corresponding decrease in their computational complexity and, therefore, an increase in the optimization-seeking efficiency of an algorithm.

Moreover, we compared and analyzed the performances under different numbers of breakdowns (β) based on the statistical results. Figure 10 shows the performance curves of the comparative algorithms. It can be seen that ARPD values increase as β gradually increases, representing that all the algorithms are affected by the frequency of breakdowns. Compared with other algorithms, ARPDs of DMA are improved by at least 88%. Moreover, ARPD values of ILS, IG, IMBO, and DTLO algorithms fluctuate more and show an increasing trend with the increase in β. Comparatively, ARPD values of DMA fluctuate the least, which indicates that the robustness of DMA under different scenarios is better than the compared algorithms.

In summary, DMA has a good performance for local reordering. Its innovation and advantages can be summarized as follows: (1) WPNEH initialization method provides a better initial population and high-quality search starting point for DMA; (2) the mutation and crossover operators based on DE provides an excellent neighborhood search capability; (3) BRRLS provides stronger local exploitation; (4) the SA-based reception criterion can help DMA jump out of the local optimum effectively. Among them, (2), (3) and (4) balance the exploration and exploitation of DMA.

Table 5. Comparison results of different algorithms by $\beta = 2$.

Initial Instance	ARPD				
$n \times m \times f$	LS	IG	IMBO	DTLO	DMA
$50 \times 5 \times 2$	2.89	1.66	1.82	1.45	0.83
$50 \times 5 \times 3$	2.06	0.41	0.52	0.28	0.25
$50 \times 5 \times 4$	0.00	0.00	0.00	0.00	0.00
$50 \times 10 \times 2$	3.04	1.77	1.84	1.50	0.98
$50 \times 10 \times 3$	2.11	0.32	0.63	0.26	0.13
$50 \times 10 \times 4$	0.22	0.00	0.00	0.00	0.00
$50 \times 20 \times 2$	3.28	1.74	1.90	1.56	0.95
$50 \times 20 \times 3$	2.04	0.36	0.61	0.33	0.21
$50 \times 20 \times 4$	1.65	0.55	0.53	0.00	0.00
$100 \times 5 \times 2$	3.55	2.76	2.87	2.51	0.68
$100 \times 5 \times 3$	3.79	1.84	1.97	1.59	0.19
$100 \times 5 \times 4$	2.74	1.71	1.69	1.28	0.84
$100 \times 10 \times 2$	3.92	2.66	2.75	2.44	0.55
$100 \times 10 \times 3$	3.72	1.93	1.93	1.64	0.31
$100 \times 10 \times 4$	2.63	1.58	1.82	1.35	0.67
$100 \times 20 \times 2$	4.73	2.85	2.92	2.67	0.41
$100 \times 20 \times 3$	3.96	1.85	2.05	1.78	0.18
$100 \times 20 \times 4$	3.04	1.66	1.92	1.20	0.23
$200 \times 5 \times 2$	6.12	4.03	4.73	4.09	0.12
$200 \times 5 \times 3$	6.58	3.04	3.67	2.82	0.05
$200 \times 5 \times 4$	4.16	2.84	3.04	2.55	0.08
$200 \times 10 \times 2$	6.49	4.39	4.54	4.15	0.00
$200 \times 10 \times 3$	6.71	3.05	3.44	2.96	0.00
$200 \times 10 \times 4$	4.39	2.65	3.45	2.91	0.13
$200 \times 20 \times 2$	6.88	4.45	4.96	4.32	0.00
$200 \times 20 \times 3$	6.33	3.58	3.70	2.71	0.00
$200 \times 20 \times 4$	4.47	2.43	3.49	2.63	0.00
Ave.	3.75	2.07	2.33	1.88	0.29

Table 6. Comparison results of different algorithms by $\beta = 3$.

Initial Instance	ARPD				
$n \times m \times f$	LS	IG	IMBO	DTLO	DMA
$50 \times 5 \times 2$	1.80	0.58	0.73	0.39	**0.20**
$50 \times 5 \times 3$	2.01	0.34	0.64	0.28	**0.15**
$50 \times 5 \times 4$	**0.00**	**0.00**	**0.00**	**0.00**	0.00
$50 \times 10 \times 2$	2.93	1.70	1.88	1.47	**0.11**
$50 \times 10 \times 3$	2.17	0.45	0.70	0.30	**0.24**
$50 \times 10 \times 4$	1.31	0.00	0.00	0.00	**0.00**
$50 \times 20 \times 2$	3.28	2.01	2.03	1.92	**0.06**
$50 \times 20 \times 3$	1.94	0.63	0.78	0.44	**0.17**
$50 \times 20 \times 4$	1.59	0.24	0.43	0.00	**0.00**
$100 \times 5 \times 2$	4.37	3.04	3.15	2.83	**0.05**
$100 \times 5 \times 3$	3.65	2.19	2.24	2.02	**0.19**
$100 \times 5 \times 4$	2.93	2.11	2.18	1.85	**0.00**
$100 \times 10 \times 2$	4.62	2.96	3.29	2.89	**0.52**
$100 \times 10 \times 3$	3.98	2.73	2.81	2.50	**0.28**
$100 \times 10 \times 4$	3.15	2.06	2.86	1.72	**0.00**
$100 \times 20 \times 2$	5.89	3.12	3.56	2.84	**0.13**
$100 \times 20 \times 3$	4.71	3.17	3.37	3.01	**0.06**
$100 \times 20 \times 4$	3.44	2.88	2.94	1.56	**0.00**
$200 \times 5 \times 2$	6.61	4.09	4.91	3.47	**0.00**
$200 \times 5 \times 3$	6.23	4.61	5.23	2.98	**0.00**
$200 \times 5 \times 4$	6.16	3.77	4.53	1.93	**0.00**
$200 \times 10 \times 2$	6.79	4.61	4.86	3.54	**0.00**
$200 \times 10 \times 3$	6.47	4.87	5.65	3.94	**0.00**
$200 \times 10 \times 4$	6.57	4.19	4.29	2.20	**0.00**
$200 \times 20 \times 2$	6.84	5.12	5.27	4.07	**0.00**
$200 \times 20 \times 3$	6.97	5.43	5.19	4.83	**0.00**
$200 \times 20 \times 4$	6.21	3.98	4.84	2.58	**0.00**
Ave.	4.16	2.63	2.94	2.04	**0.08**

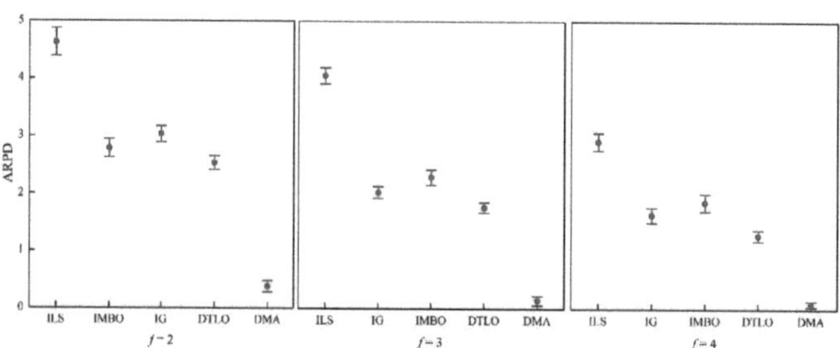

Figure 9. Mean plot with 95% confidence interval of the compared algorithms on different scenarios.

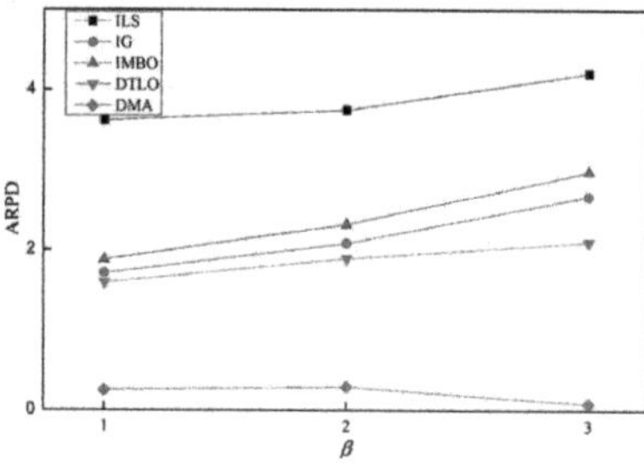

Figure 10. Performance comparison with different breakdown numbers.

6. Conclusions

Building rescheduling optimization models and designing effective optimization methods according to the characteristics of distributed manufacturing are of significance to promote the research of the dynamic scheduling theory of distributed manufacturing. This study investigated the rescheduling strategy and algorithm for DDBFSP, in which machine breakdown events are considered as the disruption in the manufacturing site. Firstly, the mathematical model of DDBFSP including the event simulation mechanism is constructed. We consider makespan and stability as the objectives. The goal of this study is to optimize the bi-objective when the stochastic breakdown occurs in any distributed factories. We apply the "event-driven" policy in response to the disruption. A two-stage "predictive-reactive" rescheduling strategy is proposed. In the first stage, a static environment (DBFSP) without machine breakdown is considered, and the global initial schedules are generated; in the second stage, after machine breakdown occurs, the initial schedule is locally optimized by a hybrid repair policy based on "right-shift repair + local reorder", and the DMA reordering algorithm based on DE is proposed for local reorder operation. For DMA, a WPNEH initialization method is designed to generate a high-quality population. In the neighborhood search phase, DE is embedded to improve the neighborhood structure and expand the target space by using mutation and crossover operators; in the local search phase, the BRRLS framework is proposed to perturb the high-quality solutions. To maintain the diversity, BRRLS has combined with the SA mechanism to receive the worse solutions with a certain probability. To obtain the best performance of DMA, the DOE method is used to calibrate three key parameters. The effectiveness of the proposed optimization strategy for DMA is verified through comparative experiments. Finally, DMA is compared with other algorithms on different test instances. The statistical analysis using ANOVA has verified the superiority of DMA.

Although the proposed rescheduling strategy has shown effectiveness, there still exist shortcomings. In this study, we only considered the breakdown event as the disruption. Real-life manufacturing suffers from far more than one disruption. The other common disruptions such as job cancellations and their interaction mechanism should be deeply investigated. Therefore, future works will concentrate on the construction of a more refined model that can manage more disruptions simultaneously.

This study attempts to explore the dynamic scheduling problem from the perspective of operational research optimization. With the development of the Industrial 4.0 network and big data, other artificially intelligent technologies play increasingly important roles in smart manufacturing. Combining data-driven technology with intelligent algorithms could adopt their respective advantages, and create more advanced optimization frameworks. For example, intelligent optimization can provide a large amount of historical scheduling data, which can be aggregated with other industrial information as a sample source for data-driven and machine learning. Therefore, the scheduling decision-making function can be deployed hierarchically and decoupled according to different scenarios and environments, thus making rational use of computing resources and improving the flexibility and stability of the system.

Author Contributions: Conceptualization, Z.L.; methodology, X.Z.; software, Y.H.; validation, Z.L., X.Z. and M.R.; formal analysis, G.K.; investigation, R.S.; resources, G.K.; data curation, X.Z.; writing—original draft preparation, X.Z. and Y.H.; writing—review and editing, Z.L.; visualization, M.R.; supervision, G.K.; project administration, R.S.; funding acquisition, Z.L. All authors have read and agreed to the published version of the manuscript.

Funding: This research is funded by the Natural Science Foundation of Xuzhou, China (KC21070), National Natural Science Foundation of China (61803192), and the Narodowego Centrum Nauki, Poland (No. 2020/37/K/ST8/02748 & No. 2017/25/B/ST8/00962).

Data Availability Statement: All data can be requested from the corresponding author.

Conflicts of Interest: The authors declare no conflict of interest.

References

1. Helu, M.; Sobel, W.; Nelaturi, S.; Waddell, R.; Hibbard, S. Industry Review of Distributed Production in Discrete Manufacturing. *J. Manuf. Sci. Eng.* **2020**, *142*, 110802. [CrossRef]
2. Cheng, Y.; Bi, L.; Tao, F.; Ji, P. Hypernetwork-based manufacturing service scheduling for distributed and collaborative manufacturing operations towards smart manufacturing. *J. Intell. Manuf.* **2020**, *31*, 1707–1720. [CrossRef]
3. Valilai, O.F.; Houshmand, M. A collaborative and integrated platform to support distributed manufacturing system using a service-oriented approach based on cloud computing paradigm. *Robot. Comput.-Integr. Manuf.* **2013**, *29*, 110–127. [CrossRef]
4. Chen, S.; Pan, Q.K.; Gao, L. Production scheduling for blocking flowshop in distributed environment using effective heuristics and iterated greedy algorithm. *Robot. Comput.-Integr. Manuf.* **2021**, *71*, 102155. [CrossRef]
5. Ribas, I.; Companys, R.; Tort-Martorell, X. Efficient heuristics for the parallel blocking flow shop scheduling problem. *Expert Syst. Appl.* **2017**, *74*, 41–54. [CrossRef]
6. Ribas, I.; Companys, R.; Tort-Martorell, X. An iterated greedy algorithm for solving the total tardiness parallel blocking flow shop scheduling problem. *Expert Syst. Appl.* **2019**, *121*, 347–361. [CrossRef]
7. Zhang, G.; Xing, K.; Cao, F. Discrete differential evolution algorithm for distributed blocking flowshop scheduling with makespan criterion. *Eng. Appl. Artif. Intell.* **2018**, *76*, 96–107. [CrossRef]
8. Zhang, X.; Liu, X.; Tang, S.; Królczyk, G.; Li, Z. Solving Scheduling Problem in a Distributed Manufacturing System Using a Discrete Fruit Fly Optimization Algorithm. *Energies* **2019**, *12*, 3260. [CrossRef]
9. Shao, Z.; Pi, D.; Shao, W. Hybrid enhanced discrete fruit fly optimization algorithm for scheduling blocking flow-shop in distributed environment. *Expert Syst. Appl.* **2020**, *145*, 113147. [CrossRef]
10. Li, W.; Li, J.; Gao, K.; Han, Y.; Niu, B.; Liu, Z.; Sun, Q. Solving robotic distributed flowshop problem using an improved iterated greedy algorithm. *Int. J. Adv. Robot. Syst.* **2019**, *16*, 1729881419879819. [CrossRef]
11. Zhao, F.; Zhao, L.; Wang, L.; Song, H. An ensemble discrete differential evolution for the distributed blocking flowshop scheduling with minimizing makespan criterion. *Expert Syst. Appl.* **2020**, *160*, 113678. [CrossRef]
12. Miyata, H.H.; Nagano, M.S. The blocking flow shop scheduling problem: A comprehensive and conceptual review. *Expert Syst. Appl.* **2019**, *137*, 130–156. [CrossRef]
13. Chen, Q.; Deng, L.-F.; Wang, H.-M. Optimization of multi-task job-shop scheduling based on uncertainty theory algorithm. *Int. J. Simul. Model.* **2018**, *17*, 543–552. [CrossRef]
14. Liu, W.; Jin, Y.; Price, M. New meta-heuristic for dynamic scheduling in permutation flowshop with new order arrival. *Int. J. Adv. Manuf. Technol.* **2018**, *98*, 1817–1830. [CrossRef]
15. Chen, J.; Wang, M.; Kong, X.-T.; Huang, G.-Q.; Dai, Q.; Shi, G. Manufacturing synchronization in a hybrid flowshop with dynamic order arrivals. *J. Intell. Manuf.* **2019**, *30*, 2659–2668. [CrossRef]
16. Zhang, B.; Pan, Q.-K.; Gao, L.; Zhang, X.-L.; Peng, K.K. A multi-objective migrating birds optimization algorithm for the hybrid flowshop rescheduling problem. *Soft Comput.* **2019**, *23*, 8101–8129. [CrossRef]
17. Moghaddam, S.-K.; Saitou, K. Predictive-Reactive Rescheduling for New Order Arrivals with Optimal Dynamic Pegging. In Proceedings of the 2020 IEEE 16th International Conference on Automation Science and Engineering, Hong Kong, China, 20–21 August 2020; pp. 710–715.
18. Han, Y.; Gong, D.; Jin, Y.; Pan, Q. Evolutionary Multiobjective Blocking Lot-Streaming Flow Shop Scheduling With Machine Breakdowns. *IEEE Trans. Syst. Man Cybern.* **2019**, *49*, 184–197. [CrossRef] [PubMed]
19. Nica, E.; Stan, C.-I.; Luțan (Petre), A.-G.; Oașa (Geambazi), R.-Ș. Internet of Things-based Real-Time Production Logistics, Sustainable Industrial Value Creation, and Artificial Intelligence-driven Big Data Analytics in Cyber-Physical Smart Manufacturing Systems. *Econ. Manag. Financ. Mark.* **2021**, *16*, 52–62.
20. Popescu, G.-H.; Petreanu, S.; Alexandru, B.; Corpodean, H. Internet of Things-based Real-Time Production Logistics, Cyber-Physical Process Monitoring Systems, and Industrial Artificial Intelligence in Sustainable Smart Manufacturing. *J. Self-Gov. Manag. Econ.* **2021**, *9*, 52–62.
21. Cohen, S.; Macek, J. Cyber-Physical Process Monitoring Systems, Real-Time Big Data Analytics, and Industrial Artificial Intelligence in Sustainable Smart Manufacturing. *Econ. Manag. Financ. Mark.* **2021**, *16*, 55–67.
22. Vieira, G.-E.; Herrmann, J.-W.; Lin, E. Rescheduling manufacturing systems: A framework of strategies, policies, and methods. *J. Sched.* **2003**, *6*, 39–62. [CrossRef]
23. Framinan, J.-M.; Fernandez-Viagas, V.; Perez-Gonzalez, P. Using real-time information to reschedule jobs in a flowshop with variable processing times. *Comput. Ind. Eng.* **2019**, *129*, 113–125. [CrossRef]
24. Katragjini, K.; Vallada, E.; Ruiz, R. Flow shop rescheduling under different types of disruption. *Int. J. Prod. Res.* **2013**, *51*, 780–797. [CrossRef]
25. Iris, Ç.; Lam, J.-S.-L. Recoverable robustness in weekly berth and quay crane planning. *Transp. Res. Part B Methodol.* **2019**, *122*, 365–389. [CrossRef]
26. Ma, Z.; Yang, Z.; Liu, S.; Song, W. Optimized rescheduling of multiple production lines for flowshop production of reinforced precast concrete components. *Autom. Constr.* **2018**, *95*, 86–97. [CrossRef]
27. Li, J.; Pan, Q.; Mao, K. A hybrid fruit fly optimization algorithm for the realistic hybrid flowshop rescheduling problem in steelmaking systems. *IEEE Trans. Autom. Sci. Eng.* **2016**, *13*, 932–949. [CrossRef]

28. Li, J.; Pan, Q.; Mao, K. A discrete teaching-learning-based optimization algorithm for realistic flowshop rescheduling problems. *Eng. Appl. Artif. Intell.* **2015**, *37*, 279–292. [CrossRef]
29. Valledor, P.; Gomez, A.; Priore, P.; Puente, J. Solving multi-objective rescheduling problems in dynamic permutation flow shop environments with disruptions. *Int. J. Prod. Res.* **2018**, *56*, 6363–6377. [CrossRef]
30. Wade, K.; Vochozka, M. Artificial Intelligence Data-driven Internet of Things Systems, Sustainable Industry 4.0 Wireless Networks, and Digitized Mass Production in Cyber-Physical Smart Manufacturing. *J. Self-Gov. Manag. Econ.* **2021**, *9*, 48–60.
31. Hamilton, S. Real-Time Big Data Analytics, Sustainable Industry 4.0 Wireless Networks, and Internet of Things-based Decision Support Systems in Cyber-Physical Smart Manufacturing. *Econ. Manag. Financ. Mark.* **2021**, *16*, 84–94.
32. Riley, C.; Vrbka, J.; Rowland, Z. Internet of Things-enabled Sustainability, Big Data-driven Decision-Making Processes, and Digitized Mass Production in Industry 4.0-based Manufacturing Systems. *J. Self-Gov. Manag. Econ.* **2021**, *9*, 42–52.
33. Pelau, C.; Dabija, D.-C.; Ene, I. What Makes an AI Device Human-Like? The Role of Interaction Quality, Empathy and Perceived Psychological Anthropomorphic Characteristics in the Acceptance of Artificial Intelligence in the Service Industry. *Comput. Hum. Behav.* **2021**, *122*, 106855. [CrossRef]
34. Richard, D. *The Selfish Gene*; Oxford University Press: New York, NY, USA, 1976.
35. Wang, J.; Zhou, Y.; Wang, Y.; Zhang, J.; Chen, C.; Zheng, Z. Multiobjective vehicle routing problems with simultaneous delivery and pickup and time windows: Formulation, instances, and algorithms. *IEEE Trans. Cybern.* **2015**, *46*, 582–594. [CrossRef] [PubMed]
36. Decerle, J.; Grunder, O.; El Hassani, A.-H.; Barakat, O. A memetic algorithm for multi-objective optimization of the home health care problem. *Swarm Evol. Comput.* **2019**, *44*, 712–727. [CrossRef]
37. Yancey, S.-K.; Tsvetkov, P.-V.; Jarrell, J.-J. A greedy memetic algorithm for a multiobjective dynamic bin packing problem for storing cooling objects. *J. Heuristics* **2019**, *25*, 1–45.
38. Zhou, Y.; Fan, M.; Ma, F.; Xu, X.; Yin, M. A decomposition-based memetic algorithm using helper objectives for shortwave radio broadcast resource allocation problem in china. *Appl. Soft Comput.* **2020**, *91*, 106251. [CrossRef]
39. Jiang, E.; Wang, L.; Lu, J. Modified multi-objective evolutionary algorithm based on decomposition for low-carbon scheduling of distributed permutation flow-shop. In Proceedings of the 2017 IEEE Symposium Series on Computational Intelligence, Honolulu, HI, USA, 27 November–1 December 2017; pp. 1–7.
40. Abedi, M.; Chiong, R.; Noman, N.; Zhang, R. A multi-population, multi-objective memetic algorithm for energy-efficient job-shop scheduling with deteriorating machines. *Expert Syst. Appl.* **2020**, *157*, 113348. [CrossRef]
41. Kurdi, M. An improved island model memetic algorithm with a new cooperation phase for multi-objective job shop scheduling problem. *Comput. Ind. Eng.* **2017**, *111*, 183–201. [CrossRef]
42. Minguillon, F.E.; Stricker, N. Robust predictive–reactive scheduling and its effect on machine disturbance mitigation. *CIRP Ann.* **2020**, *69*, 401–404. [CrossRef]
43. Deng, J.; Wang, L.; Wu, C.; Wang, J.; Zheng, X. A competitive memetic algorithm for carbon-efficient scheduling of distributed flow-shop. In *International Conference on Intelligent Computing Lanzhou, China, 2–5 August 2016*; Springer: Cham, Switzerland, 2016; pp. 476–488.
44. Liu, X.; Zhan, Z.; Lin, Y.; Chen, W.; Gong, Y.; Gu, T.; Yuan, H.; Zhang, J. Historical and heuristic-based adaptive differential evolution. *IEEE Trans. Syst. Man Cybern. Syst.* **2018**, *49*, 2623–2635. [CrossRef]
45. Boejko, W.; Grabowski, J.; Wodecki, M. Block approach-tabu search algorithm for single machine total weighted tardiness problem. *Comput. Ind. Eng.* **2006**, *50*, 1–14. [CrossRef]
46. Osman, I.; Potts, C. Simulated annealing for permutation flow-shop scheduling. *Omega* **1989**, *17*, 551–557. [CrossRef]
47. Duan, J.; Sun, W.; Li, J.; Xu, Y. A Flowshop rescheduling algorithm based on migrating birds optimization. *Control. Eng. China* **2017**, *24*, 1656–1661. (In Chinese)

Article

Coal Thickness Prediction Method Based on VMD and LSTM

Yaping Huang [1,*] , Lei Yan [1], Yan Cheng [2], Xuemei Qi [1,*] and Zhixiong Li [3]

1   School of Resources and Geosciences, China University of Mining and Technology, Xuzhou 221116, China; ts20010069a31@cumt.edu.cn
2   China National Administration of Coal Geology, Beijing 100038, China; cumt_cheng@sohu.com
3   Yonsei Frontier Lab, Yonsei University, Seoul 03722, Korea; zhixiong.li@yonsei.ac.kr
*   Correspondence: yphuang@cumt.edu.cn (Y.H.); qixuemei@cumt.edu.cn (X.Q.)

Abstract: The change in coal seam thickness has an important influence on coal mine safety and efficient mining. It is very important to predict coal thickness accurately. However, the accuracy of borehole interpolation and BP neural network is not high. Variational mode decomposition (VMD) has strong denoising ability, and the long short-term memory neural network (LSTM) is especially suitable for the prediction of complex sequences. This paper presents a method of coal thickness prediction using VMD and LSTM. Firstly, empirical mode decomposition (EMD) and VMD methods are used to denoise simple signals, and the denoising effect of the VMD method is verified. Then, the wedge-shaped coal thickness model is constructed, and the seismic forward modeling and analysis are carried out. The results show that the coal thickness prediction based on seismic attributes is feasible. On the basis of VMD denoising of the original 3D seismic data, VMD-LSTM is used to predict coal thickness and compared with the prediction results of the traditional BP neural network. The coal thickness prediction method proposed in this paper has high accuracy and basically coincides with the coal seam information exposed by existing boreholes. The minimum absolute error of the predicted coal thickness is only 0.08 m, and the maximum absolute error is 0.48 m. This indicates that VMD-LSTM has high accuracy in predicting coal thickness. The proposed coal thickness prediction method can be used as a new method for coal thickness prediction.

Keywords: VMD; LSTM; coal thickness; seismic attribute

Citation: Huang, Y.; Yan, L.; Cheng, Y.; Qi, X.; Li, Z. Coal Thickness Prediction Method Based on VMD and LSTM. *Electronics* **2022**, *11*, 232. https://doi.org/10.3390/electronics11020232

Academic Editors: Darius Andriukaitis, Yongjun Pan and Peter Brida

Received: 8 December 2021
Accepted: 10 January 2022
Published: 12 January 2022

Publisher's Note: MDPI stays neutral with regard to jurisdictional claims in published maps and institutional affiliations.

1. Introduction

In the construction and production process of large-scale mines, all aspects of coal mine safety production need to accurately calculate the change in coal thickness [1]. When the coal thickness of the working face changes dramatically, the reserves per unit length of the working face will change greatly. If the same mining speed is adopted, the monthly coal output will be unbalanced, and the mine production will be greatly affected. According to statistics, when the actual coal thickness is 10–20% thinner than the designed coal thickness, the coal output will decrease by 35–40% [2]. In general, the thicker the coal is, the more gas is formed in the coalification process, and the greater the probability of gas leakage in the mining process is. The impact risk degree is closely related to the coal thickness and its changes. A large number of field observations and in situ stress measurements have found that in the coal thickness change area, the abnormal phenomenon of in situ stress field occurs, and the stress concentration degree is high, which is easy to induce strong mine earthquakes and rock bursts [3]. Therefore, coal thickness is an essential type of data in the process of coal mine design and mining [4]. Accurately predicting coal thickness can not only improve the economic benefit of coal mines, but also provide a strong geological guarantee for the safe mining of coal mines [5].

With the development of coal mining, in the seismic exploration of coal fields, it is necessary not only to find out the structure in the mining area, but also to provide the change of the coal seam thickness [6]. However, because most coal seams are typically thin layers, their vertical resolution is very low, which is difficult to distinguish in seismic records and cannot meet the requirements of solving coal thickness, so the quantitative prediction of coal thickness has been one of the recognized problems in the geophysical field. The traditional method of coal thickness is to compare and interpolate the borehole data, and the accuracy of the coal thickness data provided is limited. In particular, when the coal seam is missed, stripped, and bifurcated, the prediction error of coal thickness is large [7].

In recent years, scholars have conducted a lot of research on coal thickness prediction and have achieved more significant results and progress. Some scholars use seismic attributes to predict coal thickness and have achieved good results. Zou et al. [8] used the staggered grid finite difference method to establish a wedge model to study coal thickness. The results show that when the coal seam is thin, the amplitude attribute is correlated with the coal thickness, and the law can be used to predict the coal thickness qualitatively. Guo et al. [9] analyzed the correlation between seismic attributes at the borehole and coal thickness. They obtained the regression equation between each attribute and coal seam thickness through optimal seismic attributes and achieved good application effects in coal thickness prediction. Lin et al. [10] used logging data combined with 3D seismic data and their interpretation results to conduct wave impedance inversion under multi-well constraints. The corresponding wave impedance layer of the coal seam was determined through horizon calibration, and then the thickness of the wave impedance layer was extracted and matched with the thickness of the coal seam exposed by the borehole. Finally, the distribution law of coal thickness in the 3D seismic survey area was obtained. Some scholars have introduced geostatistics into coal thickness prediction. Du et al. [11] used the Cokriging method to predict coal thickness by taking the thickness of drilled coal seam and seismic amplitude as regional variables, which reduced the error of coal thickness prediction and significantly improved the prediction accuracy. Cheng et al. [12] combined the variogram and the Kriging method to fully consider well data and seismic attribute data. He assigned certain weight coefficients to the two data types and performed a weighted average to calculate the thickness of the coal. The in-seam wave is a kind of guided wave propagating in the coal seam. It has the advantages of long propagation distance, carrying much information about the coal seam, roof, and floor; strong energy; and obvious dispersion characteristics. Based on these advantages, the in-seam wave can be used to study coal thickness. Liu et al. [13] found that the periodicity of the refraction P-wave in the in-seam wave is proportional to the thickness of coal, and the thickness of coal can be predicted by using the period of the refraction P-wave. Li et al. [14] chose to collect the appropriate frequency at the mining face in the transmission method of love waves with wave CT tomography, according to the wave velocity and thickness of the negative correlation and dispersion curve. Combined with the tunnel exposing coal thickness, the wave velocity contour map can be converted to a coal thickness contour map. Implementing the coal thickness quantitative prediction method proved high accuracy. With the development of artificial intelligence, more and more artificial intelligence technology is gradually applied to coal thickness prediction. The mature development of deep neural network technology urges mining workers to study the geological prediction method based on data and forms geological prediction technology driven by data [15,16]. Sun et al. [17] used non-linear BP neural network technology to predict coal thickness by extracting seismic attributes in the wavelet domain of different frequency bands according to the changing characteristics of coal thickness. Wu et al. [18] combined the least square support vector machine (LSSVM) with the Kriging method to predict coal thickness. They used the strong non-linear fitting and generalization capabilities of LSSVM to adaptively fit the experimental variogram, which overcame the disadvantages of the traditional variation function, such as difficulty in solving and strong subjectivity, and greatly improved the prediction accuracy. Guo et al. [19] proposed the coal thickness prediction method by combining genetic algorithm (GA) and simultaneous iterative reconstruction technology (SIRT), which can improved the accuracy of coal thickness inversion. However, the above methods are basically coal thickness prediction methods directly based on seismic data, and discussion on the actual situation that seismic data usually contains noise is limited. Seismic data contains a lot of noise, which is a practical problem that needs to be faced when predicting coal seam thickness.

Variational mode decomposition (VMD) is a non-recursive modal decomposition method proposed by Dragomiretskiy in 2014 [20]. The VMD algorithm decomposes the signal according to the center frequency, divides the signal around a certain center frequency into one component, and repeats this process to achieve the decomposition of the original signal. When VMD is used to process non-stationary signals, it can effectively avoid the mode-aliasing effect and the endpoint effect caused by the empirical mode decomposition (EMD) algorithm, and the intrinsic mode function (IMF) is redefined as an amplitude modulation–frequency modulation signal. It has a solid mathematical basis. More importantly, the VMD method can effectively attenuate a large amount of random noise contained in seismic data. The long short-term memory neural network (LSTM) was proposed by Hochreiter et al. in 1997 and further improved in the following years [21]. LSTM is designed to learn long-term memory, and the model can overcome the inherent problems of the recurrent neural network (RNN), namely "vanishing gradients" and "exploding gradients". LSTM is one of the most successful recurrent neural networks, and its prediction accuracy is high.

Based on the good denoising ability of VMD and good prediction ability for complex sequences of LSTM, this paper proposes VMD-LSTM method for coal thickness prediction. Firstly, the denoising effect of the VMD method is verified. Then, the coal thickness wedge model is used to illustrate that the seismic attribute can predict the coal thickness. The VMD-LSTM method is used for coal thickness prediction of actual seismic data. The prediction results have high accuracy and basically coincide with the coal seam information exposed by existing boreholes, which shows that the coal thickness prediction method proposed in this paper can be used as a new method for coal thickness prediction.

2. Basic Principle of VMD and LSTM for Predicting Coal Thickness

2.1. Basic Principles of VMD

During the late 1990s, Huang introduced the algorithm called EMD, which is widely used today to recursively decompose a signal into different modes of unknown but separate spectral bands [22]. However, due to the lack of mathematical theory and freedom, the robustness of the algorithm is very low, which leaves room for the development and improvement of EMD. VMD is an adaptive, non-recursive decomposition method that can decompose signals into the sum of finite component signals [23]. It is a new decomposition algorithm based on the Wiener filter, Hilbert transform, and heterodyne demodulation. It has a solid mathematical and theoretical foundation. VMD decomposes the signal according to the central frequency, divides the signal around a certain central frequency into a component, and repeats this process [24]. Finally, the original signal is decomposed. When VMD is used to process non-stationary signals, it can effectively avoid the mode-aliasing effect and the endpoint effect caused by the EMD algorithm, and IMF is redefined as an amplitude modulation–frequency modulation signal.

2.1.1. Construction of Variational Problems

The goal of VMD is to decompose a real-valued input signal into a discrete number of IMF. The IMF expression is as follows:

$$u_k(t) = A_k(t)\cos(\varphi_k(t)) \tag{1}$$

In Equation (1), $A_k(t)$ is the non-negative envelope, and $\varphi_k(t)$ is the phase of the signal.

For each mode, compute the associated analytic signal by means of the Hilbert transform in order to obtain a unilateral frequency spectrum:

$$[\delta(t) + \frac{j}{\pi t}] * u_k(t) \tag{2}$$

In Equation (2), $\delta(t)$ is a Dirac delta function, j is an imaginary unit, and $*$ is the convolution symbol.

For each mode, shift the mode's frequency spectrum to "baseband" by mixing with an exponential tuned to the respective estimated center frequency.

$$[(\delta(t) + \frac{j}{\pi t}) * u_k(t)]e^{-jw_k t} \tag{3}$$

In Equation (3), $e^{-jw_k t}$ is the center frequency index.

According to the estimated bandwidth of IMF, the variational constraint equation is obtained:

$$\min_{\{u_k\},\{w_k\}} \{\sum_k \|\partial_t[(\delta(t) + \frac{j}{\pi t}) * u_k(t)]e^{-jw_k t}\|_2^2\}$$
$$s.t.\sum_k u_k = f \tag{4}$$

$\{u_k\} := \{u_1, u_2 \cdots, u_K\}$ and $\{w_k\} := \{w_1, w_2 \cdots, w_K\}$ are shorthand notations for the set of all modes and their center frequencies, respectively. $\sum_k := \sum_{k=1}^{K}$ is understood as the summation over all modes.

2.1.2. Solution of Variational Problem

In order to obtain the optimal solution, the augmented Lagrange expression is obtained by introducing the quadratic penalty factor α and the Lagrange multiplication operator γ. The purpose of introducing α is to ensure the signal reconstruction accuracy under the influence of noise, and the purpose of introducing γ is to maintain strict constraints in the solving process.

$$L(\{u_k\}, \{w_k\}, \gamma) = \alpha\left\{\sum_{k=1}^{K} \partial_t[(\delta(t) + \frac{j}{\pi t}) * u_k(t)]e^{-jw_k t}\right\}\|^2 +$$
$$\sum(f(t) - \sum_{k=1}^{K} u_k(t))\|_2^2 - \left\langle\gamma(t), f(t) - \sum_{k=1}^{K} u_k(t)\right\rangle \tag{5}$$

The alternating direction method of multipliers (ADMM) is used to solve the formula. In order to achieve the purpose of complete signal decomposition, the central frequency of each IMF is divided according to the frequency domain characteristics of the original signal, and then the central frequency of IMF is updated. The updated IMF expression is as follows:

$$u_k^{n+1} = \underset{u_k \in X}{\arg\min}\left\{\alpha\|\partial_t[(\delta(t) + \frac{j}{\pi t}) * u_k(t)]e^{-jw_k t}\|^2 + \|f(t) - \sum_i u_i(t) + \frac{\gamma(t)}{2}\|_2^2\right\} \tag{6}$$

Fourier transform is used to transform the formula from the time domain to the frequency domain, and then the non-negative frequency domain interval is integrated to obtain the optimal solution. Among them, the frequency domain calculation criterion expression of each IMF is as follows:

$$\hat{u}_k^{n+1}(\omega) = \frac{\hat{f}(\omega) - \sum_{i\neq}\hat{u}_i(\omega) + \frac{\hat{\gamma}(\omega)}{2}}{1 + 2\alpha(\omega - \omega_k)^2} \tag{7}$$

The updated iteration formula of IMF center frequency is:

$$\omega_k^{n+1} = \frac{\int_0^\infty \omega \left| \hat{u}_k(\omega) \right|^2 d\omega}{\int_0^\infty \left| \hat{u}_k(\omega) \right|_2 d\omega} \tag{8}$$

In Equation (8), $\hat{u}_k^{n+1}(\omega)$ is the Wiener filter of $\hat{f}(\omega) - \sum_{i \neq} \hat{u}_i(\omega)$, and ω_k^{n+1} is the frequency spectrum center of each IMF.

2.2. Basic Principles of LSTM

Hochreiter et al. proposed LSTM on the basis of RNN. RNN takes the sequence data as the input of the model, calculates the influence on the final output by learning the changes of a series of data, and modifies the weight matrix to optimize the network by calculating the gradient through positive feedback and negative feedback so as to realize the learning of time series. However, RNN systems are problematic in practice, despite their stability. During training, they suffer from well-documented issues, known as "vanishing gradients" and "exploding gradients" [25]. These difficulties become pronounced when the dependencies in the target subsequence span a large number of samples, requiring the window of the unrolled RNN model to be commensurately wide in order to capture these long-range dependencies [26]. LSTM is a good solution to the problem of "vanishing gradients" and "exploding gradients" and can be efficiently learned through memory cells and "gates" and for long-term memory of useful information. The structure of LSTM is similar to a traditional neural network, which has one input layer, one or more hidden layers, and one output layer. The hidden layer of the LSTM neural network contains many neurons called memory cells, and each of these memory cells has three "gates", whose function is to maintain and adjust the state of the memory cells. These three "gates" are called the forget gate, input gate, and output gate, respectively. The essence of the control gate of LSTM is a fully connected layer. The input is a vector x, which is calculated with weight vector W and bias item b and filtered by the activation function. The output will obtain a real vector of 0–1, whose calculation formula is as follows:

$$g(x) = \sigma(wx + b) \tag{9}$$

The activation function of the gates in the hidden layer of LSTM is the sigmoid function with a range of (0, 1). The output of the gates is multiplied by the input vector to determine how much information is passed. When the output of the gate is 0, the product of the output vector and the input vector results in 0, and the information cannot pass. When the output is 1, the input vector is unaffected, and the gate is open. The gate is always ajar because of the sigmoid function range properties.

LSTM controls the unit state c with the forget gate and the input gate. The forget gate determines how much of the unit state c_{t-1} at the previous moment is retained to the current moment c_t, and the input gate determines how much of the network input x_t at the current moment is saved to the unit state c_t. LSTM uses the output gate to control how much output the unit state c_t has to the current output of LSTM h_t.

Updating a cell can be broken down into the following steps [24].

(1) Calculate the value of the forget gate:

$$f_t = \sigma(W_f \cdot [h_{t-1}, x_t] + b_f) \tag{10}$$

In Equation (10), W_f is the weight matrix of the forget gate, $[h_{t-1}, x_t]$ is the joining of two vectors into a longer vector, b_f is the offset of the forget gate, and σ is the sigmoid function.

(2) Calculate the value of the input gate:

In Equation (11), W_i is the weight matrix of the forget gate, and b_i is the offset of the forget gate.

$$i_t = \sigma(W_i \cdot [h_{t-1}, x_t] + b_i) \tag{11}$$

(3) Calculate the value of $\tilde{c}_t$. It is used to describe the unit state of the current input, calculated from the previous output and the current input:

$$\tilde{c}_t = \tanh(W_c \cdot [h_{t-1}, x_t] + b_c) \tag{12}$$

(4) Calculates the cell state at the current time c_t:

$$c_t = f_t \odot c_{t-1} + i_t \odot \tilde{c}_t \tag{13}$$

In Equation (13), $\odot$ is the symbol for multiplying matrices by bits, and c_t is the result of combining current and long-term memories.

(5) Calculate the value of the output gate:

$$o_t = \sigma(W_o \cdot [h_{t-1}, x_t] + b_o) \tag{14}$$

In Equation (14), W_o is the weight matrix of the output gate, and b_o is the offset of the forget gate.

(6) Calculate the current cell output of LSTM:

$$h_t = o_t \odot \tanh(c_t) \tag{15}$$

2.3. Coal Thickness Prediction Process of VMD-LSTM

Firstly, the wedge model of coal thickness is constructed and simulated by seismic forward modeling, and the sensitive seismic attributes of coal thickness such as instantaneous amplitude, instantaneous frequency, and relative wave impedance are extracted, then the feasibility of coal thickness prediction using seismic attributes is analyzed. Secondly, the seismic data were decomposed into IMFs with different characteristics and frequencies by using the VMD method, the correlation coefficients between seismic attributes of seismic data IMF1 and coal thickness were calculated, and the appropriate IMFs were selected in order to effectively remove random noise and highlight effective signals. Finally, the LSTM network is used to predict coal thickness. The specific flow chart is shown in Figure 1.

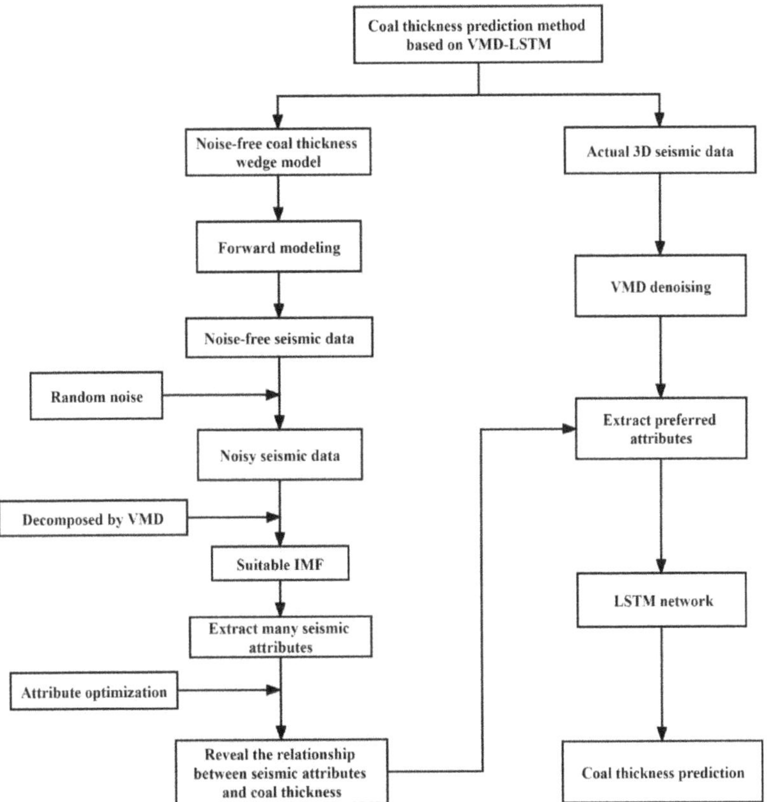

Figure 1. Flow chart of coal thickness prediction by VMD and LSTM.

3. Numerical Calculation

3.1. Simple Signal Test

Formula (16) is adopted to synthesize the signal, and Gaussian white noise with a signal-to-noise ratio (SNR) of 30 dB is added to the signal to form a signal containing noise (Figure 2). The EMD and VMD methods are used to denoise simple signals to test their denoising ability. The denoising results are shown in Figures 3 and 4. It can be seen from Figure 3 that there is an obvious mode-aliasing problem in EMD decomposition, which cannot effectively decompose the random noise information in simple signals. Figure 4 shows that VMD has a better decomposition effect on random noise, IMF1 corresponds to the main components of the signal, and the correlation coefficient with the noise-free signal reaches 0.9998. The result shows that VMD can be used in signal denoising, and the denoising effect is good.

$$\begin{cases} y_1 = \cos(30\pi * t) \\ y_2 = 0.5 * \sin(30\pi * t) \\ y = y_1 + y_2 \end{cases} \tag{16}$$

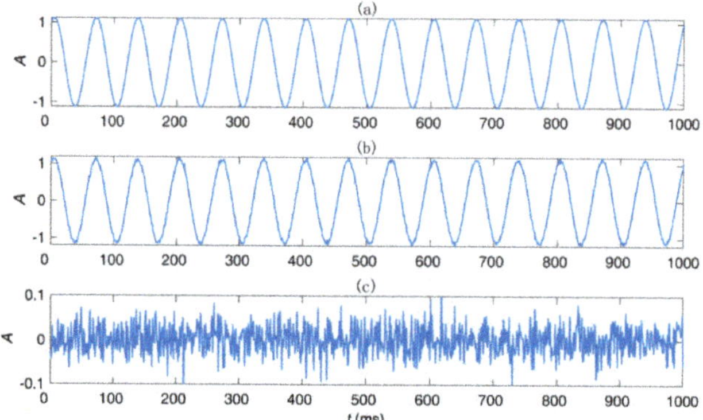

Figure 2. Simple signal and noisy signal: (**a**) simple signal; (**b**) noisy signal; (**c**) noise.

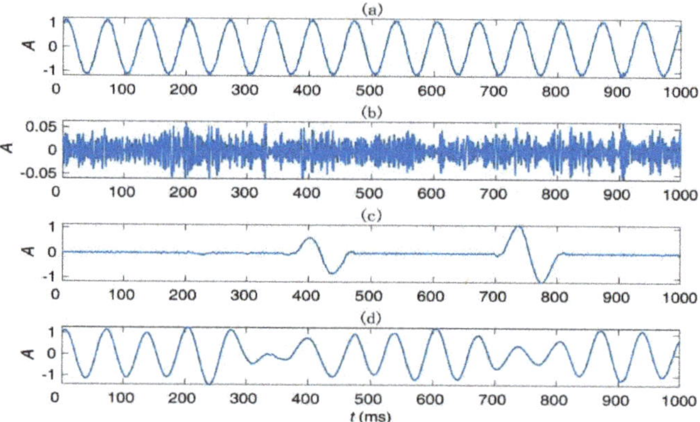

Figure 3. EMD decomposition results of noisy signal: (**a**) noisy signal; (**b**) IMF 1; (**c**) IMF 2; (**d**) IMF 3.

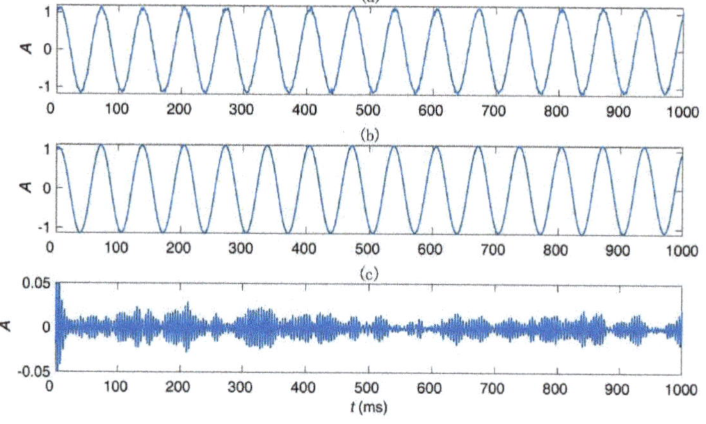

Figure 4. VMD decomposition results of noisy signal: (**a**) noisy signal; (**b**) IMF 1; (**c**) IMF 2.

The instantaneous amplitude and frequency of IMF1 denoised by VMD are calculated (Figure 5). As shown in Figure 5, the correlation coefficients of the instantaneous amplitude and instantaneous frequency between the noise-free signal and IMF1 are high, reaching 0.9861 and 0.9709.

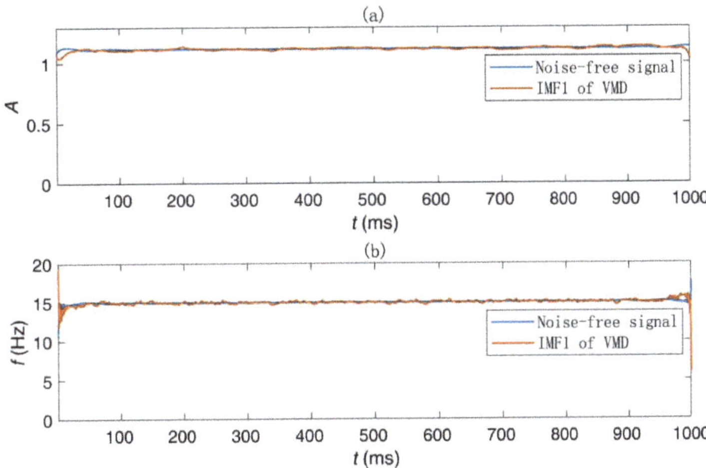

Figure 5. VMD decomposition results of noisy signal: (**a**) instantaneous amplitude of noise-free signal and IMF1 of VMD; (**b**) instantaneous frequency of noise-free signal and IMF1 of VMD.

The application results of VMD in the simple signal show that VMD has good denoising ability, the noisy signal processed by VMD can accurately contain the effective components of the noise-free signal, and the instantaneous amplitude and instantaneous frequency extracted from IMF1 have a high correlation with the instantaneous frequency and instantaneous amplitude of the noise-free signal.

3.2. VMD Decomposition of Coal Thickness Wedge Model

Firstly, a wedge model, which is commonly used in the study of coal thickness, is constructed. The model mainly consists of three layers: sandstone, mudstone, and coal seam. The coal thickness varies from 0 m to 40 m, and the velocity, density, and thickness of each layer are obtained from [5]. A Ricker wavelet with a main frequency of 50 Hz was used for forward modeling, and the distance between the receiving channels was 10 m, with a total of 40 receiving channels. The forward modeling section is shown in Figure 6a. Random noise of 40 dB was added to the forward record of the wedge model, as shown in Figure 6b. As can be seen from Figure 6, seismic records are significantly affected by noise, and the continuity of the in-phase axis becomes poor, which is not conducive to the subsequent prediction of coal thickness. VMD was carried out for noisy signals. The experiment found that when the decomposed IMF number K was greater than 2, the center frequency was very close, so K was set to 2. The results of the seismic records of the wedge model of the coal seam decomposed by VMD are shown in Figure 6c,d. The correlation coefficients between the noise-free signal and IMF1 seismic tracks are calculated, with the maximum value reaching 0.9659 and the average value 0.9258. It shows that VMD has strong denoising ability, and IMF1 decomposed by VMD is basically consistent with the noise-free synthetic seismic record.

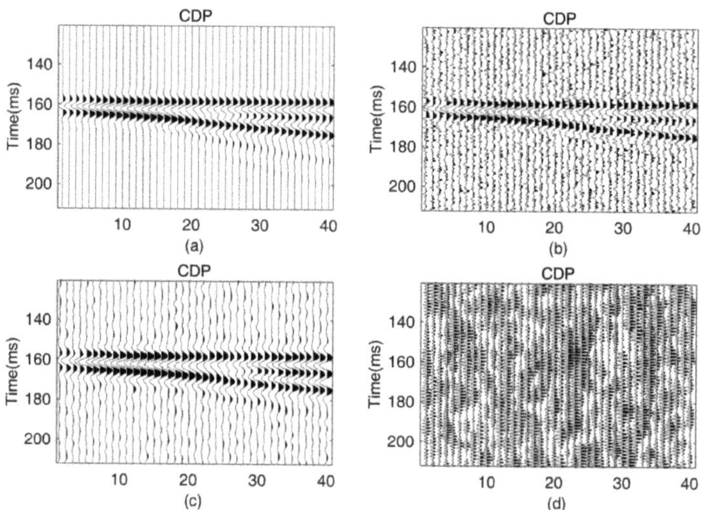

Figure 6. Wedge model seismic record and VMD decomposition result: (**a**) forward seismic record of the wedge-shaped model; (**b**) seismic record of the model after adding random noise; (**c**) IMF1 after VMD denoising of noisy signals; (**d**) IMF2 after VMD denoising of noisy signal.

3.3. Seismic Attributes Analysis of Wedge Model Seismic Records

According to the existing research, seismic attributes such as instantaneous amplitude, instantaneous frequency, and relative wave impedance have a certain correlation with the thickness of coal. Therefore, the three seismic attributes mentioned above are extracted, and their relationship with the thickness of coal is analyzed. Figure 7 shows the relationship between the instantaneous amplitude, instantaneous frequency, and relative wave impedance of seismic records in the noise-free wedge model and coal thickness. When the coal seam thickness is less than 10 m, the instantaneous amplitude attribute shows an obvious increasing trend with the increase in the coal seam thickness; in other words, the instantaneous amplitude attribute has a good positive correlation with the coal thickness. When the coal thickness is greater than 10 m, the instantaneous amplitude attribute gradually decreases and tends to change gently. When the thickness of coal is less than 20 m, the instantaneous frequency attribute gradually decreases and tends to the minimum value with the increase of coal thickness. The instantaneous frequency attribute has a good negative correlation with the coal thickness. With the continuous increase in coal thickness, the instantaneous frequency attribute gradually increases and tends to be stable. When the thickness of coal is less than 20 m, the relative wave impedance attribute increases gradually with the increase in the coal thickness and has a good positive correlation with the coal thickness. When the thickness of coal is greater than 20 m, the relative wave impedance attribute decreases and tends to be stable with the change in coal thickness. In conclusion, when the thickness of coal is small, the relationship between the attributes of the instantaneous amplitude, instantaneous frequency and relative wave impedance, and coal thickness is simple, and the coal thickness can be predicted by using seismic attributes.

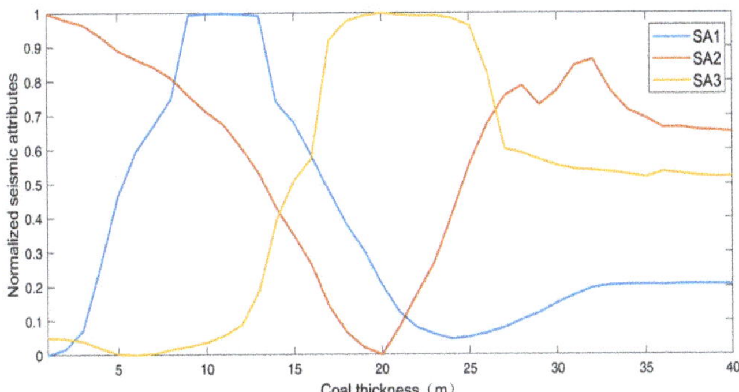

Figure 7. Relationship between seismic attributes and coal thickness of noise-free wedge model.

VMD was used to decompose the seismic record of the wedge model with noise, extract instantaneous amplitude, instantaneous frequency, and relative wave impedance of IMF and draw the relationship between the three attributes and coal seam thickness (Figure 8). It can be seen from Figure 8 that the variation trend of seismic attributes extracted after denoising from the wedge model with noise is basically consistent with that extracted from the model without noise.

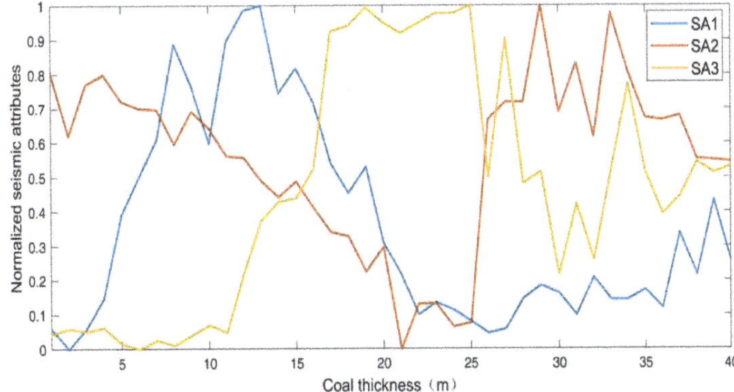

Figure 8. Relationship between the instantaneous amplitude, instantaneous frequency, and relative wave impedance of IMF and coal thickness.

4. Application of VMD-LSTM Method in Coal Thickness Prediction

4.1. Geological Survey of the Working Area

The working area is located in Ordos Basin. The coal-bearing strata in the mine field are the Carboniferous Taiyuan Formation and the Lower Permian Shanxi Formation. The total thickness of the coal-bearing strata revealed by drilling is 118.26–192.88 m, with an average of 142.74 m. The #4 coal seam is located in the upper part of the second rock member of the Lower Permian Shanxi Formation. The natural thickness of the coal seam is 0–5.65 m, with an average thickness of 3.82 m. The thickness of the working area is 0.86–3.79 m, 2.44 m on average. The structure of the coal seam is complex. There are 0–9 layers of partings, generally three layers. Most partings are located in the middle of the coal seam. The coefficient of variation of coal thickness is 32%. The coal seam gradually thickens from west to east. The #4 coal seam is a relatively stable coal seam with reliable comparison and most of the mining area. The thickness of the #4 coal seam in the working area varies from 2.30 to 4.70 m.

4.2. Coal Thickness Prediction and Result Analysis

Seismic data processing mainly adopts conventional processes such as preprocessing, filtering, deconvolution, velocity analysis, normal moveout correction and static correction, stack, and migration. The focus of this paper is to use the VMD method to attenuate the noise of post-stack migration 3D seismic data at any time so as to provide seismic data with a high SNR for coal thickness prediction. The VMD-LSTM method was used to predict the thickness of the #4 coal seam in this area. The parameters of LSMT are as follows: the solver is 'adam', the gradient threshold is 1, the maximum number of rounds is 500, and the execution environment is specified as 'cpu'.

Figure 9 shows the original seismic record and the seismic record after VMD denoising of the #4 coal seam. It can be seen that the #4 coal seam is between 240 ms and 290 ms of the seismic record. It can be seen from the comparison between Figure 9a,b that the original seismic record is greatly affected by noise, and the seismic event of the #4 coal seam is not clear. After VMD denoising, the seismic event of the #4 coal seam is clearer, and the energy is significantly enhanced, which is conducive to the subsequent coal thickness prediction.

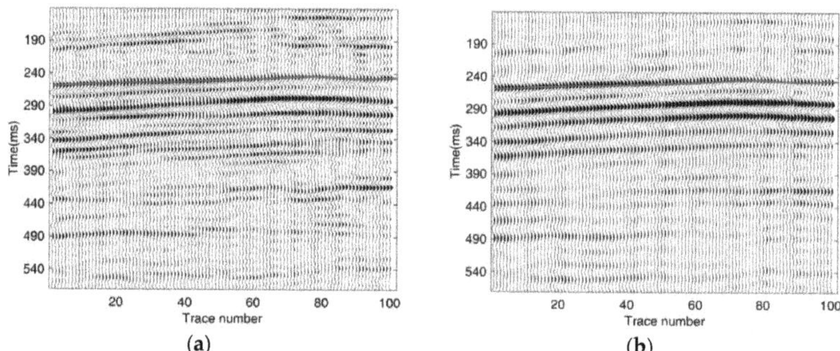

Figure 9. Original seismic record and seismic record after VMD denoising of #4 coal seam: (**a**) original seismic record of #4 coal seam; (**b**) seismic record after VMD denoising of #4 coal seam.

Instantaneous amplitude, instantaneous frequency, and relative wave impedance attributes of the seismic record after VMD denoising were extracted (Figures 10–12). The relationship between the coal thickness and instantaneous amplitude attribute derived from the wedge model shows that the coal seam in the western part of the area may be thinner, while the coal seam in the eastern part may be thicker. The instantaneous frequency and relative wave impedance attributes also have similar characteristics, but the response characteristics of each seismic attribute are quite different. The results of a single seismic attribute reflect the possible distribution characteristics of the coal seam, but it is impossible to quantitatively predict the coal thickness.

In order to study the variation of coal thickness in this area more accurately, based on the coal thickness information and seismic attributes of 10 known boreholes in this area, the instantaneous amplitude, instantaneous frequency, and relative wave impedance attributes of the borehole side channel were used as the input of the training samples, and the known coal thickness was used as the output of the training samples. The coal thickness predicted by the traditional BP neural network method is shown in Figure 13. The LSTM network was also used for training, which has six layers: an input layer, an output layer, and four hidden layers. In order to prevent over-fitting, the dropout layer was introduced into the model, and the dropout value was set to 0.1. The root mean square error is used for the loss function, and the optimizer is Adam. After obtaining satisfactory training results, the coal thickness of the whole area was predicted, as shown in Figure 14. Comparing Figures 13 and 14, it can be seen that the LSTM prediction results have better regularity. Figure 14 shows that the area of the southeast coal seam is thick, between 3 and 5 m, and the southwest coal seam is thin, about 3 m. The predicted coal thickness information is basically consistent with the trend that the coal seam in this area gradually thickens from west to east and from south to north. This regularity is not obvious in Figure 13. The A10 borehole reveals a coal thickness of 4.7 m. Figure 14 shows that the coal seam in the area where A10 is located is thicker and the surrounding area is stable, but there is a certain deviation in the prediction results in Figure 13. The A9 borehole reveals a coal thickness of 2.3 m. Figure 14 also shows that the coal seam in the area where the borehole is located is thinner, and the coal seam is thinner in a certain range around. The analysis of the prediction results of coal thickness at 10 boreholes shows that the minimum absolute error of the BP neural network coal thickness prediction is 0.35 m, and the maximum absolute error is 1.27 m. However, the minimum absolute error of the LSMT coal thickness prediction is only 0.08 m, and the maximum absolute error is 0.48 m. The application results of actual data show that the coal thickness prediction method proposed in this paper has high accuracy, and the predicted coal thickness results have important guiding significance and reference value for later coal mining.

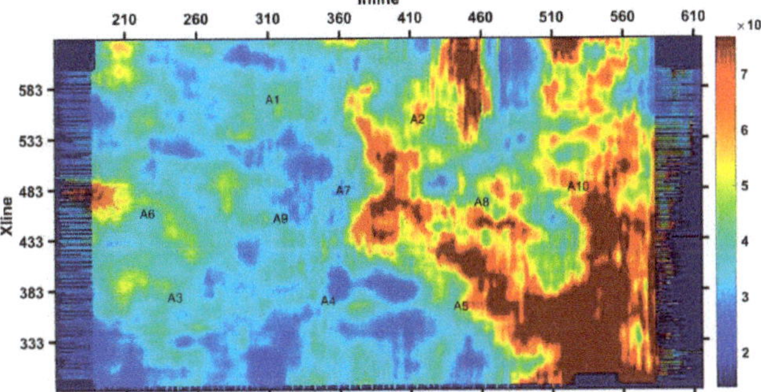

Figure 10. Instantaneous amplitude attribute of #4 coal seam.

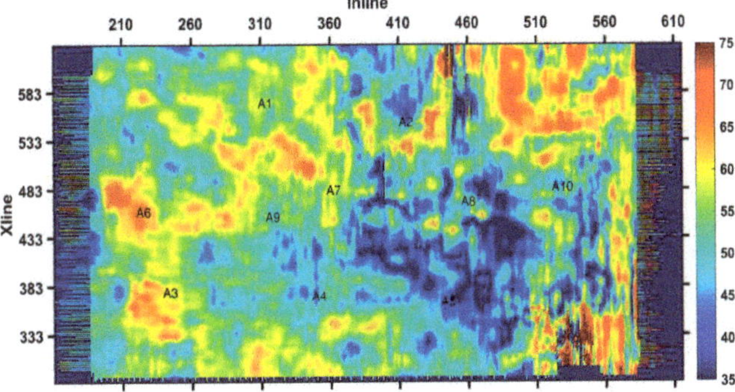

Figure 11. Instantaneous frequency attribute of #4 coal seam.

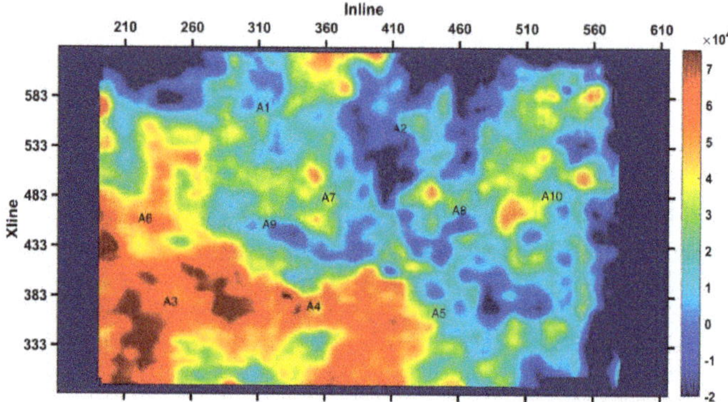

Figure 12. Relative wave impedance amplitude attribute of #4 coal seam.

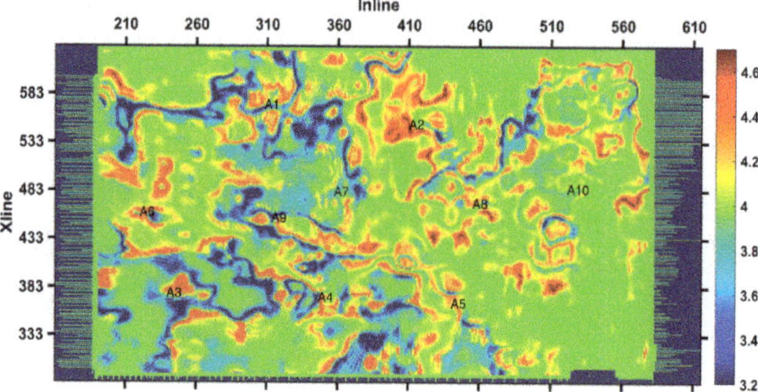

Figure 13. Coal thickness prediction results based on BP neural network.

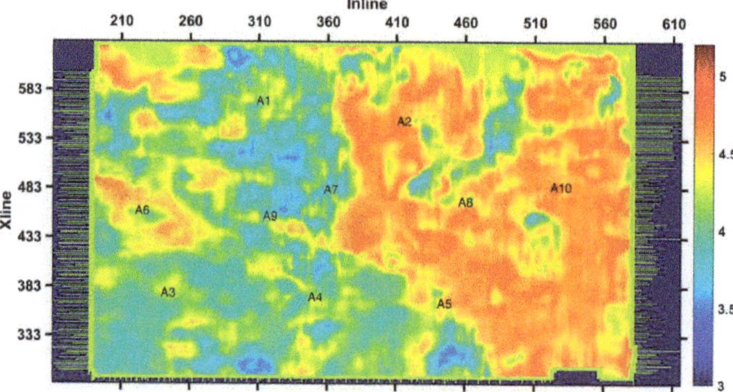

Figure 14. Coal thickness prediction results based on VMD-LSMT method.

5. Conclusions

(1)　EMD and VMD were used to denoise simple signals. There is an obvious mode-aliasing problem in EMD decomposition, which cannot effectively decompose the random noise. VMD can be used in signal denoising, and the denoising effect is good.

(2)　It can be seen from the forward simulation of the coal thickness wedge model that there is a good positive correlation between the instantaneous amplitude attribute, the relative wave impedance attribute, and the coal thickness, while the instantaneous frequency attribute has a good negative correlation with the coal thickness, which indicates that the seismic attribute is feasible for coal thickness prediction.

(3) It can be seen from the comparison with traditional BP neural network coal thickness prediction results that the VMD-LSTM method has higher prediction accuracy. The prediction results are in good agreement with the coal seam information exposed by the existing boreholes, which can be used as a new method for coal thickness prediction.

(4) The influence of different seismic attributes on coal thickness prediction needs to be further explored. In the process of using LSTM to predict, different weights can be assigned to each seismic attribute. This will help improve the accuracy of coal thickness prediction. Further in-depth research is needed in the future.

Author Contributions: Y.H., L.Y., and X.Q. conceived and designed the algorithms; Y.H., X.Q., Z.L., and L.Y. performed the algorithms; Y.H., X.Q., and Y.C. analyzed the data; Y.H. and L.Y. wrote the paper. All authors have read and agreed to the published version of the manuscript.

Funding: This research was funded by the Fundamental Research Funds for the Central Universities (Grant Number 2018QNA45) and a project funded by the Priority Academic Program Development of Jiangsu Higher Education Institutions (PAPD).

Institutional Review Board Statement: Not applicable.

Informed Consent Statement: Not applicable.

Data Availability Statement: The data presented in this study are available on request from the corresponding author.

Conflicts of Interest: The authors declare no conflict of interest.

References

1. Sun, Z.W. Numerical simulation of geostress field distribution in local variation area of coal seam thickness. *Ground Press. Strata Control* **2003**, *5*, 95–97.
2. Zhong, Q.T. Research and Application of Coal Seam Thickness Inversion Method. Ph.D. Thesis, China University of Mining and Technology, Xuzhou, China, 2001.
3. Shang, X.Q.; Yang, H.Y.; Ai, G. Study on the evolution and impact of mining-induced stress in coal thickness variation area. *China Min. Mag.* **2020**, *29*, 148–151+157.
4. Cui, W.X.; Wang, B.L.; Wang, Y.H. High-precision inversion method of coal seam thickness based on transmission channel wave. *J. China Coal Soc.* **2020**, *45*, 2482–2490.
5. Zeng, A.P.; Zhang, J.W.; Ren, E.M.; Liu, T.; Jiang, F.; Liu, X.J.; Jiang, F. Research on coal thickness prediction method based on VMD and SVM. *Coal Geol. Explor.* **2021**, *49*, 243–250.
6. Li, Q.H.; Zhang, C.Z.; Li, K.X. Analysis of influencing factors of mining surface subsidence under huge thick unconsolidated strata. *Coal Sci. Technol.* **2020**, *49*, 191–199.
7. Dong, S.H.; Xu, Y.Z. Inversion of coal seam thickness from seismic data by spectral moment method. *J. Liaoning Tech. Univ.* **2005**, *24*, 38–40.
8. Zou, G.G.; Peng, S.P.; Hao, X.X.; Zhang, J.; Wang, L. The relationship between coal thickness and seismic amplitude is analyzed based on wedge model. *Coal Sci. Technol.* **2014**, *42*, 88–91+100.
9. Guo, Y.X.; Meng, Z.P.; Yang, R.Z.; Zhang, L.H.; Liu, Y.C.; Sun, X.Y.; Zhao, G.P. Seismic attribute and its application to coal seam thickness prediction. *J. China Univ. Min. Technol.* **2004**, *33*, 67–72.
10. Lin, J.D.; Huo, Q.M.; Wu, Y.F. Application of multi-well constrained 3D seismic inversion technique in coal thickness prediction. *Coal Geol. China* **2003**, *15*, 47–49.
11. Yu, S.J.; Wang, Z.S.; Liu, Y.X. Application of wavelet multiscale analysis in coal thickness detection. *Coal Geol. Explor.* **2005**, *33*, 73–75.
12. Du, W.F.; Peng, S.P. Use geostatistics to predict coal seam thickness. *Chin. J. Rock Mech. Eng.* **2010**, *29*, 2762–2767.
13. Cheng, Y. Geological statistics coal thickness prediction method and application analysis. *China Min. Mag.* **2019**, *28*, 245–249.
14. Liu, Z.L.; Wang, J. Periods of refracted P-waves in coal seams and their applications in coal thickness estimations. *Acta Geophys.* **2020**, *68*, 1753–1762. [CrossRef]
15. Brosig, A.; Knobloch, A.; Barth, A.; Legler, C.; Hielscher, P.; Noack, S.; Etzold, S.; Dickmayer, E.; Kaufmann, H.; Franke, D. Mineral predictive mapping in 2D, 2.5D and 3D using Artificial Neural Networks—Case study of Sn and W deposits in the Erzgebirge, Germany. *Miner. Prospect. Curr. Approaches Future Innov. Orléans* **2017**, *1*, 24–26.

16. Noack, S.; Barth, A.; Irkhin, A.; Bennewitz, E.; Schmidt, F. Spatial modeling of natural phenomena and events with Artificial Neural Networks and GIS. *Int. J. Appl. Geospat. Res.* **2012**, *3*, 1–20. [CrossRef]
17. Sun, Y.; Zhang, L.; Zhu, J.; Xue, T.; Ding, J. Application of seismic attribute parameters in coal seam thickness prediction. *Coal Geol. Explor.* **2008**, *36*, 58–60.
18. Wu, W.W.; Yang, Y.G.; Chen, Y.K. Prediction of coal thickness Change by Kriging Method based on LSSVM Optimization. *Coal Technol.* **2015**, *34*, 89–91.
19. Guo, C.F.; Zhen, Y.; Shuai, C.; Ren, T.; Yao, W.L. Precise Identification of Coal Thickness by Channel Wave Based on a Hybrid Algorithm. *Appl. Sci.* **2019**, *9*, 1493. [CrossRef]
20. Dragomiretskiy, K.; Zosso, D. Variational mode decomposition. *IEEE Trans. Signal Process* **2014**, *62*, 531–544. [CrossRef]
21. Hochreiter, S.; Schmidhuber, J. Long short-term memory. *Neural Comput.* **1997**, *9*, 1735–1780. [CrossRef]
22. Huang, N.E.; Shen, Z.; Long, S.R.; Wu, M.C.; Shih, H.H.; Zheng, Q.; Yen, N.C.; Tung, C.C.; Liu, H.H. The empirical mode decomposition and the Hilbert spectrum for nonlinear and non-stationary time series analysis. *Proc. R. Soc.* **1998**, *454*, 903–995. [CrossRef]
23. Lian, J.; Liu, Z.; Wang, H.; Dong, X. Adaptive variational mode decomposition method for signal processing based on mode characteristic. *Mech. Syst. Signal Process.* **2018**, *107*, 53–77. [CrossRef]
24. Hochreiter, S.; Yoshua, B.; Paolo, F.; Schmidhuber, J. Gradient flow in recurrent nets: The difficulty of learning long-term dependencies. In *A Field Guide To Dynamical Recurrent Neural Networks*; IEEE Press: Piscataway Township, NJ, USA, 2001.
25. Sherstinsky, A. Fundamentals of Recurrent Neural Network (RNN) and Long Short-Term Memory (LSTM) Network. *Phys. D Nonlinear Phenom.* **2020**, *404*, 1–45. [CrossRef]
26. Cao, C.F. Research and Application of Coal Mine Floor Water Inrush Warning Model Based on LSTM. Master's Thesis, Xi'an University of Architecture and Technology, Xi'an, China, 2017.

 *electronics*

MDPI

Article

A Sensor-Based Drone for Pollutants Detection in Eco-Friendly Cities: Hardware Design and Data Analysis Application

Roberto De Fazio [1], Leonardo Matteo Dinoi [1], Massimo De Vittorio [1,2] and Paolo Visconti [1,*]

[1] Department of Innovation Engineering, University of Salento, 73100 Lecce, Italy; roberto.defazio@unisalento.it (R.D.F.); leonardomatteo.dinoi@studenti.unisalento.it (L.M.D.); massimo.devittorio@unisalento.it (M.D.V.)
[2] Center for Biomolecular Nanotechnologies, Italian Technology Institute IIT, 73010 Arnesano, Italy
* Correspondence: paolo.visconti@unisalento.it; Tel.: +39-0832-297334

Abstract: The increase in produced waste is a symptom of inefficient resources usage, which should be better exploited as a resource for energy and materials. The air pollution generated by waste causes impacts felt by a large part of the population living in and around the main urban areas. This paper presents a mobile sensor node for monitoring air and noise pollution; indeed, the developed system is installed on an RC drone, quickly monitoring large areas. It relies on a Raspberry Pi Zero W board and a wide set of sensors (i.e., NO_2, CO, NH_3, CO_2, VOCs, $PM_{2.5}$, and PM_{10}) to sample the environmental parameter at regular time intervals. A proper classification algorithm was developed to quantify the traffic level from the noise level (NL) acquired by the onboard microphone. Additionally, the drone is equipped with a camera and implements a visual recognition algorithm (Fast R-CNN) to detect waste fires and mark them by a GPS receiver. Furthermore, the firmware for managing the sensing unit operation was developed, as well as the power supply section. In particular, the node's consumption was analysed in two use cases, and the battery capacity needed to power the designed device was sized. The onfield tests demonstrated the proper operation of the developed monitoring system. Finally, a cloud application was developed to remotely monitor the information acquired by the sensor-based drone and upload them on a remote database.

Keywords: air pollution; drone; land survey; visual recognition algorithm; fast R-CNN; dust sensors; microcontroller

Citation: De Fazio, R.; Dinoi, L.M.; De Vittorio, M.; Visconti, P. A Sensor-Based Drone for Pollutants Detection in Eco-Friendly Cities: Hardware Design and Data Analysis Application. *Electronics* 2022, 11, 52. https://doi.org/10.3390/electronics11010052

Academic Editors: Darius Andriukaitis, Yongjun Pan and Peter Brida

Received: 17 November 2021
Accepted: 22 December 2021
Published: 24 December 2021

Publisher's Note: MDPI stays neutral with regard to jurisdictional claims in published maps and institutional affiliations.

1. Introduction

Human civilization and globalization are the primary culprits of the constant change in the global environment in the current scenario, mainly due to air and water pollution, global warming, ozone depletion, acid rain, natural resource depletion, overpopulation, waste disposal, deforestation, and biodiversity loss. Almost all of these processes result from the unsustainable use of natural resources; waste is a growing environmental, social, and economic problem for all modern economies. The air pollution generated by waste causes impacts felt by a large part of the population living in and around the main urban areas. Landfill management is a deeply felt problem by government authorities, given the enormous environmental impact of the pollutants due to naturally evaporated gases or substances generated by self-combustion or man-made fires, potentially dangerous for human health being diffused into the atmosphere.

According to the World Health Organization (WHO), air pollution is the most considerable environmental risk to health in the European Union (EU) [1]. Each year in the EU, it causes about 400,000 premature deaths and hundreds of billions of euros in health-related external costs. People in urban areas are particularly exposed; particulate matter, nitrogen dioxide, and ground-level ozone are the air pollutants responsible for most early deaths. These concepts are summarized in the initial section of the Special Report 23/2018, called "Air pollution: Our health still insufficiently protected" and published by the European

Court of Auditors, which stresses the importance of the pollution problem that can no longer be ignored. To solve this problem, the European Parliament and European Union Council have adopted the directive 2016/22884 on reducing national emissions of certain atmospheric pollutants, amending directive 2003/35/E.C. and repealing directive 2001/81/EC [2]. This last act establishes the emission reduction commitments for the member states' anthropogenic atmospheric emissions of sulfur dioxide (SO_2), nitrogen oxides (NO_x), non-methane volatile organic compounds (NMVOC), ammonia (NH_3), and fine particulate matter ($PM_{2.5}$). Additionally, the directive imposes that national air pollution control programs are drawn up, adopted, and implemented and that pollutant emissions and their impacts are monitored and reported. For air pollution measurements, a sampling height between 3–10 m must be considered; at this altitude, the vertical mixing is homogeneous, and representative of pollutants transported from neighbouring sources. Several monitoring methods and approaches are available for air pollution, including diffusion tubes [3,4], bubbler sampler [5], gas chromatography (GC) analyzers [6,7], remote optical/long path analyzers [8,9], photochemical and optical sensor systems [10,11], etc. However, the sampling method must be selected according to several parameters and requirements, including analyte typology, sampling duration and frequency, portability, maintenance need, costs, etc.

The sensors networks, including multiple sensor nodes distributed in strategic points, represent a valid approach for monitoring air quality with high precision in a relatively short time interval [12]. Dam et al. developed a wearable air quality sensor that analyzes the personal exposition to pollutions [13]. The designed device, called *EnviroSensor*, was a low-cost, open-source, mobile air-quality monitor that gathered real-time air quality data (ozone—O_3, PM and CO concentrations). A dashboard was designed to display and analyse the air quality data and a mobile application for connecting the sensors and enabling real-time data sharing. Additionally, Dhingra et al. presented a three-phase air pollution monitoring system featured by high sensitivity and precision [14]. The proposed IoT device comprises multiple gas sensors, an Arduino board, and a Wi-Fi module to collect environmental parameters and send them to a cloud server. This last one stores the incoming data, accessible using a custom Android application called *IoT-Mobair*. Wearable and portable devices are lastly founding applications in the environmental monitoring field. Specifically, Teriús-Padrón et al. worked on a wearable device to acquire PM concentration and send collected data to other devices using Wi-Fi and Bluetooth Low Energy (BLE) connection [15]. The device is placed into a case with three caps to fix it to a belt, pants, bags, or an armband. Moreover, a custom user interface shows the real-time Air Quality Index (AQI) level. This last is calculated from the received data according to the PM level, using the EPA (Environmental Protection Agency) formula reported in [16]:

$$I_p = \frac{I_{Hi} - I_{Lo}}{BP_{Hi} - BP_{Lo}}(C_p - BP_{Lo}) + I_{Lo}, \tag{1}$$

where I_p is the index for the pollutant p, C_p its truncated average concentration over 24 h based on 1 h measurements, BP_{Hi} and BP_{Lo} the concentration breakpoints greater or equal and lower or equal than C_p, and I_{Hi} and I_{Lo} the corresponding AQI levels.

Similarly, P. Arroyo et al. developed a portable system for air quality measurements in outdoor environments, detecting the NO_2, NO, CO, O_3, $PM_{2.5}$ and PM_{10} concentrations, along with the temperature, humidity and location [17]. The device comprises electrochemical gas sensors and optical particulate sensors, as well as a GSM module to transmit the acquired data, MQTT-Message Queue Telemetry Transmission communication protocol, to a cloud platform for storing and processing them. In [18], a novel Wireless Sensor Network (WSN) was introduced to detect pollutants, such as CO, NO_2, PM_{10}, PM_1, $PM_{2.5}$, and O_3, produced by urban transport and domestic heating systems. The proposed WSN, based on sensor nodes called *AIRBOX*, has been installed in several hotspots and on public buses [18]. In [19], the authors described an air quality monitoring system for urban scenarios applied to the citizens' garments and their bikes. In particular, the sensor node, equipped with

CO, SO_2, and NO_2 electrochemical gas sensors, was based on a 16-bit MSP430BT5190 microcontroller and a CC256x Bluetooth chip. The tests demonstrated good accuracy in temperature, humidity, and pressure measurements with accuracy within 1 °C, 2% RH, and 2 hPa; for the electrochemical sensors, 0.6 ppm accuracy was obtained [19].

F. Tsow et al. developed a portable/wearable volatile organic toxicants sensor equipped with Bluetooth connectivity [20]. It is based on an innovative tuning fork sensor for detecting VOC concentration. A dedicated oscillator drives each fork sensor; the resulting oscillations are integrated for a set number of cycles and integrated with a high-frequency clock generated by a high-frequency crystal oscillator. The sensor outputs are digitalized and transmitted to a host device such as smartphones and/or laptops.

Several scientific works are proposed in the literature that have as main elements a drone for acquiring air quality data. For instance, in [21], the authors introduced a novel drone for detecting air quality parameters in a given location, constructing a 3D map of the air quality measurements. Furthermore, in [22], the authors developed an Environmental Drone (E-drone) to gather information about concentrations of air pollutants (i.e., CO, CO_2, SO_2, NH_3, PM, O_3, and NO_2) in a specific place. The device implements onboard pollution abatement solutions. The drone comprises a 500 mL tank containing a solution for decreasing the NO_2 level, dispersed when an excessive NO_2 concentration is detected. A custom software gathered the data from multiple E-drones, drawing an Air Quality Health Index (AQHI) map for environmental analysis purposes. Similarly, Q. Gu et al. described an unmanned aerial vehicle (UAV) for air monitoring applications to obtain high-resolution and punctual profiling of air pollution [23]. The drone was equipped with low-cost microsensors, measuring the air concentrations of particulate matter and NO_2. A fusion servlet software, implemented by the NanoPI Neo Air board, fuses the data from PM and NO_2 sensors and flight controller, providing an aggregated data output.

Traffic is a critical issue in urban areas which are densely populated, requiring careful supervision to avoid areas with high congestion zones, resulting in high levels of pollutants with consequent risks to human health [24]. The scientific community has addressed this problem, finding new solutions for monitoring traffic based on environmental parameters. Notably, the noise level (NL) is a good indicator for forecasting the traffic level since it depends on traffic volume and speed, vehicle content, road surface and structure of the surrounding area. By gathering data related to the noise level and correlating them with other environmental parameters, a good estimation of the traffic level can be inferred.

In this paper, a monitoring system based on a low-cost drone is presented, equipped with a series of photochemical and optical sensors for monitoring the primary sources of pollution present in an urban scenario. The monitoring system is based on a Raspberry Pi Zero W board, which acquires and processes the data from sensors and stores them into the internal memory (SD card) along with alarm flags related to overcoming specific threshold values. Furthermore, the sensing unit is equipped with a wide set of sensors to detect concentrations of dangerous gaseous species and particulates (i.e., NO_2, CO, NH_3, CO_2, VOCs, $PM_{2.5}$, and PM_{10}) as well as the noise levels [25]. Additionally, the drone comprises an IR camera supported by a visual recognition algorithm to detect fires of hazardous materials and simultaneously capture their location by an onboard GNSS (Global Navigation Satellite System) receiver [26]. The presented device is thought to monitor the pollutant species in urban or suburban environments, along with the traffic level correlating it with the noise level detected by an onboard sound sensing module. Thus, the smart drone can easily monitor the traffic load in restricted city areas, allowing real-time management of moving vehicles on different city streets. Specifically, we employed a simple data fusion algorithm to combine the noise level acquired by the onboard microphone module and the air concentrations of gaseous species strictly correlated to vehicular traffic, such as NO_2 and $PM_{2.5}$. The developed solution offers numerous advantages over fixed monitoring systems available on the market, including portability, low cost, customization, and a wide operating range.

The proposed paper aims to develop a low-cost sensor-based drone for pollutant detection and visual recognition of waste fires, enabling quick supervising large areas and identifying pollution sources. Additionally, the consumption analysis is presented to design the sensing unit's supply section. Furthermore, characterization and testing of the proposed sensor-based drone in different operating conditions were carried out, demonstrating the correct operation of the developed system. Finally, a cloud-based application for remote monitoring of the historical data acquired by the sensor-based drone is introduced; it relies on a local application, enabling the operator to upload the data on the remote database and a mobile application to monitor the acquired measurements.

2. Materials and Methods

In this section, we first describe the architecture of the proposed sensing unit is introduced, along with its integration inside the Phantom 3 drone. Afterwards, the specifications of each component constituting the sensing unit onboard the drone are discussed, along with the threshold values set by the regulations for each pollutant. Lastly, a simple data fusion technique is presented to determine the traffic load in an urban scenario combining the noise level and pollutants measurements.

2.1. Architecture of the Developed Pollution Monitoring Drone

Figure 1 represents the 3D picture of the proposed mobile monitoring system constituted by the Phantom 3 drone (manufactured by DJI, Shenzen, China) hooked the sensing section to acquire the environmental parameters (1 in Figure 1). This last one is placed into a plastic box with air intakes (5 mm diameter), enabling the gas sensors to be exposed to the incoming air forced inside the case by the drone movement (2 in Figure 1). Moreover, an IR camera is placed on the box front section supported by a recognition algorithm to detect fires of hazardous materials and simultaneously capture their location by an onboard GNSS receiver (3 in Figure 1) [27]. Moreover, the sensing section is equipped with an electret microphone for detecting the noise pollution level in the overflown area. The monitoring system core is the mobile sensor node mounted on the previously chosen drone (4 in Figure 1). A Raspberry Pi Zero board acquires and processes the sensors' signals and implements a visual recognition algorithm to recognise waste fires.

Figure 1. 3D image of the proposed mobile monitoring system.

The performances of every gas sensor typology are affected by both analyte flow rate and direction, causing variation of their properties, and thus inducing measurement errors if not compensated [28,29]. Relatively to the developed sensor-based drone, the airflow changes due to variations of drone speed and the turbulence created by the rotation of the drone blades. Two precautions were used to mitigate the possible negative effects on the parameters acquisition. The first one is the application of a 40 μm stainless steel mesh layer behind the air inlets placed in front of the plastic case. This last one acts as a diffusion barrier, reducing the air velocity into the case, thus protecting the stability of the sensing part; also, it avoids the entry of foreign bodies into the plastic box. The latter is to keep relatively low and stable drone velocity (<10 km/h) during measurements.

Specifically, the system is equipped with two particulate detection sensors (ZH03A/ZH03B, manufactured by Winsen Electronics Technology, Zhengzhou, China) based on optical technology able to detect the concentrations of fine dust $PM_{2.5}$ and PM_{10}. Since these sensors provide analog outputs, the motherboard must be equipped with two external ADC (Analog-to-Digital Converter) (model ADS1115, manufactured by Texas Instruments, Dallas, TX, USA) interfaced by an I^2C bus. To make the system polyvalent, it was decided to equip it with sensors capable of detecting a wide range of gaseous species. Notably, the system includes a MOx technology sensor (CCS811, manufactured by AMS Technologies, Premstätten, Austria) to detect TVOC (Total Volatile Organic Compounds) and the CO_2 emitted in all applications that require combustion of fossil fuels.

Moreover, the system is equipped with an additional sensor based on MOx technology (MiCS6814, manufactured by SGX Sensortech, Neuchâtel, Switzerland) for detecting nitrogen hydroxide, ammonia, and carbon monoxide, gaseous species attributable to activities in an urban or industrial environment. The mobile sensor nodes are also provided with an audio detection system (model MAX4466, manufactured by Maxim Integrated, San Jose, CA, USA) based on an electric microphone and an adjustable low power gain stage to detect the level of noise pollution present in the overflown area. Once an area with pollution levels beyond the thresholds defined by current regulations or the presence of fires have been identified, the system stores its position using a low consumption GNSS receiver (model NEO M8N, manufactured by U-Blox, Thalwil, Switzerland). Additionally, the drone is equipped with an IR camera (model OV5647, manufactured by OmniVision Technologies, Santa Clara, CA, USA) to detect waste fires and take photos of areas where an abnormal level of pollutants has been detected to improve detection reliability (Figure S1).

The connections between the Raspberry motherboard and the various used components are depicted in Figure 2. The two ADS1115 ADC are interfaced with the microcontroller board using the I^2C bus, configuring them with different addresses. The analog signals supplied by the microphone sensor (MAX4466) and the PM_{10} laser dust sensor (ZH03A) is connected to the analog inputs of the first ADC (ADC1). In contrast, the second ADC (ADC2) converts the analog signals supplied by the MOx sensor (MiCS 6814) and the $PM_{2.5}$ sensor (ZH03B). Furthermore, the CCS811 MOX sensor is interfaced with the Raspberry board through the I^2C bus, whereas the GNSS receiver (NEOM8) uses the UART (Universal Asynchronous Receiver Transmitter) interface, sending NMEA packets containing the drone position. The GNSS receiver is used to acquire the coordinates of the locations where the pollutant concentrations are greater than WHO limits, as detailed in Section 3.1. Finally, the IR Camera (OV5647) is connected to the Raspberry board with the Camera Serial Interface Type 2 (CSI-2).

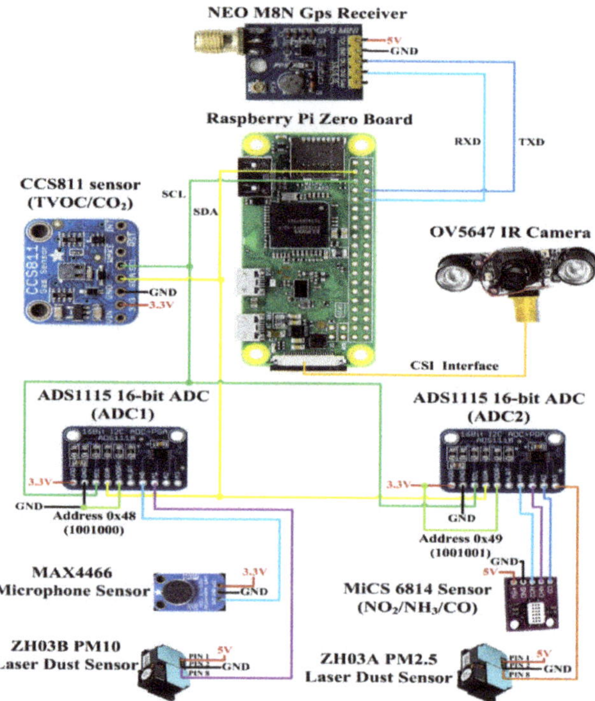

Figure 2. Connections of devices equipping the sensing section with the Raspberry Pi Zero board.

The developed sensor-based drone can operate in a wide range of environmental conditions, excluding those with strong wind (>35 km/h) and heavy rain, for avoiding drone instability. Additionally, water vapour represents the main disturbing factor in the performances of gas sensors, affecting the sensitivity and calibration of employed gas sensors, thus inducing measurement errors if not compensated [30]. At present, no compensation has been applied to the measurements acquired from gas sensors; as future development, a firmware compensation of the acquired gas concentrations based on air humidity measurements will be implemented.

Furthermore, the Phantom 3 drone is featured by a transmission distance up to 0.5 miles (1 km), depending on the environmental conditions. This distance allows simply reaching inaccessible and dangerous places (i.e., the central area of a landfill) while keeping the operator at a safe distance. As aforementioned, a relatively low sampling height (between 3 to 10 m) has been considered since a vertical homogeneity of gas species is obtained. Moreover, the presented device is extremely resistant to environmental agents (humidity, temperature and chemical species, etc.) and mechanical stress, both from the point of view of the RC drone and sensing section, suitably protected by a proper plastic cover.

Table S1 summarizes the threshold values, provided by the WHO (World Health Organization), of each pollutant component referred to a specific exposition time. The threshold values represent the maximum concentrations above which pollutants are highly harmful to humans and the environment [31].

2.2. Description of Used Devices and Sensors: Technical Features and Functionalities

The core of the sensing unit equipping the drone is the Raspberry Pi Zero W board; it is the smaller board of the Raspberry line, featured by Wi-Fi and Bluetooth capabilities. Specifically, the board includes Broadcom BCM2835 SoC, complying with the ARM11 architecture running at 1 GHz clock, and featured by 512 KB RAM. Moreover, the board is equipped with 40 GPIO (General Purpose Input-Output) and a wide range of interfaces (UART, I^{2}C, USB, CSI), enabling a high use versatility.

Two external ADCs are employed to acquire the analog signals provided by sensors and modules; the ADS1115 is a higher precision 16-bit ADC with four multiplexed channels [32]. It has a four multiplexed input, usable as four single-ended or two fully differential inputs, and programmable gain from $2/3\times$ to $16\times$ to amplify small signals and acquire them with higher resolution (Figure 3). Additionally, the sensing unit is equipped with two laser dust sensors for monitoring the environmental particulate ($PM_{2.5}$ and PM_{10}) concentrations. The ZH03A/ZH03B laser dust sensors can measure the number of a particular matter in a unit volume of air. In particular, the ZH03B sensor can detect coarse particles with a diameter lower or equal to 10 μm (PM_{10}), whereas the ZH03A module can detect fine particles with a diameter of 2.5 μm or less ($PM_{2.5}$) [33]. They are featured by 5V working voltage, current absorption lower than 120 mA, and response time (T_{90}) lower than 90 s, making the concentration data available using the UART interface.

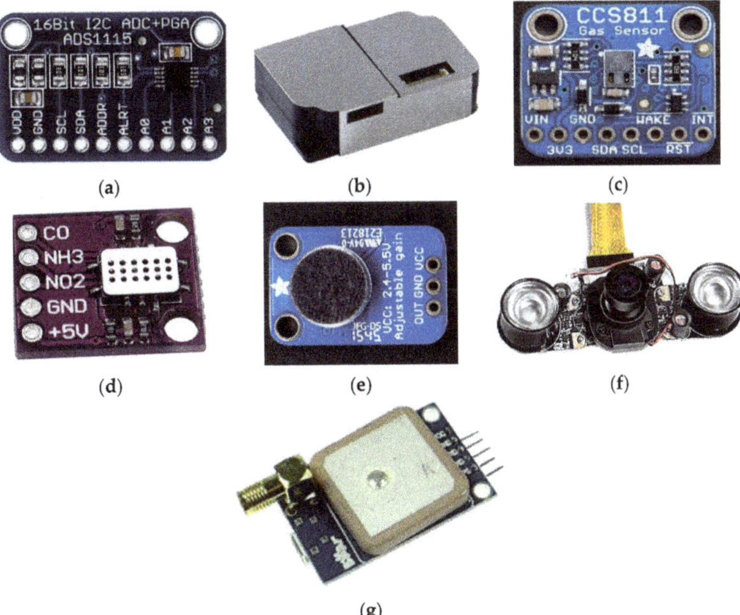

(**a**) (**b**) (**c**)

(**d**) (**e**) (**f**)

(**g**)

Figure 3. Modules constituting the sensing unit integrated inside the drone: AD1115 ADC module (**a**), ZH03A/B laser dust sensors (**b**), CCS811 gas sensor (**c**), MiCS-6814 sensor (**d**), Microphone sensor module (**e**), IR camera module (**f**), and NEO M8 GNSS receiver module (**g**).

Additionally, the system comprises a CCS811 gas sensor to monitor the air concentration of TVOC and CO_2; it is an ultra-low-power digital sensor that integrates a metal oxide (MOx) gas sensor to detect CO_2 in the 400–8192 ppm and Volatile Organic Compounds (VOC) in the range 0–1187 ppb [34]. The sensor has a small microcontroller to manage heater power, acquire the sensor voltage, and provide them through an I^{2}C interface; it is featured by 1.8–3.6 V supply voltage, 30 mA maximum supply current, and 80 mW maximum power consumption. Similarly, a MiCS-6814 sensor is comprised inside the developed drone; it is a compact MOS sensor with three metal oxide semiconductor sensors (Red sensor, Ox sensor, NH$_3$ sensor) able to detect several gas species [35]. The detectable gases are Nitrogen dioxide (NO_2), in the range of 0.05–10 ppm, Carbon monoxide (CO), in the range of 1–1000 ppm, Ammonia (NH_3) in the range of 1–500 ppm. The device has a 5 V supply voltage, 88 mW maximum heater power dissipation, and 8 mW maximum sensitive layer power dissipation (Figure 3).

The drone is equipped with a microphone sensor module based on a capsule microphone and a MAX4466-based amplifier, featured by low operating voltage and an excellent power supply noise rejection. The module has a 2.5–5.5 supply voltage, 24 μA supply current, 600 kHz operating bandwidth, and adjustable gain. Furthermore, an IR camera module is mounted in front of the plastic case containing the sensing unit; it is based on OV5647 CMOS (Complementary Metal Oxide Semiconductor) image sensor, providing 2592 × 1944 video output using Omni BSI (back-illuminated sensor) technology [36]. In addition, the camera is integrated with an IR filter, sensitive to Infrared radiation, making the camera operational similar to a night vision camera with external light sources such as IR LED illuminators. The device is interfaced with the Raspberry Pi board using a short ribbon cable to the board processor via the CSI bus, a higher bandwidth link that carries pixel data from the camera to the processor.

Lastly, a NEO M8 GNSS receiver is integrated inside the drone for acquiring the coordinates of zones where anomalous values of detected parameters or a waste fire are detected. The module has 2.5 m horizontal position accuracy, 10 Hz maximum update rate, −166 dBm sensitivity, and 29 s (cold start) time to first fix.

As shown in Figure 1, the sensing section is installed inside a plastic case realized in ABS (Acrylonitrile Butadiene Styrene) by 3D printing, having dimensions 7.8 cm × 6 cm × 6 cm and weight 15 g. The case cover includes a flange, enabling its connection to the joint used for mounting the onboard camera (Figure 1). The entire sensing unit has 294 g weight, including the power supply section and the battery. The drone with the sensing section installed is characterized by a total weight of 1350 g, allowing for easy transport.

As previously discussed, the proposed work aims to develop a low-cost tool for monitoring environmental concentrations of pollutants; considering this goal, the total cost of the developed sensor-based drone is about EUR 780, where most of the cost is attributable to the drone, whereas the cost of sensing unit is around EUR 80. Table 1 shows the breakdown of costs on the various components that make up the developed device. Compared with commercial monitoring systems with similar capabilities [37], the designed device is surely cheaper and more functional, but ensuring easier portability.

Table 1. Table reporting the breakdown of costs on the various components constituting the sensor-based drone.

Component	Indicative Cost (EUR)
Phantom 3 Drone	699
Raspberry Pi Zero W Board	12
ADS1115 Breakout board	3
ZH03A/B Sensor module	8
CCS811 Breakout board	10
MiCS6814 Breakout board	16
MAX4466 Microphone module	3
OV5647 IR Camera module	12
NEO 8M GNSS receiver	5
Battery	5
MP 207 DC/DC Converter	0.5
Battery protection board	0.3

2.3. Data Fusion Technique for Monitoring the Traffic Level from Noise and Pollutants Data

The presented air monitoring system represents a valid solution for the rapid and punctual monitoring of the traffic level in smart city scenarios through a suitable data fusion algorithm. In particular, the acquisition section measures the voltage signal provided by the microphone module over a one-minute interval; the microphone module is set to provide a +15 gain on the signal generated by the microphone. The voltage level is converted in a corresponding dB-level, using an empirical function deduced by a characterization of the used module.

$$\mathrm{NL_{dB}} = \frac{\overline{V} + 0.2674}{0.03534 \frac{V}{dB}},$$ (2)

Because the noise measurements are performed to an altitude different than zero, the noise level measurement is compensated using a corrective factor derived by Stokes' law, supposing a reference attenuation coefficient $\alpha = 0.005 \frac{dB}{m}$ (@ 70% relative humidity, 1000 Hz frequency, 18 °C temperature) and considering the altitude measurement provided by the onboard GNSS module. Thus, the NL at zero quotas measured by the drone is given by:

$$\mathrm{NL_{dB}}(0) = \mathrm{NL_{dB}}(d) + \alpha \cdot d,$$ (3)

Furthermore, the collected noise level (NL) is combined with the NO_2 and $PM_{2.5}$ concentration acquired by the onboard sensors, defining a Traffic Level Index (TLI) defined as:

$$\mathrm{TLI} = k \cdot [a \cdot \mathrm{NL_{dB}} + b \cdot PM_{2.5} + c \cdot NO_2],$$ (4)

where $a[dB^{-1}]$, $b[\frac{\mu g}{m^3}^{-1}]$, and $c[ppm^{-1}]$ are combination coefficients, as well as k a normalization parameter. These parameters are set to $0.25\ dB^{-1}$ (a), $24\ \frac{\mu g}{m^3}^{-1}$ (b), $825\ [ppm^{-1}]$ (c), and 0.5 (k), respectively.

The following classification rule is employed to discern the traffic condition according to TLI value:

$$\begin{cases} \mathrm{TLI} \leq 60 \rightarrow \text{LOW TRAFFIC} \\ 60 < \mathrm{TLI} \leq 105 \rightarrow \text{MODERATE TRAFFIC} \\ 105 < \mathrm{TLI} \leq 200 \rightarrow \text{INTENSE TRAFFIC} \\ \mathrm{TLI} > 200 \rightarrow \text{EXCESSIVE TRAFFIC} \end{cases},$$ (5)

where the threshold values are determined according to the limitations imposed by the European or international regulations on the parameters above (Table S1).

According to the acquired TLI level, the drone can discern in real-time the traffic load in four different conditions (LOW, MODERATE, INTENSE, and EXCESSIVE), combining the noise level and the pollutant measurements and providing a quick indicator to traffic managers to balance the traffic load on a wider area.

3. Results

In this section, at first, the sensing unit's firmware for coordinating the acquisition of environmental and traffic conditions is introduced; then, the results of onfield tests on the presented sensor-based drone are reported to verify the correct operation of all the systems functionalities.

3.1. Description of the Sensing Unit's Firmware for Managing the Data's Acquisition and Processing

This section describes the flowchart of the firmware implemented by the Raspberry Pi Zero board integrated into the proposed mobile sensor node to manage the acquisition of environmental parameters and detect possible waste fire. The sensing unit's firmware was developed with Python (third version) programming language, fully compatible with the used computational platform.

The first step is the declaration and initialization of all the useful variables, among which three arrays are fundamental:

1. THRESHOLD array, containing the threshold values of the monitored pollutants, defined by the European and national regulations, as described in Section 2.1. Additionally, a noise threshold value equal to 85 dB is set, corresponding to the maximum noise value tolerated by the human ear [38].
2. DATA array, containing the values measured by all the sensors present in the drone, compared later with the threshold values.

3. FLAGS array, which, unlike the two previous arrays, is a boolean array, i.e., it can contain only the values 0 and 1. It contains the comparison results between the measured value and threshold; a "1" in the i-th position indicates that the value of the i-th pollutant component is above the threshold, whereas "0" indicates a value below the threshold.

Afterwards, two nested cycles are started; the first one lasts 60 s used to evaluate the traffic level and the latter with 5 s period for synchronizing the acquisition and storing of the environmental parameters (Figure 4). Particularly, every 5 s, an acquisition cycle is started; the first step is resetting the DATA and FLAGS arrays since they store the previous measurements if not the first acquisition. Afterwards, the acquisition step begins by measuring gas concentrations and noise level from the eight sensors (ZH03A, ZH03B, CCS811, MiCS 6814, and MAX 4466 sensors). Specifically, each iteration provides for selecting the i-th channel and acquiring the corresponding signal, followed by storing this information in the i-th element of the DATA array. Then, the measurements (DATA[i]) are individually compared with the corresponding thresholds (THRESHOLD[i]) implemented by a for loop constituted by eight iterations and moving on DATA[] and THRESHOLD[] array elements; if the i-th value is greater than the limits, it is annotated with a "1" in the boolean FLAGS array.

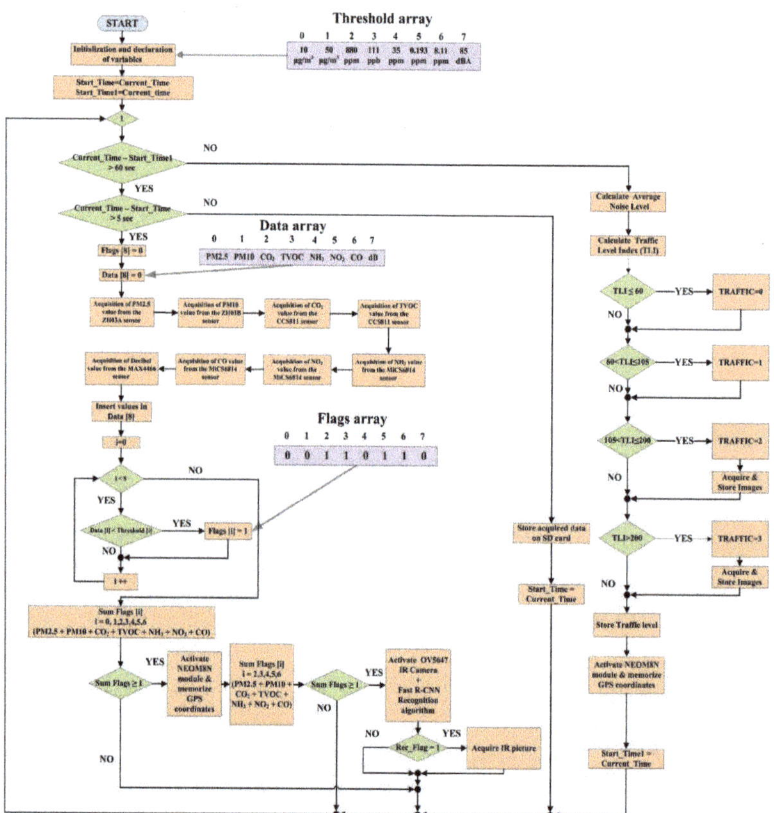

Figure 4. Flowchart of the firmware executed the sensor-based drone for pollutants detection and waste fire localization.

Later, the sum of all flags is calculated, and if a single species is over-threshold, the drone acquires the GPS coordinates. Then, the system looks for waste fires; in this case, only the data from two sensors are considered to detect and localize possible waste fires by monitoring the combustion gases produced. In general, the so-called micro-pollutants are produced from waste combustion, such as acid gases, nitrogen oxides, and unburnt gases, and the so-called micro-pollutants, such as heavy metals and organochlorine compounds. Carbon monoxide (CO) is formed in combustion reactions in the absence of sufficient oxygen and represents the primary combustion indicator. Other compounds generated during the fire event are nitrogen oxides, particularly the NO_2, toxic gas with a reddish color, irritating, with a pungent odour, and produced following the combustion of nitrogen materials, such as nitrocellulose and organic nitrates.

Furthermore, in the presence of materials containing nitrogen (wool, silk, acrylic materials, etc.), the formation of ammonia can be detected from combustion. Finally, the so-called "greenhouse gases" cannot fail to be mentioned, in particular carbon dioxide (CO_2), methane (CH_4) and nitrous oxide (N_2O). Therefore, the two considered sensors are the CCS811 for monitoring CO_2 and TVOC and the MiCS6814 for monitoring CO, NO_2, and NH_3 [39]. If at least one of the five values above mentioned are greater than the corresponding threshold values, or in other words, if the sum of the FLAGS array values is greater or equal than one, the IR camera is activated, followed by the launching of a recognition algorithm (Figure 4).

Specifically, the Fast Region-based Convolutional Network (Fast R-CNN) method is employed for object detection [40]; the method accepts the input picture and a set of proposal objects. At first, the method calculates multiple convolutions of the input picture and maximum polling layers generating a feature map. Later, a fixed-length vector is derived for each proposal object from the feature map. A sequence of fully connected layers processes each vector, deriving two sibling layers; the first generates a softmax probability estimation relative to the K object classes and the overall background class. The latter produces four real numbers for each K object classes, defining a bounding box over each of them. In particular, a Python implementation of the Fast R-CNN was implemented inside the realized firmware, trained using a set of annotated images containing fires as well as corresponding negatives. Finally, the iteration is terminated by updating the start_time variable with the current instant (start_time = current_time) to restart the described process and repeat the polluting quantities' sampling every 5 s (Figure 4).

Additionally, every 60 s, the algorithm calculates the mean noise level over the considered observation interval and, thus, the TLI value according to Equation (3). According to the obtained TLI, the traffic load is classified into four categories following the classification rule (6), storing the corresponding GPS coordinates and a timestamp. Furthermore, in the conditions of intense ($105 < \mathrm{TLI} \leq 200$) and excessive ($\mathrm{TLI} > 200$) traffic, the drone acquires two pictures with different levels of zoom ($1\times$ and $20\times$), allowing direct observation of the traffic state.

Once the drone has finished patrolling, the user can extract the SD card mounted into the Raspberry Pi Zero board, download the acquired data in csv files for numerical values and JPG for the caught picture, process them to detect any anomalies, followed by subsequent field checks. Additionally, the acquired data was uploaded on a cloud application realized on the IBM Cloud platform, allowing users remote monitoring of acquired environmental parameters and traffic conditions, as detailed in Section 3.2.

Before each measurement, the drone's gas sensors were tested by comparing the environmental concentration values of the gases and particulates acquired on the ground with those from calibrated portable instruments. Specifically, the multi-parametric detector RS-9680 (manufactured by RS Pro, Corby, UK) was used to verify the $PM_{2.5}$, PM_{10}, and TVOC concentrations, whereas the detector KANE101 (manufactured by Kane International Limited Inc., Welwyn Garden City, UK) was used to verify CO and CO_2 levels. Furthermore, the handheld detectors model GAXT-D-DL (manufactured by Frontline Safety Inc., Glasgow, UK) and NH3 Responder (manufactured by CTI Inc., Columbia, MO, USA) are used for detecting the NO_2 and NH_3 air concentrations, respectively. A good agreement between the measurements obtained with the portable detectors and those obtained from the sensors onboard the drone is obtained, with a maximum deviation of $\pm 5\%$, demonstrating the proper operation of all integrated sensors.

3.2. Power Consumption Measurements and Power Supply Section Design

In the following section, the consumption characterisation of the sensing section is presented to size the battery capacity, given a fixed energy autonomy; moreover, the power supply section of the proposed mobile sensor node is developed.

Table 2 summarizes the current values absorbed by each component obtained during the laboratory tests, depending on the operative modality (i.e., active mode or sleep/dormancy mode). The obtained values agree with the data reported in the datasheets of the respective components.

Table 2. Summarizing table with supply current values of the different components included in the developed mobile sensor node.

Component	Operative Modality	Supply Current
Raspberry Pi Zero W	Idling	120 mA
	Loading LXDE	160 mA
	Shoot 1080p Video (Rapberry + Pi Camera)	230 mA
ADS1115	Active mode ($T_A = 25\ ^\circ C$)	150 µA (TYP)/200 µA (MAX)
	Power-down ($T_A = 25\ ^\circ C$)	0.5 µA (TYP)/2 µA (MAX)
MiCS6814	Heating current	32 mA (RED sensor)
		26 mA (OX sensor)
		30 mA (NH_3 sensor)
CCS811	During measuring	26 mA
	Sleep mode	19 µA
ZH03A/ZH03B	Working current	<120 mA
	Dormancy current	<10 mA
MAX4466	Active mode ($T_A = 25\ ^\circ C$)	24 µA (TYP)/48 µA (MAX)
	Shutdown ($T_A = 25\ ^\circ C$)	5 nA (TYP)/50 nA (MAX)
NEO M8N	Acquisition	32 mA
	Tracking mode	30 mA
	Power save mode	13 mA
OV5647	Dormancy current	20 µA
	Working current	110 mA

After identifying the current consumption of each component, to perform adequate battery sizing, it is necessary to calculate the total current consumed during the drone operation. In particular, two specific cases must be considered; the first one represents the best case, whereas the second is a more realistic use case.

The first case refers to an optimal situation; particularly, after an entire session of monitoring the gases in flight, the acquired values has never been detected above the threshold values, defined in Section 2.1. The overall power consumptions of the sensing section during active and sleep modes are 487.08 mA (I_{ACTIVE}) and 241.04 mA (I_{SLEEP}), respectively. Therefore, the weighted average ($\bar{I}_{Patrol\ flight}$) is calculated according to the time duration of the two phases to obtain an average current value during the drone operation; specifically, the acquisition period lasts about 20 ms, and the sleep period 4.98 s, as demonstrated by the experimental tests.

$$\bar{I}_{Patrol\ flight} = a_1 \times I_{ACTIVE} + a_2 \times I_{SLEEP} = \frac{20\ ms}{10\ s} \times 487.08\ mA + \frac{9.98\ s}{10\ s} \times 241.04\ mA = 242.02\ mA, \qquad (6)$$

The second case refers to a more realistic use case. In particular, during a patrol flight lasting 30 min, equivalent to the autonomy of the designated drone, the various sensors detected ten times the value of at least one gas concentration above the set threshold value. In response, the system promptly activated the GNSS module to acquire the GPS coordinates and took a photo by the onboard camera module. In detail, the current absorbed by the entire system activates the GNSS module during the tracking phase is 260.56 mA (I_{GPS}), as well as during the picture acquisition 370.56 mA (I_{Photo}) (Figure 5).

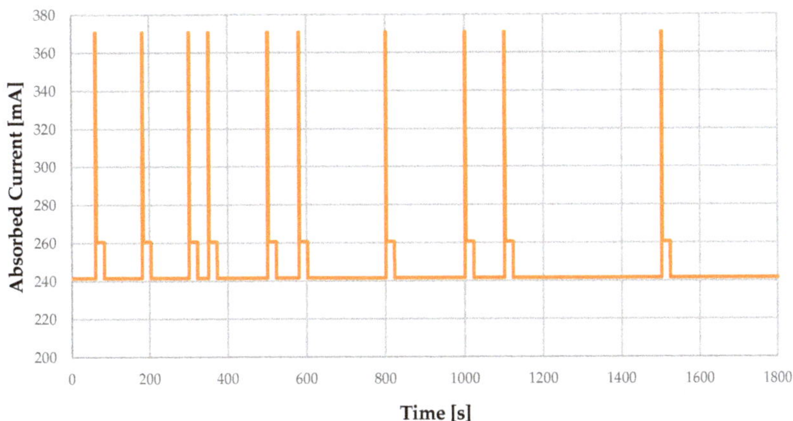

Figure 5. Time trend of the absorbed current during a patrol flight using camera and GPS.

In this second case, the mean current consumption ($\bar{I}_{Patrol\ flight\ camera\ \&\ GPS}$) is calculated, knowing the mean current during the drone patrolling ($\bar{I}_{Patrol\ flight}$) and the current values during the GPS tracking (I_{GPS}) and picture acquisition (I_{Photo}) and the corresponding time duration of each phase. Particularly, the acquisition of the GPS coordinates lasts 20 s, and the capturing and storing of a picture by the onboard camera requires 1.9 s. The three weights, multiplied by current values, called b_1, b_2, and b_3, are obtained by dividing the time in which the consumption refers by the autonomy time declared by the drone manufacturer, i.e., 30 min.

$$\bar{I}_{Patrol\ flight\ camera\ \&\ GPS} = b_1 \times \bar{I}_{Patrol\ flight} + b_2 \times I_{Photo} + b_3 \times I_{GPS}$$
$$= \tfrac{1581\ s}{1800\ s} \times 242.02\ mA + \tfrac{19\ s}{1800\ s} \times 370.56\ mA + \tfrac{200\ s}{1800\ s} \times 260.56\ mA = 245.43\ mA, \tag{7}$$

In conclusion, the last step is the sizing of the battery to guarantee the autonomy of the mobile sensor node equal to that of the drone (i.e., 30 min). The two quantities required to select and size the battery used to power the mobile sensor nodes are nominal and capacity. The nominal voltage must be at least 5 V, as all components used are powered either at 5 V or 3.3 V; therefore, a two-cell Li-Po battery was chosen, featured by 7.4 V nominal voltage. The overall charge required in the worst-case scenario previously discussed is given:

$$Q_{Patrol\ flight\ camera\ \&\ GPS} = \frac{\bar{I}_{Patrol\ flight\ camera\ \&\ GPS} \times \Delta t}{(1 - M)} = \frac{245.43\ mA\ \times 0.5\ h}{0.7} = 175.30\ mAh, \tag{8}$$

where a 30% discharge margin (M) on the overall battery capacity has been considered. We can conclude that a two-cell Li-Po battery with a minimum capacity of 175.30 mAh is required to ensure 0.5 h autonomy to the developed sensing unit for monitoring environmental pollutants. For this purpose, a two-cell Li-Po battery was selected, featured by a 380 mAh capacity and 7.4 V nominal voltage (model VZKT5088, manufactured by YuanHan Co., Jiangsu, China), as below described (Figure 6).

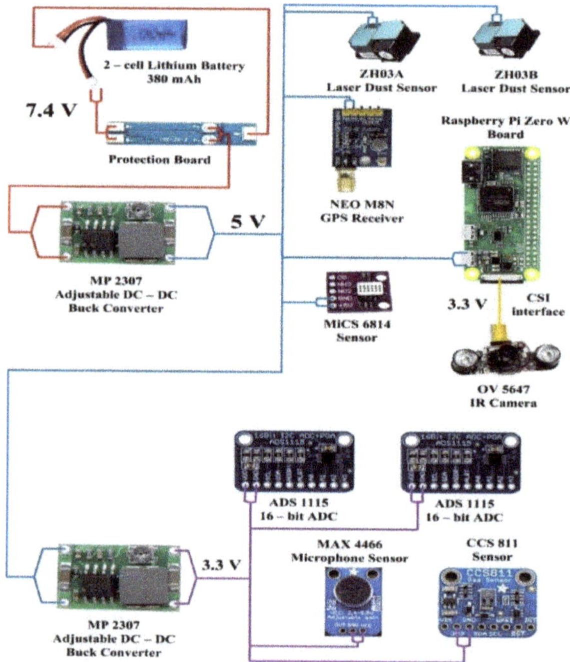

Figure 6. Physical connection scheme for supplying the various peripheral devices included in the developed sensing section.

It is connected to two adjustable buck DC-DC converters (model MP2307, manufactured by Monolithic Power Systems Co., Kirkland, WA, USA); the first one is used to scale down the battery voltage from 7.4 V to 5 V to power supply the MiCS6814 sensor, the NEOM8N GPS receiver, the ZH03A and ZH03B laser dust sensors, and the Raspberry Pi Zero W board, to which the OV5647 camera is connected via the CSI interface. On the other hand, the second buck converter reduces the 5 V provided by the first converter to 3.3 V used to feed the CCS811 sensor, the MAX4466 microphone sensor, and the two ADS1115 16-bit ADC (Figure 6).

3.3. Onfield Tests of Developed Sensor-Based Drone for Monitoring Environmental Parameters and Traffic Conditions

Figure 7 depicts two application scenarios describing how and where the proposed mobile monitoring system can be applied. In particular, the first figure shows the drone that identifies a fire in a suburban area (Figure 7a), whereas the second one shows a fire in a landfill (Figure 7b).

To test the presented sensor-based drone, we carried out different measurement campaigns to evaluate the correct operation of the entire detection system. The measurement campaigns were carried out considering a height from the ground of the drone between 8 m and 10 m and maintaining a cruising speed below 10 km/h, for the reasons detailed in Section 2.1. To carry out the tests, the monitored area was divided according to a grid with a 20 m pitch. The drone lingered within the quadrant for about 1 min, making it follow a circular trajectory; at the end of the rest period, the drone was brought to the adjacent quadrant till covering the entire monitored area. Relatively to the height range from 3 m to 10 m, significant variation of the parameters acquired with the acquisition altitude was not observed, confirming the hypothesis of homogeneous distribution of the pollutants within the aforementioned height range. Nevertheless, a gradual reduction in the environmental concentration of CO_2, NO_2 and particulate was observed for acquisitions carried out beyond the indicated range (>10 m), probably due to their specific weight greater than 1 with respect to the air, resulting in the accumulation of these species at low altitude. Furthermore, no significant variations were observed in the sensor measurements with the trajectory described by the drone.

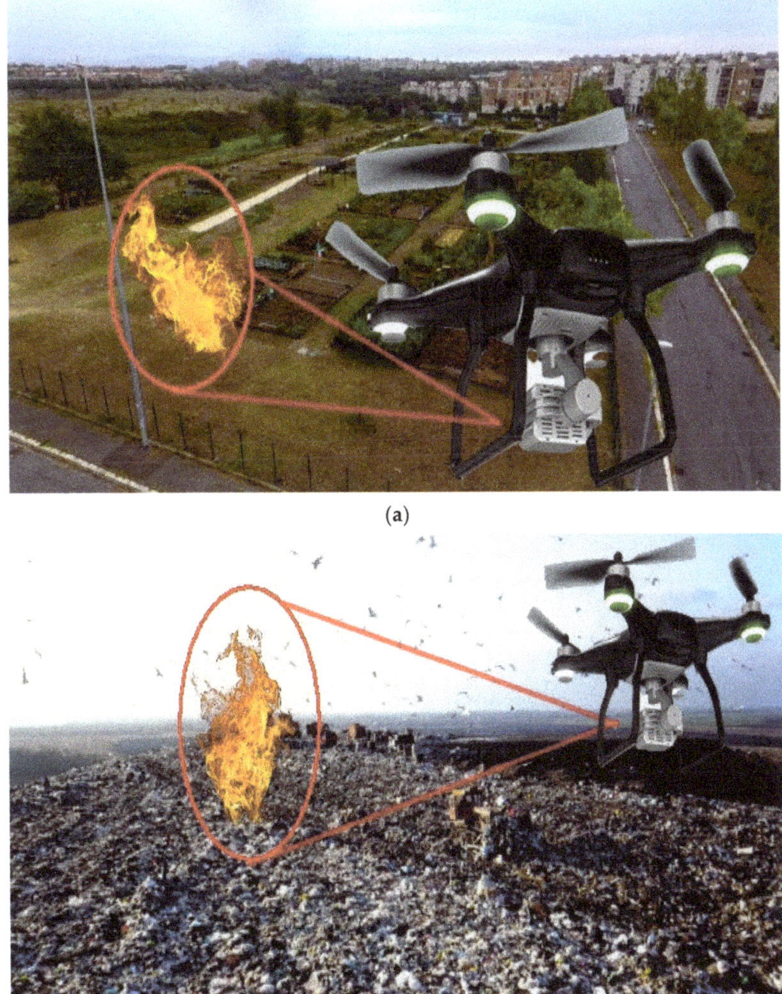

(a)

(b)

Figure 7. Pictures describing two application scenarios: in a suburban area (**a**); in a landfill (**b**).

Specifically, the first measurement campaign was carried in a suburban area (Ecotekne Campus, Lecce, Italy), away from possible pollutants sources. Figure 8a depicts the data acquired during the measuring campaign for the seven considered gaseous and particulate quantities. The drone flew over the area for 20 min, using a 5 s sampling period, as described in Section 3.1; later, the data acquired by the drone were downloaded and analyzed. As evident, the detected values are all below the set thresholds (Table S1); thus, the drone does not acquire any picture and GPS coordinates.

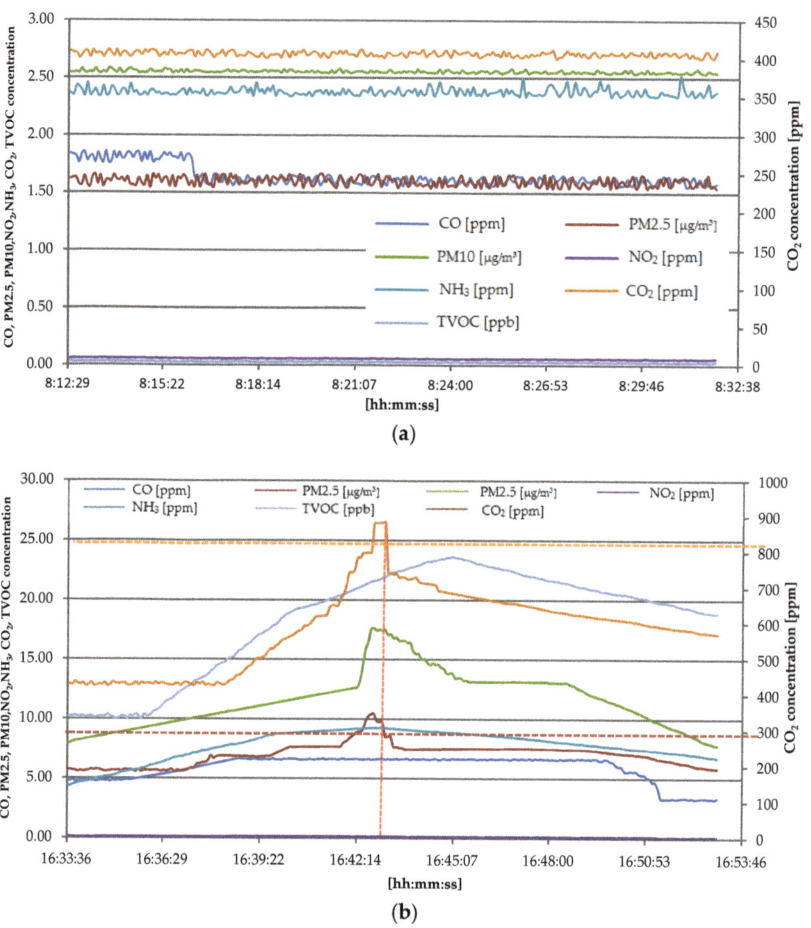

Figure 8. Time trend of the air concentrations of gaseous and particles species acquired by developed drone, using a sampling period of 5 s: first test campaign in a suburban area (**a**) and close a very busy state road (**b**).

The second test campaign was carried near a particularly busy state road (SS694, Lecce, Italy), using the previously described modalities. As evident, a rapid increase in CO_2, $PM_{2.5}$, and PM_{10} concentrations occurred when the drone flew near the road. Specifically, the CO_2 and $PM_{2.5}$ concentrations overcame the threshold set according to the WHO directives, reaching the peak values of 882 ppm and 10.23 µg/m^3, respectively (Figure 8b). In these conditions, as previously described, the drone acquires the specific position of the place where the thresholds were exceeded (40.335011N, 18.130470E). Since the sum of CO_2, TVOC, NH_3, NO_2, and CO flags equals 1, the drone acquires a picture and launches the visual recognition algorithm, not detecting any fire.

Furthermore, another test campaign was carried out to prove the capability of the developed sensor-based drone to detect waste fire. We prepared a setup constituted by a camping stove placed in an outdoor environment and modified the detection firmware for triggering the position and visual acquisition when only the CO_2 concentration exceeds a reduced threshold value of 500 ppm. The tests were performed in the evening to test the detection effectiveness of the IR camera and recognition algorithm. The drone was flown over the stove at a low altitude and verified the correct detection of the stove position and the acquisition IR picture of the area. Figure 9 depicts the trend of the CO_2 concentration acquired by the drone during the third test campaign. As evident, the CO_2 concentration overcame the set threshold (500 ppm) when the drone passed over the camping stove, acquiring the IR picture of the observed area.

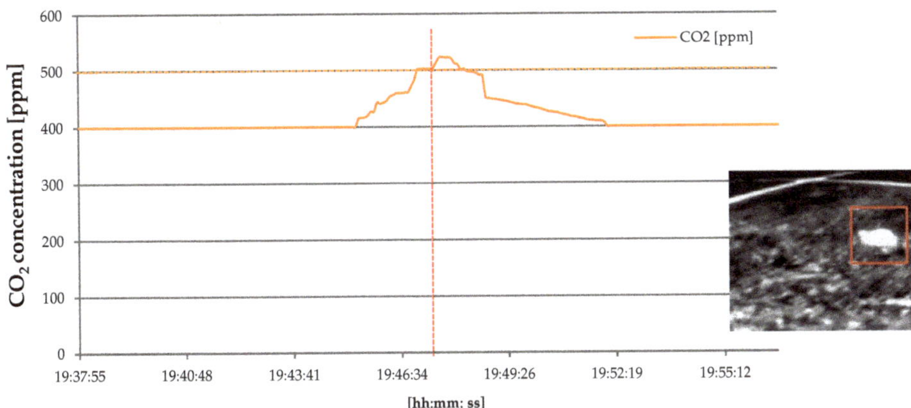

Figure 9. CO_2 trend acquired during the third test campaign used to test the functionality of the IR camera and visual recognition algorithm.

4. Discussion

This section analyses the experimental results previously reported, highlighting novelties and potentialities of developed air and land monitoring systems. Later, a discussion on the power consumption of the developed system is presented, designing the corresponding power supply section.

The previously presented results show the correct operation of the different modules and sensors included inside the developed air monitoring system. Specifically, the three test campaigns demonstrated the effectiveness of implemented firmware that carries out a crosscheck on the acquired gas species, triggering an alarm routine of only more than two parameters overcomes the corresponding thresholds. In Figure 8b, it is clear the fast response of sensors when the drone flew over the busy road, allowing identifying the areas with high pollutant levels. The sensing unit has stored the GPS coordinates when CO_2 and $PM_{2.5}$ concentrations overcome the set threshold values (880 ppm and 10 $\mu g/m^3$). In this condition, the drone has acquired a picture and launched a visual recognition algorithm, not detecting any fire.

The Fast R-CNN algorithm has been successfully tested in conjunction with the IR module for detecting waste-fire when anomalous environmental parameters are detected. As evident from Figure 9, once the detected CO_2 overcome the set threshold (i.e., 500 ppm), the visual recognition algorithm is launched for detecting the presence of waste-fire. When a fire is detected, the system marks the presence of fire by setting a flag and storing the acquired picture inside the SD card.

Additionally, a cloud application was developed for uploading the data provided by the drone patrolling. Specifically, it is based on a local application realized by Microsoft Visual Studio® used by the operator to upload the data on the IBM Cloud platform, exploiting the MQTT client package. The GPS coordinates, timestamps, and environmental data of places with anomalous environmental parameters are uploaded in the form of.csv file and stored into a remote database (Cloudant NoSQL Database). Figure 10a reports the loading of sample data inserted inside a. csv file on the remote database. The places with anomalous values are marked on a map, allowing the operator to check the correct loading of the data.

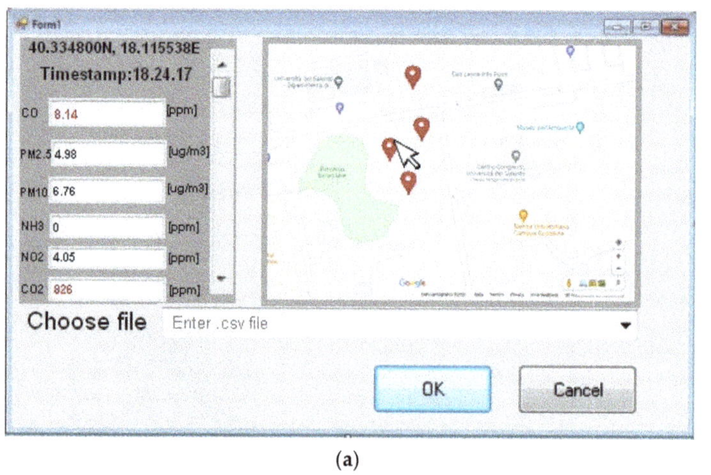

(a)   (b)

Figure 10. Screenshots of the local application used to upload the drone's data on cloud platform (**a**) and mobile application for monitoring the measurements acquired by drone (**b**).

Furthermore, by clicking on the marks, a list view is opened showing the acquired parameters, along with the corresponding timestamp. A mobile application allows users to access the data stored in the remote database and display them on an interactive map (Figure 10b). After login, the user can monitor the places with anomalous parameters, consulting the corresponding environmental parameters and timestamp, allowing a deep understanding of the polluting phenomena in the considered places.

5. Conclusions

This manuscript presents the development of a sensor-based drone for monitoring air and noise pollution. In particular, a Phantom 3 drone is equipped with a Raspberry Pi Zero W board and a wide set of sensors (i.e., NO_2, CO, NH_3, CO_2, VOCs, $PM_{2.5}$, and PM_{10}) to acquire environmental data. When a detected value exceeds the fixed threshold, the drone stores the GPS coordinates and a timestamp, enabling tracing historical pollution data in the considered area. Additionally, a proper classification algorithm was developed to estimate the traffic level according to the noise level acquired by the integrated microphone. Furthermore, an IR camera supported by a Fast R-CNN algorithm allows the recognition of waste fire even in dark conditions. Additionally, power consumption measurements were carried out in two different use cases to suitably design the sensing unit's supply section. Onfield tests were performed to verify all functionalities of developed monitoring systems, proving it in different operative scenarios. The tests demonstrated the proper operation of the developed sensor-based drone in all considered scenarios, thus representing a useful tool for performing environmental monitoring in a non-invasive and economical way. Finally, a cloud application was developed for uploading and monitoring the data regarding places and times where excessive levels of pollutants have occurred. In this way, a deeper understanding of pollutant trends can be obtained, enabling the implementation of countermeasures to reduce pollution levels.

Supplementary Materials: The following supporting information can be downloaded at: https://www.mdpi.com/article/10.3390/electronics11010052/s1, Figure S1: Block diagram of the electronic section for the environmental pollutants monitoring; Table S1: Summarizing table with threshold values of the main polluting species, issued by provided by the WHO (World Health Organization) [28].

Author Contributions: Conceptualization, R.D.F. and P.V.; methodology, R.D.F. and L.M.D.; software, L.M.D. and R.D.F.; validation, M.D.V.; data curation, R.D.F. and P.V.; writing—original draft preparation, R.D.F. and L.M.D.; writing—review and editing, P.V. and R.D.F.; supervision, M.D.V. All authors have read and agreed to the published version of the manuscript.

Funding: This research received no external funding.

Informed Consent Statement: Informed consent was obtained from all subjects involved in the study.

Data Availability Statement: Data of our study are available upon request.

Conflicts of Interest: The authors declare no conflict of interest.

References

1. Air Pollution: Our Health Still Insufficiently Protected. Available online: https://www.eca.europa.eu/Lists/ECADocuments/SR18_23/SR_AIR_QUALITY_EN.pdf (accessed on 9 April 2021).
2. Ddirective (EU) 2016/2284 of the European Parliament and of the Council. Available online: https://eur-lex.europa.eu/legal-content/EN/TXT/PDF/?uri=CELEX:32016L2284&from=ITA (accessed on 9 April 2021).
3. Cape, J.N. The Use of Passive Diffusion Tubes for Measuring Concentrations of Nitrogen Dioxide in Air. *Crit. Rev. Anal. Chem.* **2009**, *39*, 289–310. [CrossRef]
4. Rowell, A.; Terry, M.E.; Deary, M.E. Comparison of diffusion tube–measured nitrogen dioxide concentrations at child and adult breathing heights: Who are we monitoring for? *Air Qual. Atmos. Health* **2021**, *14*, 27–36. [CrossRef]
5. Filho, J.P.; Costa, M.A.M.; Cardoso, A.A. A Micro-impinger Sampling Device for Determination of Atmospheric Nitrogen Dioxide. *Aerosol Air Qual. Res.* **2019**, *19*, 2597–2603. [CrossRef]
6. Soo, J.-C.; Lee, E.G.; LeBouf, R.F.; Kashon, M.L.; Chisholm, W.; Harper, M. Evaluation of a portable gas chromatograph with photoionization detector under variations of VOC concentration, temperature, and relative humidity. *J. Occup. Environ. Hyg.* **2018**, *15*, 351–360. [CrossRef]
7. Watson, N.; Davies, S.; Wevill, D. Air Monitoring: New Advances in Sampling and Detection. *Sci. World J.* **2012**, *11*, 2582–2598. [CrossRef]
8. Geiko, P.P.; Smirnov, S.S.; Samokhvalov, I.V. Long Path Detection of Atmospheric Pollutants by UV DOAS Gas-Analyzer. In Proceedings of the 25th International Symposium on Atmospheric and Ocean Optics: Atmospheric Physics, Novosibirsk, Russia, 1–5 July 2019; SPIE: Novosibirsk, Russia, 2019; Volume 11208, pp. 596–600.
9. Yoshii, Y.; Kuze, H.; Takeuchi, N. Long-path measurement of atmospheric NO_2 with an obstruction flashlight and a charge-coupled-device spectrometer. *Appl. Opt.* **2003**, *42*, 4362–4368. [CrossRef]
10. Perevoshchikova, M.; Sandoval-Romero, G.E.; Argueta-Diaz, V. Developing an optical sensor for local monitoring of air pollution in México. *J. Opt. Technol.* **2009**, *76*, 274–278. [CrossRef]
11. Kuula, J.; Kuuluvainen, H.; Rönkkö, T.; Niemi, J.V.; Saukko, E.; Portin, H.; Aurela, M.; Saarikoski, S.; Rostedt, A.; Hillamo, R.; et al. Applicability of Optical and Diffusion Charging-Based Particulate Matter Sensors to Urban Air Quality Measurements. *Aerosol Air Qual. Res.* **2019**, *19*, 1024–1039. [CrossRef]
12. De Fazio, R.; Cafagna, D.; Marcuccio, G.; Minerba, A.; Visconti, P. A Multi-Source Harvesting System Applied to Sensor-Based Smart Garments for Monitoring Workers' Bio-Physical Parameters in Harsh Environments. *Energies* **2020**, *13*, 2161. [CrossRef]
13. Dam, N.; Ricketts, A.; Catlett, B.; Henriques, J. Wearable sensors for analyzing personal exposure to air pollution. In Proceedings of the 2017 Systems and Information Engineering Design Symposium (SIEDS), Charlottesville, VA, USA, 28 April 2017; pp. 1–4.
14. Dhingra, S.; Madda, R.B.; Gandomi, A.H.; Patan, R.; Daneshmand, M. Internet of Things Mobile–Air Pollution Monitoring System (IoT-Mobair). *IEEE Internet Things J.* **2019**, *6*, 5577–5584. [CrossRef]
15. Teriús-Padrón, J.G.; García-Betances, R.I.; Liappas, N.; Cabrera-Umpiérrez, M.F.; Waldmeyer, M.T.A. Design, Development and Initial Validation of a Wearable Particulate Matter Monitoring Solution. In *How AI Impacts Urban Living and Public Health, Proceedings of the 17th International Conference on Smart Homes and Health Telematics, ICOST 2019, New York, NY, USA, 14–16 October 2019*; Pagán, J., Mokhtari, M., Aloulou, H., Abdulrazak, B., Cabrera, M.F., Eds.; Springer International Publishing: Cham, Switzerland, 2019; pp. 190–196.
16. U.S. Environmental Protection Agency. Office of Air Quality Planning and Standards a Guide to Air Quality and Your Health. Available online: https://www.airnow.gov/sites/default/files/2018-04/aqi_brochure_02_14_0.pdf (accessed on 9 December 2021).
17. Arroyo, P.; Gómez-Suárez, J.; Suárez, J.I.; Lozano, J. Low-Cost Air Quality Measurement System Based on Electrochemical and PM Sensors with Cloud Connection. *Sensors* **2021**, *21*, 6228. [CrossRef]
18. Penza, M.; Suriano, D.; Pfister, V.; Prato, M.; Cassano, G. Urban Air Quality Monitoring with Networked Low-Cost Sensor-Systems. *Proceedings* **2017**, *1*, 573. [CrossRef]
19. Oletic, D.; Bilas, V. Design of sensor node for air quality crowdsensing. In Proceedings of the 2015 IEEE Sensors Applications Symposium (SAS), Zadar, Croatia, 13–15 April 2015; Volume 2012, pp. 1–5.

20. Tsow, F.; Forzani, E.; Rai, A.; Wang, R.; Tsui, R.; Mastroianni, S.; Knobbe, C.; Gandolfi, A.J.; Tao, N.J. A Wearable and Wireless Sensor System for Real-Time Monitoring of Toxic Environmental Volatile Organic Compounds. *IEEE Sens. J.* **2009**, *9*, 1734–1740. [CrossRef]

21. Wivou, J.; Udawatta, L.; Alshehhi, A.; Alzaabi, E.; Albeloshi, A.; Alfalasi, S. Air quality monitoring for sustainable systems via drone based technology. In Proceedings of the 2016 IEEE International Conference on Information and Automation for Sustainability (ICIAfS), Galle, Sri Lanka, 16–19 December 2016; pp. 1–5.

22. Rohi, G.; Ejofodomi, O.; Ofualagba, G. Autonomous Monitoring, Analysis, and Countering of Air Pollution Using Environmental Drones. *Heliyon* **2020**, *6*, e03252. [CrossRef]

23. Gu, Q.; Jia, C. A Consumer UAV-based Air Quality Monitoring System for Smart Cities. In Proceedings of the 2019 IEEE International Conference on Consumer Electronics (ICCE), Las Vegas, NV, USA, 11–13 January 2019; pp. 1–6.

24. Matz, C.J.; Egyed, M.; Hocking, R.; Seenundun, S.; Charman, N.; Edmonds, N. Human Health Effects of Traffic-Related Air Pollution (TRAP): A Scoping Review Protocol. *Syst. Rev.* **2019**, *8*, 1–5. [CrossRef]

25. Visconti, P.; de Fazio, R.; Velázquez, R.; Del-Valle-Soto, C.; Giannoccaro, N.I. Development of Sensors-Based Agri-Food Traceability System Remotely Managed by a Software Platform for Optimized Farm Management. *Sensors* **2020**, *20*, 3632. [CrossRef]

26. Visconti, P.; Iaia, F.; De Fazio, R.; Giannoccaro, N.I. A Stake-Out Prototype System Based on GNSS-RTK Technology for Implementing Accurate Vehicle Reliability and Performance Tests. *Energies* **2021**, *14*, 4885. [CrossRef]

27. Visconti, P.; de Fazio, R.; Costantini, P.; Miccoli, S.; Cafagna, D. Innovative complete solution for health safety of children unintentionally forgotten in a car: A smart Arduino-based system with user app for remote control. *IET Sci. Meas. Technol.* **2020**, *14*, 665–675. [CrossRef]

28. Sedlák, P.; Kuberský, P. The Effect of the Orientation Towards Analyte Flow on Electrochemical Sensor Performance and Current Fluctuations. *Sensors* **2020**, *20*, 1038. [CrossRef]

29. Dong, L.; Xu, Z.; Xuan, W.; Yan, H.; Liu, C.; Zhao, W.-S.; Wang, G.; Teh, K.S. A Characterization of the Performance of Gas Sensor Based on Heater in Different Gas Flow Rate Environments. *IEEE Trans. Ind. Inform.* **2020**, *16*, 6281–6290. [CrossRef]

30. Krivec, M.; Mc Gunnigle, G.; Abram, A.; Maier, D.; Waldner, R.; Gostner, J.M.; Überall, F.; Leitner, R. Quantitative Ethylene Measurements with MOx Chemiresistive Sensors at Different Relative Air Humidities. *Sensors* **2015**, *15*, 28088–28098. [CrossRef]

31. Ambient (Outdoor) Air Pollution. Available online: https://www.who.int/news-room/fact-sheets/detail/ambient-(outdoor)-air-quality-and-health (accessed on 16 June 2021).

32. ADS1113 Datasheet. Available online: https://www.ti.com/lit/ds/symlink/ads1113.pdf?ts=1616751793744&ref_url=https%253A%252F%252Fwww.google.com%252F (accessed on 28 March 2021).

33. ZH03-Series-Laser-Dust-Module Datasheet. Available online: https://www.winsen-sensor.com/d/files/air-quality/zh03-series-laser-dust-module-v2_0.pdf (accessed on 30 March 2021).

34. CCS811_Datasheet. Available online: https://cdn.sparkfun.com/assets/learn_tutorials/1/4/3/CCS811_Datasheet-DS000459.pdf (accessed on 31 March 2021).

35. MICS-6814 Datasheet. Available online: https://sgx.cdistore.com/datasheets/sgx/1143_datasheet%20mics-6814%20rev%208.pdf (accessed on 30 March 2021).

36. OV5647 Datasheet. Available online: https://cdn.sparkfun.com/datasheets/Dev/RaspberryPi/ov5647_full.pdf (accessed on 29 March 2021).

37. Synetica Enlink Air Wireless Air Quality Monitoring Datasheet. Available online: https://synetica.net/wp-content/uploads/2019/03/enLink-Air-V1_3.pdf (accessed on 14 December 2021).

38. Guidance & Regulations on Reducing Noise Exposure|NIOSH|CDCace Safety and Health Topic. Available online: https://www.cdc.gov/niosh/topics/noise/reducenoiseexposure/regsguidance.html (accessed on 16 June 2021).

39. Waste Fires: Combustion Gases. Available online: https://www.ingenio-web.it/28737-i-gas-di-combustione-generati-dagli-incendi-di-rifiuti-e-la-loro-tossicita (accessed on 17 June 2021).

40. Girshick, R. Fast R-CNN. In Proceedings of the IEEE International Conference on Computer Vision, Santiago, Chile, 7–13 December 2015; pp. 1440–1448.

 *electronics*

Article

Beacon-Based Hybrid Routing Protocol for Large-Scale Unmanned Vehicle Ad Hoc Network

Weiwei Mu [1], Guang Li [2], Yulin Ma [3,*], Rendong Wang [4,*], Yanbo Li [2] and Zhixiong Li [3]

[1] Fifth Team of Cadets, Army Military Transportation University, Tianjin 300161, China; mvv1130@163.com
[2] Tianjin 712 Communication Broadcasting Co., Ltd., Tianjin 300462, China; smileangellg@163.com (G.L.); liyanbo@712.cn (Y.L.)
[3] Suzhou Automotive Research Institute, Tsinghua University, Suzhou 215134, China; zhixiong.li@yonsei.ac.kr
[4] Institute of Military, Army Transportation, Military Transportation University, Tianjin 300161, China
* Correspondence: mayulin@tsari.tsinghua.edu.cn or mayulin1983@tsari.tsinghua.edu.cn (Y.M.); wrd1992@163.com (R.W.)

Abstract: In this paper, we designed a beacon-based hybrid routing protocol to adapt to the new forms of intelligent warfare, accelerate the application of unmanned vehicles in the military field, and solve the problems such as high maintenance cost, path failure, and repeated routing pathfinding in large-scale unmanned vehicle network communications for new battlefields. This protocol used the periodic broadcast pulses initiated by the beacon nodes to provide synchronization and routing to the network and established a spanning tree through which the nodes communicated with each other. An NS3 platform was used to build a dynamic simulation environment of service data to evaluate the network performance. The results showed that when it was used in a range of 5 ~ 35 communication links, the beacon-based routing protocol's PDR was approximately 10% higher than that of AODV routing protocol. At 5 ~ 50 communication links, the result was approximately 20% higher than the DSDV routing protocol. The routing load was not related to the number of nodes and communication link data and the protocol had better performance than traditional AODV and DSDV routing protocol, which reduced the cost of the routing protocol and effectively improved the stability and reliability of the network. The protocol we designed is more suitable for the scenarios of large-scale unmanned vehicle network communication in the future AI battlefield.

Keywords: communication technology; beacon; large-scale; unmanned vehicle; ad hoc network; hybrid routing protocol

Citation: Mu, W.; Li, G.; Ma, Y.; Wang, R.; Li, Y.; Li, Z. Beacon-Based Hybrid Routing Protocol for Large-Scale Unmanned Vehicle Ad Hoc Network. *Electronics* **2021**, *10*, 3129. https://doi.org/10.3390/electronics10243129

Academic Editor: Darius Andriukaitis

Received: 17 November 2021
Accepted: 13 December 2021
Published: 16 December 2021

Publisher's Note: MDPI stays neutral with regard to jurisdictional claims in published maps and institutional affiliations.

1. Introduction

A modern battlefield considers large-scale unmanned vehicle network (LUVN) operations, with a single unmanned vehicle (UV) as the basic task unit. That is, the individual low intelligence is coupled into swarm intelligence, which could execute wide area target searches, target continuous mission coverage, and perform other complex tasks that could not be achieved with traditional vehicles [1]. An unmanned vehicle network (UVN) is a cluster network composed of multiple UV communication systems in which the vehicle is capable of information sharing and dynamic allocation. In large-scale specific application scenarios, the network could stably and reliably execute information interaction [2,3]. Instructions for UVs and information exchange among them mainly rely on wireless networks. Considering LUVN communication in future military applications, it is particularly important to design a network routing protocol that can satisfy the requirements of a large-scale network of 100 or more nodes while maintaining good network performance with the topology of the network changing rapidly. Figure 1 illustrates the scheme diagram of LUVN.

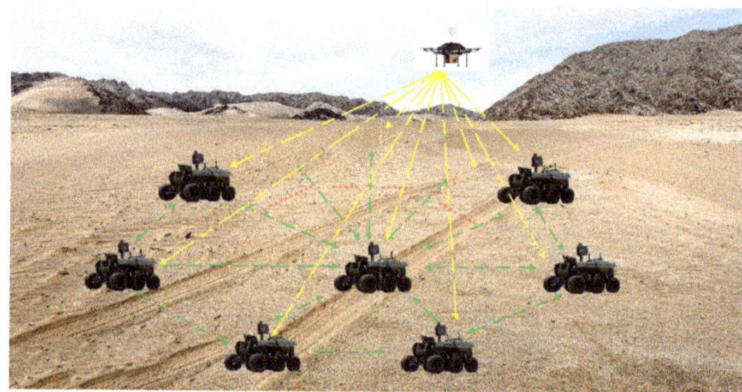

Figure 1. Schematic diagram of large-scale unmanned vehicle networking.

Currently, according to the network routing discovery policy, there are active routes, on-demand routes, and hybrid routes [4]. In the active routing protocol, all nodes maintain network topology information. Whether or not the network topology changes, its updated information will be transmitted throughout the entire network. In addition, the active routing protocol would generate a large amount of control overhead in the network, which consumes more bandwidth resources in the data transmission process. The nodes of an on-demand routing protocol trigger the path search process only when there is demand for service data. Route maintenance is executed on demand in the process of data transmission, that is, once the demand of service data stops, route maintenance will also be terminated, and the control cost can be reduced to some extent [5]. With the application of traditional protocols such as destination sequence distance vector (DSDV) and ad hoc on-demand distance vector (AODV) routing protocols in more scenarios, more obvious problems are exposed, mainly in the network overhead of active routing protocols and the long end-to-end transmission delay of data packets of on-demand routing protocols [6]. These problems are not suitable for LUVN scenarios.

In network routing protocols, literature [7] comprehensively analyzed and simulated typical AODV, DSDV, and OLSR protocols, but has not proposed a routing protocol with higher efficiency according to the differences obtained by simulation. Literature [8] introduced a clustering algorithm into the DSDV protocol and modified the criterion of link quality in the protocol so as to improve the network throughput. However, this method still used the method of flooding in the process of maintenance and path discovery, and the problems of transmission delay were not solved. Another study [9] adopted the frequency-based policy to maintain routing to improve the AODV protocol, which effectively reduced the transmission delay of the protocol. Literature [10] used ant colony algorithm to discover paths, which solved the single path problem in AODV protocol and improved the network stability. Literature [11] modified the AODV routing protocols to reduce the number of route request (RREQ) and route reply (RREP) messages by adding direction parameters and two-step filtering. The two-step filtering process reduced the number of RREQ and RREP packets, reduced the packet overhead, and helped select the stable route. Literature [12] improved the MPR algorithm of the OLSR protocol and selected the MPR path to the left or right of the source node according to the destination node location, so as to reduce the network overhead and improve the network throughput. Literature [13] proposed a service-based multipath routing protocol; this protocol was based on the quality of service (QoS) requirements of various services, considering hop count, average connection, and minimum bandwidth as comprehensive reference conditions, and the optimal path based on the QoS requirements of this service was selected.

The routing protocols in the above studies mainly reduced the routing cost to a certain extent. However, the broadcast mode was used to discover routes and maintain routing tables. As the number of network nodes increased, the network maintenance load increased exponentially. Therefore, it was difficult to support large-scale networks. In the multi-unmanned vehicle network, with each unmanned vehicle as a node, the task required strong mobility performance, resulting in the network topology changing dramatically. The node moving speed was fast, therefore data transmission had strong interference, and the communication link stability and reliability were low. As shown in Figure 2, the original 0-1-2-3 path was interrupted due to node movement and a new 0-4-3 path was generated. However, the routing protocol in the above literature could not actively perceive the change of routing path. When the path failed, the failed path was still used for communication or repeated routing path finding, which eventually led to data loss and seriously affected the communication quality.

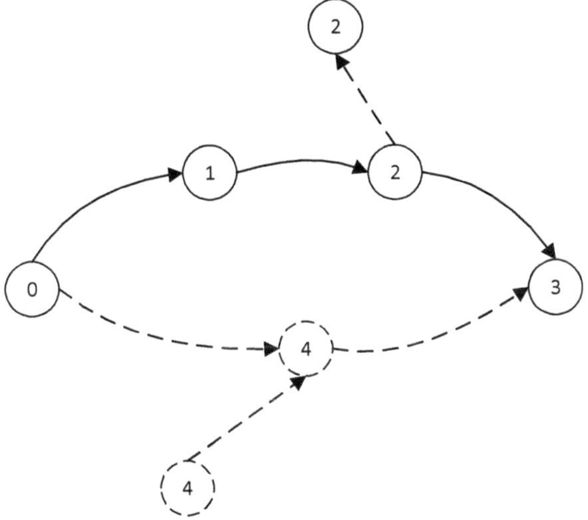

Figure 2. Schematic diagram of node path changes.

Based on the above analysis, a more efficient beacon-based hybrid routing protocol was proposed for the LUVN scenarios. In this protocol, a periodic beacon pulse was used to actively discover and maintain the path tree of the node, and the path was built from the source node to the destination on demand. At the same time, active paths were maintained under the guidance of periodic beacon pulses, so as to reduce the cost of routing protocols and improve the stability and reliability of routing network paths. An NS3 platform was used to build a dynamic service data simulation environment to verify the network performance. The simulation results showed that this protocol can solve the problems of path failure, repeated path finding, and excessive cost in route maintenance when using traditional AODV and DSDV routing protocols, which effectively improved the stability and reliability of the network. The protocol was more suitable for future AI battlefield LUVN scenarios.

The major contributions of this paper follow:

- In this protocol, a periodic beacon pulse was used to actively discover and maintain the path tree of the node, and the path was built from the source node to the destination on demand, which reduced the cost of routing protocols.
- Active paths were maintained under the guidance of periodic beacon pulses, so as to improve the stability and reliability of routing network paths.
- In the beacon-based hybrid routing protocol for LUVAN, a pulse cycle was the basic unit. A pulse cycle was divided into four phases, each of which was executed in a fixed time and the cost controlled by the routing protocol was only related to the time allocated in the first three phases.

This paper is organized as follows:

- In Section 1, the implementation method of beacon-based hybrid routing protocol is introduced.
- In Section 2, the effectiveness of beacon-based hybrid routing protocol is analyzed.
- In Section 3, an NS3 network simulation platform is used to compare and analyze the performance of beacon-based hybrid routing protocol, DSDV protocol and AODV protocol.
- Finally, the conclusion of this paper is in Section 4.

2. Beacon-Based Multi-Agent Network Hybrid Routing Algorithm

In a beacon-based hybrid routing protocol for a large-scale unmanned vehicle ad hoc network (LUVAN), the beacon node was the control center of the whole network; other nodes in the network took the pulse message sent by the beacon node as the time origin of the routing operation and established the optimal path tree to the beacon node. According to the optimal path tree, path discovery and maintenance and data transmission services were then completed.

2.1. Pulse Message

In the beacon-based hybrid routing protocol for LUVAN, a pulse cycle was the basic unit. As shown in Figure 3, a pulse cycle was divided into four phases, each of which was executed in a fixed time and the cost controlled by the routing protocol was only related to the time allocated in the first three phases.

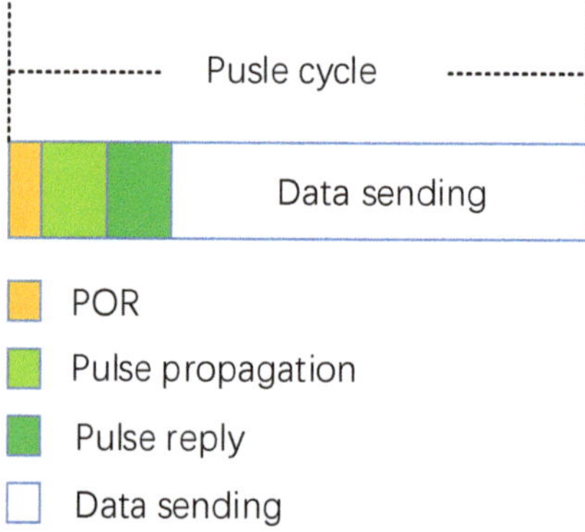

Figure 3. Schematic diagram of pulse unit.

Power on Reset (POR): In order to avoid the time error of different nodes, nodes can start receiving in advance and wait for receiving pulse messages.

Pulse Propagation: Pulse node periodically broadcasted pulse messages, other nodes received pulse messages and established the optimal path tree (OPT) to the pulse node.

Pulse Reply: For the routing request node, after receiving the pulse message, it would reply to the message in the form of unicast pulse along the OPT to the pulse node and request the path to the pulse node.

Data Sending: After the optimal path from pulse node to the destination node was established, the pulse node sent data along the OPT and sent the data to the destination node through forwarding to the nodes along the OPT.

In the beacon-based routing protocol, the non-beacon node would actively maintain the OPT pointing to the beacon node and establish the path to the destination node through the OPT.

When the non-beacon nodes received the pulse messages sent by the beacon node, they would create an OPT from non-beacon nodes to the beacon node. Because the beacon node sent pulse messages periodically, the non-beacon nodes could maintain an OPT to the beacon node in real time to ensure the real-time effectiveness of the OPT.

The beacon node sent pulse messages in a fixed period. Pulse messages included time information for time synchronization of non-beacon nodes.

The non-beacon node determined the time of the next POR state according to the pulse interval received last time and entered the receiving state at the specified time while remaining in the inactive state at other times. This node could enter the hibernation state to save power.

After receiving the pulse message, the non-beacon node would judge it according to the serial number and link metric of the pulse message. When the pulse number was the latest or the pulse number was the same as the stored pulse number but the link measure was small, the path was saved and the pulse message was forwarded at the same time. At this time, the node would establish an optimal path data to reach the beacon node.

After the beacon node sent a pulse message, all nodes in the network would create a path tree to the beacon source node, as shown in Figure 4.

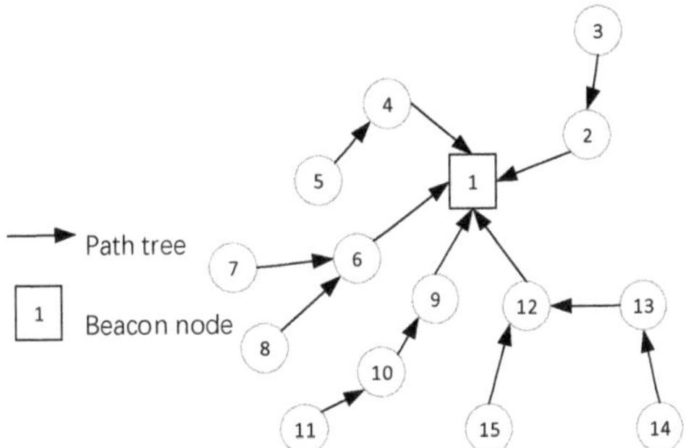

Figure 4. Schematic diagram of spanning tree.

2.2. Path Discovery and Maintenance

2.2.1. Path Discovery

When the data node had a data sending request, if there was no routing path to the destination node, the path establishment process as shown in Algorithms 1 and 2.

Algorithm 1. Source node path request process.

---

1   Tp: Pulse Propagation time
2   Tr:Pluse Reply time
3   Td:Data Sending time
4   **If** T > Tp and T < Tr
5   Source node send a pulse reply message along the OPT.
6   OPT node unicast reply message.
7   **Elif** T > Tr and T < Td
8   Source node send data along the path tree.
9   **Else**
10   Wait until the time reaches Tr.
11   **End if**

Algorithm 2. Beacon node path request process.

1   Pt: Path information table
2   Pd: Path to the destination address
3   Tp: Pulse Propagation time
4   Becaon node received the pulse reply message:
5   **If** Pd is in Pt:
6   Becaon node created a reverse path to the source node.
7   **Else:**
8   Becaon node created a reverse path to the source node.
9   **If** T < Tp:
10
11   Becaon node send a pulse message with the destination address.
12 **Else:**
13 Wait until next pulse cycle.
14 **End if**
15 **End if**

---

(1) Path creation process of data node:

① The node judged whether it was in the pulse reply state. If it was, it would send the pulse reply along the OPT, reaching the beacon node.

② If it was not in the pulse reply state, the node then judged whether it was in the data transmission state. If it was, data would be sent directly along the OPT reaching the beacon node. If not, it would wait until the data sending state.

(2) Node of OPT:

① The node in the OPT would create a reverse path to the source node after receiving the pulse reply message with the destination address, as shown in red arrows 6–7 in Figure 5a.

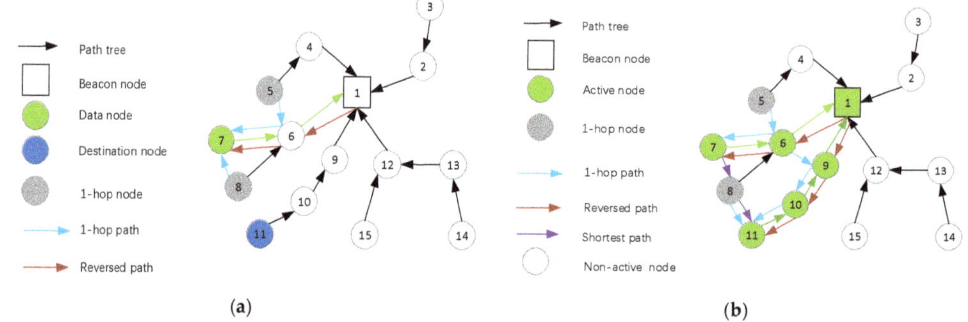

Figure 5. (a) Schematic diagram of source node path establishment process. (b) Schematic diagram of destination nodepath establishment process.

② The node in the OPT forwarded the pulse reply message with the destination address until it reached the beacon node.

(3) 1-hop node around the OPT:

When 1-hop node around the OPT received a pulse reply message with the destination address, a reverse path to the source node was established, as shown in the blue arrows 5-6-7 and 8-7 in Figure 5a.

The 1-hop node around the OPT did not forward the pulse reply message to avoid unnecessary routing cost.

(4) Beacon node:

① After the beacon node received a pulse reply request from the source node, a reverse path to the source node would be created, as shown by the red arrow 1-6-7 in Figure 5a.

② The beacon node judged if there was a path to the destination node. If not, it would send a pulse message with the destination address in the next pulse cycle for path addressing.

Subsequently, the beacon node, the source nodes in the OPT from the source node to the beacon source node, and the 1-hop nodes around the OPT would all create a reverse path to the source node.

When the destination node received the pulse message with the destination address, the node would send the pulse reply message to the beacon node along the OPT pointing to the beacon source node. The process for each type of node is illustrated Algorithm 1 and Algorithm 2. After the message reached the beacon node, the path 1-9-10-11 from the beacon node to the destination node as shown in Figure 5b was established. Node 8 around the path tree established reverse paths from 8 to 11 to the destination node.

2.2.2. Optimal Path Creation

As shown in Figure 5b, the source node would send data to the destination node along the newly established path tree. The initial possible path was 7-6-9-10-11, which was obviously not the optimal path.

During the path creation process, 1-hop nodes (nodes 5 and 8) of the path tree would monitor the pulse reply message sent by the source node and the destination node. If the node found a fast path existed, it would send a message to the corresponding node and declared that a fast path existed. The optimal path 7-8-11 shown in the purple arrow in Figure 5b would then be created.

The source node and destination node would actively respond to the pulse message broadcast by the beacon node to ensure the real-time validity of the communication link. The non-source node and destination node would not respond to the pulse message to reduce the route cost.

2.3. Beacon-Based Hybrid Routing Protocol

In active DSDV, the cost of maintaining N paths per node is $O\ (N)$. The maintenance cost of each link is $O\ (N/T)$, where N is the number of nodes and T is the lifetime of a link. Therefore, the total cost of maintaining links for N nodes in the network is $O\ (N^2/T)$.

In on-demand routing protocols, such as AODV and dynamic source routing (DSR), the routing cost is reduced by maintaining active links on demand; the cost of maintaining active links for N nodes is $O\ (NF)$, where F is the number of active links on each node. However, the on-demand routing protocol still uses the method of flooding in the process of path transmission, and the broadcast cost of node flooding is $O\ (N)$. Therefore, the total number of maintenance links of N nodes in the network is $O\ (FN^2/T)$.

The beacon-based hybrid routing protocol uses single path tree to discover and maintain links; only the beacon node in the network periodically send a flooding. Therefore, the maintenance cost of the link is $O(N/T)$, and T is the pulse period of the beacon-based protocol.

In active and on-demand routing protocols, when the number of nodes increases in a large scale, the cost of routing protocols increases exponentially, which is not suitable for large-scale networking scenarios. The beacon-based routing protocol uses the single path tree method to effectively solve the problem of flooding in the network. The routing cost and the number of nodes increase linearly, which reduces the cost of network maintenance.

3. Results

3.1. Simulation Environment

The NS3 simulation platform was used to build a dynamic simulation environment of service data to evaluate the routing protocol we designed. The MAC layer in the NS3 simulation environment uses the IEEE802.11 standard. Table 1 shows the basic simulation parameters. Table 2 shows the beacon-based hybrid routing protocol parameters.

Table 1. Simulation parameters.

Parameters	Values
Number of nodes	50
Bandwidths (Mbps)	5
Cover range (m^2)	1000 × 1000
Data package (Byte)	512
Active link data	5, 10, 15, 20, 25, 30, 35, 40, 45, 50
Communication speed rate/Kbps	10
Simulation time (s)	180
communication start time/s	80
Link active time (s)	10
Node velocity (m·s^{-1})	5
Pulse cycle (s)	1.5

Table 2. Protocol parameters.

Parameter	Value
Pulse cycle/sec	1.5
POR/msec	10
Pulse Propagation/msec	70
Pulse Reply/msec	70
Data Sending/sec	1.35

In this simulation, in order to verify the performance of multi-node concurrent network communication, the node access time is limited. The node access time is required to be randomly selected between simulation start time (communication start time/s) and active link data. That is, the network access time in this experiment was between 80 s + 5 s ~ 50 s, so that it can guarantee concurrent communication with multiple links at the same time in the active link time.

According to the above simulation parameters, 50 nodes were randomly distributed in the simulation environment with a range of 1000 m × 1000 m, moving randomly at the speed of 5 m/s, and 5 to 50 groups of active communication links were randomly generated in the network. AODV and DSDV were selected as comparisons to the proposed beacon-based routing protocol. The performance of the routing protocol was estimated by comparing the simulation results.

3.2. Results and Analysis

This paper mainly compared and analyzed the packet delivery rate and routing load. In order to ensure the randomness of node positions during the test, the nodes were in the stage of random movement before the 80 s of simulation, then communication links were established randomly from the 80 s to 170 s, in which the number of active links varied from 5 to 50. The results are discussed as follows.

(1) Packet Delivery Rate (PDR)

PDR is the ratio of packets received by the target node to packets sent by the source node. It is a statistical measure of correctly transmitted packets. This index mainly examines the reliability and communication quality of the network. The higher the successful packet delivery rate, the higher the network reliability and the better the communication quality.

The PDR of the beacon-based routing protocol and the AODV and DSDV protocols varied with the number of active links in the network, as shown in Figure 6. When the beacon-based routing protocol was used in a range 5–35 communication links, the PDR was approximately 10% higher than that of AODV routing protocol. At 5–50 communication links, the PDR was approximately 20% higher than the DSDV routing protocol. The overall PDR of the beacon-based routing protocol was better than that of the two traditional routing protocols. The beacon-based routing protocol adopted the active maintenance path tree and active link mode, which was more suitable for sensing path changes in advance and avoiding path failures and repeated path finding. Therefore, it had high reliability after networking.

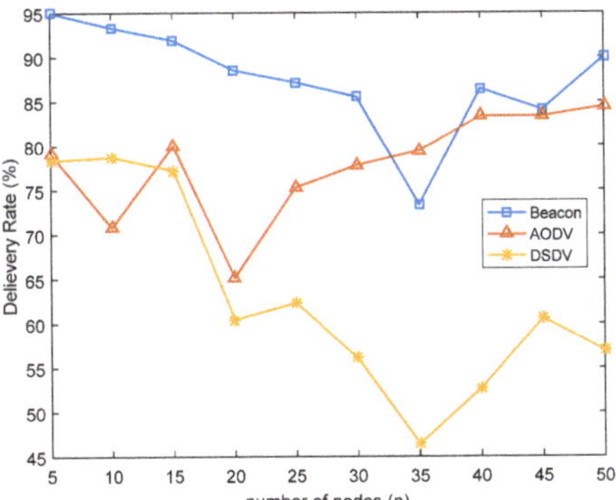

Figure 6. Delivery ratios of beacon routing protocol and AODV, DSDV protocols.

In Figure 6, the packet transmission rate of beacon-based routing protocol suddenly drops at node 35, mainly because the simulation startup time of six active communication links is basically the same, and the source node and destination node addresses of these six links are the same. The path query and maintenance of beacon-based routing protocol is limited to the first three stages of 140 ms.

As a result, when there are multiple data requests from the same node at the same time, the path query and maintenance cannot be completed within 140 ms. To solve this problem, we conducted the following experiment. In the experiment, the time of node access to the network was not limited and the time of node access to the network was randomly distributed within 80 s ~ 180 s (Beacon 2) and 80 s ~ 300 s (Beacon 3). The comparison between the delivery rate and the network entry time distribution of the original experiment in the 80 s + 5 s ~ 50 s (Beacon 1) is shown in Figure 7.

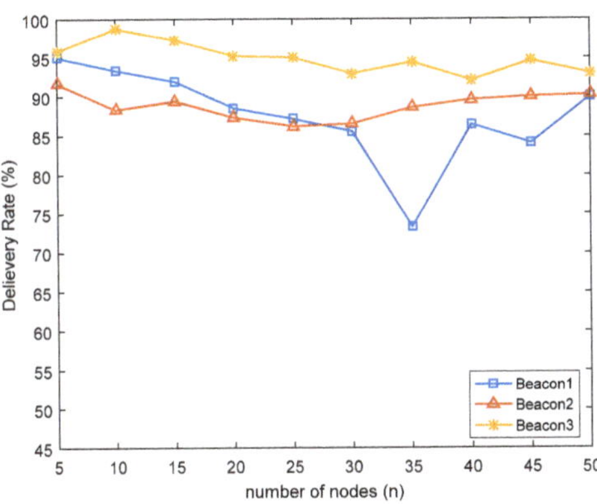

Figure 7. Delivery ratios of beacon routing protocol at different inbound times.

As shown in the figure above, the more dispersed the network time distribution of nodes is, the better the protocol delivery rate is. At the same time, there is no concurrent communication between multiple nodes at node 35, so there is no problem as shown in Figure 6.

(2) Routing Load (RL)

RL indicates the number of frames or packets borne by a communication device within a unit time. Ideally, the forwarding capability/rate should increase linearly with the increase in load. In practice, as the load increases, the processing capability decreases. Packet congestion leads to packet loss. As a result, the forwarding rate drastically decreases. Therefore, the RL is closely related to the system stability after networking.

Figure 8 showed the changes of RL of the beacon-based routing protocol and the AODV and DSDV protocols with the number of active links in the network. The result show that the cost of the RL of the beacon-based routing protocol had always been in a stable state and was unrelated to the data of the active communication link in the link, which was consistent with the original intention of the design.

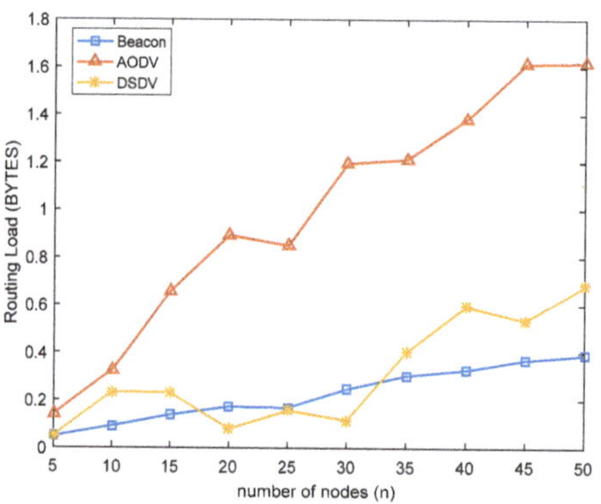

Figure 8. Routing load of beacon-based routing and the AODV and DSDV protocols.

The RL of the beacon-based routing protocol was only related to the time allocated to the first three phases of the pulse unit and unrelated to the number of nodes and communication link data. Therefore, using the beacon-based routing protocol for network unmanned equipment was more stable and suitable for the LUVN scenario.

(3) Effect of Pulse Cycle on the Network

In the beacon-based routing protocol, periodic flooding was used to create and maintain the OPT for nodes in the network to reach the beacon node. At the same time, nodes in the network completed path establishment and data transmission according to the guidance of the path tree. The sending period of the pulse message of the beacon node determined the network entry time and path update maintenance time of nodes in the network. The sending period of the pulse message should be selected according to the number of nodes in the network, the network access time, communication bandwidth, and other parameters required by the network. Figures 9 and 10 were the simulation comparison of PDR and RL when the pulse cycle was 1.5 s, 2 s, 2.5 s, and 3 s, respectively, under the configuration parameters in Table 1.

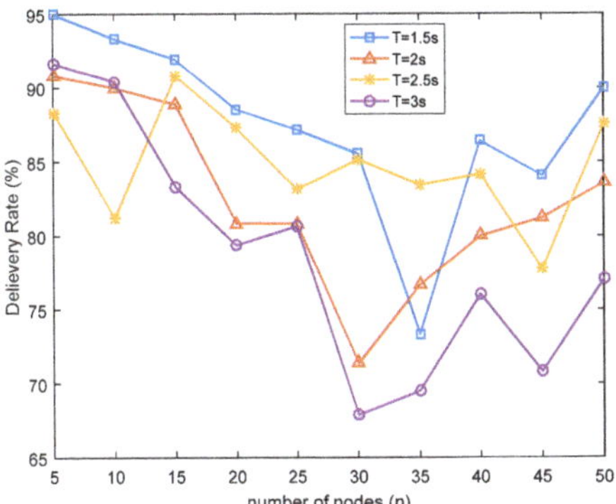

Figure 9. Comparison of pulse cycle delivery rate of beacon routing protocol.

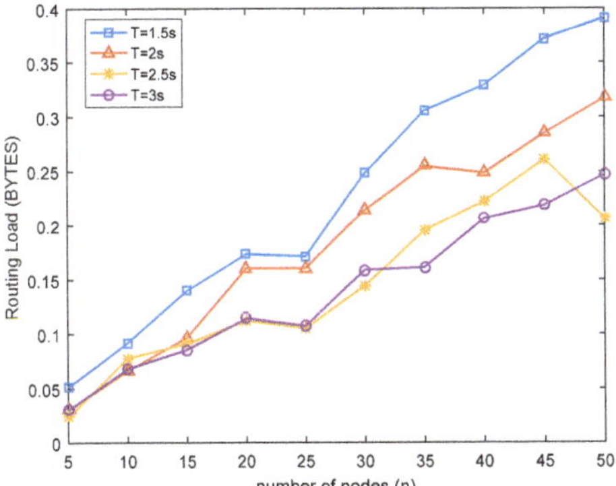

Figure 10. Comparison of pulse cycle routing load of beacon routing protocol.

As shown in Figure 9, the four curves represented the PDR of beacon-based routing protocols with pulse cycle of 1.5 s, 2 s, 2.5 s, and 3 s. The result showed that the PDR of the beacon-based routing protocol was the most stable when the pulse cycle was 1.5 s, and the performance would gradually decrease with increasing time.

The simulation result showed that the active link continuously sent data in the active time; due to constant motion of the node in the path, path failure would occur. Therefore, the shorter the time interval of the pulse cycle, the better the perception of the change of link path failure can be. Therefore, the delivery rate is the best when the pulse cycle is 1.5 s, and the delivery rate decreases gradually with the increase in time. The beacon-based routing protocol made full use of the periodic pulse message, completing the path reconstruction and path addressing operation within a pulse cycle, which effectively avoided the problems of path failure and repeated path finding.

As shown in Figure 10, the four curves represented the RL of beacon-based routing protocols with pulse periods of 1.5 s, 2 S, 2.5 s, and 3 s. The simulation result showed that with the increase in pulse cycle, the RL decreased gradually, which was also consistent with the design idea of this paper. Theoretically, the RL of the beacon-based routing protocol is only related to the time allocated to the first three phases of the pulse unit and unrelated to the number of nodes and communication link data. The simulation verified this: the larger the pulse cycle, the smaller the proportion of time occupied by the route load; the smaller the route load, the larger the effective data bandwidth.

4. Conclusions

Aiming at the problems of path failure, repeated routing path finding, and excessive cost of route maintenance in the networking of large-scale unmanned equipment using traditional routing protocol, combining the characteristics of active and on-demand routing protocols, we proposed a beacon-based hybrid routing protocol for LUVAN. In this beacon-based routing protocol, the path tree of each node pointing to the beacon node was used as guidance. Nodes were periodically maintained for OPT. Source nodes used unicast mode to create paths as required. Active nodes actively maintained routing paths in the network, maximized routing control commands, and limited the route load to a fixed time. Therefore, the cost of route maintenance in unmanned equipment network is reduced and the problems of path failure and repeated path finding are solved.

The simulation results showed that the PDR and RL performance of the beacon-based routing protocol were better than the network with traditional AODV and DSDV routing protocols. When they were used in a range 5–35 communication links, the beacon-based routing protocol PDR was approximately 10% higher than that of AODV routing protocol. At 5 ~ 50 communication links, result was approximately 20% higher than the DSDV routing protocol. The beacon-based routing protocol with different pulse cycles were compared. The RL of the beacon-based routing protocol were unrelated to the number of nodes and communication link data in the LUVN, which effectively improved the stability and reliability of the network. The beacon-based routing protocol we designed is more suitable for future LUVN battlefields.

Author Contributions: Conceptualization, Y.M. and Z.L.; methodology, W.M.; software, G.L.; validation, W.M., R.W. and Z.L.; formal analysis, W.M.; investigation, Y.L.; resources, Y.M.; data curation, Z.L.; writing—original draft preparation, W.M. and Y.M.; writing—review and editing, Z.L.; visualization, R.W.; supervision, G.L.; project administration, Y.L.; funding acquisition, Y.M. All authors have read and agreed to the published version of the manuscript.

Funding: This research was funded by the National Key Research and Development Program under Grant 2018YFE0204302, This research was funded by NSFC, grant number 51979261 and Intelligent Unmanned System Key Technology Cutting-Edge Tracking Research, grant number 4142Z17.

Data Availability Statement: All data can be requested from the corresponding author.

Acknowledgments: Authors acknowledge the technical support given by Youchun Xu.

Conflicts of Interest: The authors declare no conflict of interest.

References

1. Tahir, A.; Böling, J.; Haghbayan, M.H.; Toivonen, H.; PLoSila, J. Swarms of unmanned aerial vehicles—A survey. *J. Ind. Inf. Integr.* **2019**, *16*, 100106. [CrossRef]
2. Adiguzel, F.; Mumcu, T.V. Robust discrete-time nonlinear attitude stabilization of a quadrotor UAV subject to time-varying disturbances. *Elektron. Elektrotechnika* **2021**, *27*, 4–12.
3. Ulku, E.; Dogan, B.; Demir, O.; Bekmezci, I. Sharing location information in multi-UAV systems by common channel multi-token circulation method in FANETs. *Elektron. Elektrotechnika* **2019**, *25*, 66–71. [CrossRef]
4. Ramanathan, R.; Redi, J. A brief overview of ad hoc networks: Challenges and directions. *IEEE Commun. Mag.* **2002**, *40*, 20–22. [CrossRef]
5. Basci, A.; Can, K.; Orman, K.; Derdiyok, A. Trajectory tracking control of a four rotor unmanned aerial vehicle based on continuous sliding mode controller. *Elektron. Elektrotechnika* **2017**, *23*, 12–19. [CrossRef]
6. Zafar, S.; Tariq, H.; Manzoor, K. Throughput and delay analysis of AODV, DSDV and DSR routing protocols in mobile ad hoc networks. *Int. J. Comput. Netw. Appl.* **2016**, *3*, 1–7.
7. Shafiq, M.; Ashraf, H.; Ullah, A.; Masud, M.; Azeem, M.; Jhanjhi, N.; Humayun, M. Robust cluster-based routing protocol for IoT-assisted smart devices in WSN. *Comput. Mater. Contin.* **2021**, *67*, 3505–3521. [CrossRef]
8. Godfrey, D.; Kim, B.; Miao, H.; Shah, B.; Hayat, B.; Khan, I.; Kim, K. Q-learning based routing protocol for congestion avoidance. *Comput. Mater. Contin.* **2021**, *68*, 3671. [CrossRef]
9. Del-Valle-Soto, C.; Mex-Perera, C.; Aldaya, I.; Lezama, F.; Nolazco-Flores, J.; Monroy, R. New detection paradigms to improve wireless sensor network performance under jamming attacks. *Sensors* **2019**, *19*, 2489. [CrossRef] [PubMed]
10. Sharma, A.; Dongsoo, S. Energy efficient multipath ant colony based routing algorithm for mobile ad hoc networks. *Ad Hoc Netw.* **2021**, *113*, 102396. [CrossRef]

11.   Ahamed, A.; Vakilzadian, H. Impact of direction parameter in performance of modified AODV in VANET, J.J. *Sens. Actuator Netw.* **2020**, *9*, 40. [CrossRef]
12.   Kumar, P.; Verma, S. Implementation of modified OLSR protocol in AANETs for UDP and TCP environment. *J. King Saud Univ. -Comput. Inf. Sci.* **2019**. [CrossRef]
13.   Wahid, I.; Ullah, F.; Ahmad, M.; Khan, A.; Uddin, M.; Alharbi, A.; Alosaimi, W. Quality of service aware cluster routing in vehicular Ad Hoc Networks. *Comput. Mater. Contin.* **2021**, *67*, 3949–3965. [CrossRef]

 *electronics*

Article

Design Thinking as a Framework for the Design of a Sustainable Waste Sterilization System: The Case of Piedmont Region, Italy

Ivonne Angelica Castiblanco Jimenez [1,*], Stefano Mauro [2], Domenico Napoli [3], Federica Marcolin [1], Enrico Vezzetti [1,*], Maria Camila Rojas Torres [4], Stefania Specchia [5] and Sandro Moos [1]

1 Department of Management and Production Engineering, Politecnico di Torino, Corso Duca degli Abruzzi 24, 10129 Turin, Italy; federica.marcolin@polito.it (F.M.); sandro.moos@polito.it (S.M.)
2 Department of Mechanical and Aerospace Engineering, Politecnico di Torino, Corso Duca degli Abruzzi 24, 10129 Turin, Italy; stefano.mauro@polito.it
3 Technologies for Waste Management, Via Assietta 27, 10128 Turin, Italy; domenico.napoli@twmproject.it
4 Industrial Engineering, Escuela Colombiana de Ingenieria Julio Garavito, Bogota AK 45 205-59, Colombia; maria.rojas@mail.escuelaing.edu.co
5 Department of Applied Science and Technology, Politecnico di Torino, Corso Duca degli Abruzzi 24, 10129 Turin, Italy; stefania.specchia@polito.it
* Correspondence: ivonne.castiblanco@polito.it (I.A.C.J.); enrico.vezzetti@polito.it (E.V.)

Citation: Castiblanco Jimenez, I.A.; Mauro, S.; Napoli, D.; Marcolin, F.; Vezzetti, E.; Rojas Torres, M.C.; Specchia, S.; Moos, S. Design Thinking as a Framework for the Design of a Sustainable Waste Sterilization System: The Case of Piedmont Region, Italy. *Electronics* **2021**, *10*, 2665. https://doi.org/10.3390/electronics10212665

Academic Editor: Darius Andriukaitis

Received: 16 September 2021
Accepted: 28 October 2021
Published: 31 October 2021

Publisher's Note: MDPI stays neutral with regard to jurisdictional claims in published maps and institutional affiliations.

Abstract: The development of new methods for the correct disposal of waste is unavoidable for any city that aims to become eco-friendly. Waste management is no exception. In the modern era, the treatment and disposal of infectious waste should be seen as an opportunity to generate renewable energy, resource efficiency, and, above all, to improve the population's quality of life. Northern Italy currently produces 66,600 tons/year of infectious waste, mostly treated through incineration plants. This research aims to explore a more ecological and sustainable solution, thereby contributing one more step toward achieving better cities for all. Particularly, this paper presents a conceptual design of the main sterilization chamber for infectious waste. The methodology selected was Design Thinking (DT), since it has a user-centered approach which allows for co-design and the inclusion of the target population. This study demonstrates to the possibility of obtaining feasible results based on the user's needs through the application of DT as a framework for engineering design.

Keywords: waste management; sustainability; eco-friendly city; design thinking; infectious waste; resource efficiency

1. Introduction

Exponential human population growth, rapid industrialization, and urbanization have led to massive waste production [1]. However, Brundtland's report "Our Common Future" defined sustainable development as: "the development that meets the needs of the present without compromising the ability of future generations to meet their own needs" [2]. This indicates how, for decades, the international community's concern is outlined in economic aspects as well as environmental and social ones.

In the European Union, Italy has been one country committed to mitigating adverse environmental and social impacts. An example of Italian awareness regarding this challenge is the regulation DPR n.254 07/15/2003. It defines sanitary waste as the waste derived from public and private structures that carry out medical and veterinary activities for prevention, diagnosis, treatment, rehabilitation, and research, and all waste produced by sanitary activities regardless of its nature. [3,4].

Within the sanitary wastes, there is hazardous waste and infectious waste, which refers to waste that could be harmful to humans and the environment [5]. It is crucial to pay attention to the entire process, from the appropriate handling to the treatment and final disposition and consider that the total elimination of infectious risk can only be achieved using incineration or sterilization.

The typical treatments and disposal systems for handling infectious waste are incineration, sterilization and subsequent destruction, and in some cases, land disposal [6]. Despite its high cost and effects on the environment, incineration is the most popular method in many contexts [7–9]. However, open burning and waste incineration have contributed to a potential risk to human health, since, during the incineration process, dioxins and furans (PCDD/Fs) are released [10]. These organic pollutants are considered toxic due to their adverse effects on humans and the environment [11]. On the other hand, traditional steam sterilization facilities usually require two separate phases for the process, one for sterilization and the second to render the wastes unrecognizable [12], leading to increased precautionary measures during the waste transport between the two phases.

In Italy particularly, the report presented by Istituto Superiore per la Protezione e la Ricerca Ambientale (Higher Institute for Environmental Protection and Research) highlights that infectious waste production in Northern Italy is about 66,600 tons/year, equivalent to 46.4% of the national total, of which, the Piedmont region was responsible for the production of approximately 9040 tons/year [13].

The primary method of managing infectious waste in Italy is incineration. Nationwide, approximately 66.7% of infectious waste, equivalent to 95,815 tons/year, was handled at incineration plants; the other 33.3% was treated using sterilization [13,14]. Additionally, there is the problem concerning the lack of incineration plants for infectious waste or sterilization systems in Piedmont. None of the Italian sterilization plants are located in Piedmont, and there are no incineration facilities in the region authorized by the Italian legislation 152/2006 [15] to treat the infectious waste directly [3]. Consequently, all of the waste produced is exported outside Piedmont to other regions, which implies moving infectious waste at least 400 km, increasing health and safety risks.

Given the environmental and social concerns that governments and industries are experiencing, the need to develop methods, systems, and processes that allow for correct and effective waste disposal is unavoidable for any city that aims to become smart and eco-friendly. New technologies focused on sustainability and resource efficiency would counteract the negative social impact of this context. In modern days, while experiencing the effects of COVID-19, the treatment and disposal of infectious waste should be seen as an opportunity to explore ways to generate sustainable solutions, use renewable energy, and improve the population's quality of life.

This study aims to develop a conceptual design of the main treatment process inside a new infectious waste management system in the Piedmont region. To achieve this, we will use the Design Thinking (DT) framework methodology, a validated, human-centered, and iterative problem solving approach that involves stakeholders from various backgrounds [16]. DT has been validated as a methodology that can foster sustainability-oriented innovations [17] by focusing on the users through a flexible development.

This work is part of a larger research project that attempts to build an automatic process (Figure 1) for infectious waste management (EWC 180103), characterized by the use of innovative and sustainable technologies, with the challenge to obtain, as a final result, the recovery and possible reuse of infectious waste as a source of renewable energy. Figure 1 depicts the general scenario of future infectious waste facility management, it should begin with the collection of the infectious waste directly from hospitals and other health facilities, then it must be safely and efficiently transported to a storage center for treatment. After verification of the conditions for acceptance (for example, correct handling and proper packaging) the waste should be taken to the main process chamber for sanitization and the complete elimination of infectious risks. In order to execute this operation, the future management system must include a proper vapor and liquid control system. After the sanitization process, the treated material should be discharged to later be transformed using a compression process into a dry, homogeneous, and compact material. The final process of the future infectious waste management system foresees the recovery and possible reuse of this material as potential refuse-derived fuel.

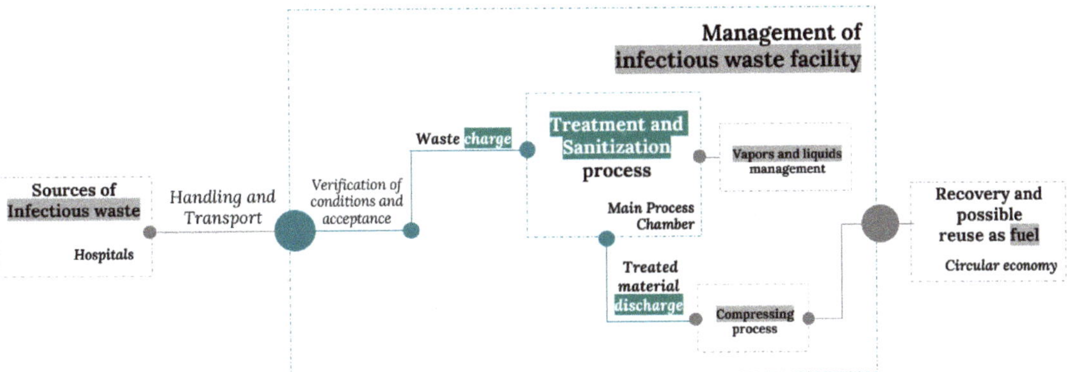

Figure 1. Infectious Waste Management Process.

In the present paper, we focus specifically on the conceptual design of the main process chamber which is responsible for waste treatment and sanitization. Following the waste management system elements proposed by Blackman [12], which include waste segregation, packaging, transport and handling, treatment, and disposal, this study focuses on the treatment technique. We use the conceptual design definition proposed by Pahl et al. [18] as the part of the engineering design process where, by identifying the fundamental problems through abstraction and establishing function structures, ideas are concretized into preliminary solutions.

The study is structured as follows: first, we explain the methodology adopted and related works. Second, there is the application of each one of the methodology phases, beginning by describing the problem, context, and stakeholders in the empathize phase. Then we proceed to the define and ideate phases for continuing to the prototyping and testing phases. Finally, we report our results and conclusions.

2. Methodology and Tools

Recently, Design Thinking methodology has been increasingly used by researchers since it shows a new perspective on the challenges posed by interacting with all stakeholders through multidisciplinary groups, formulating a user-centered solution to these problems. Additionally, DT provides a formal process that captures the needs of users [19]. Its approach is based on creativity, simplicity, and technological innovation, making DT an agent of change [20]. Together with other methods and tools, it could create resilient, significant, and sustained interventions [21]. The interest in DT has continued to grow in the last few decades. In recent years, this methodology has been gradually implemented in different companies to create products and services that tend to meet users' needs better, making them an active part of the creation process [22]. DT is presented as a methodology to develop human-centered innovation, offering a lens through which challenges can be observed, their needs detected, and be solved. It is a human-centered innovation process that emphasizes observation, collaboration, rapid learning, idea visualization, rapid concept creation prototyping, and simultaneous business analysis, ultimately influencing innovation and strategy [23].

DT consists of a collection of standard methods in engineering design [24], but unlike other methodologies from innovation management, it involves the activities of both designers and engineers [25]. Features such as problem framing, user focus, visualization, experimentation, and iteration allow DT to be an appropriate method for solving complex problems in engineering, such as creating new (or improved) products, services, processes, or technologies), obtaining environmental and social benefits in addition to economic profits [26]. Through the application of the more than 160 methods and tools associated with DT [27], these design tools help generate and incubate ideas and, therefore, create innovative solutions [28].

From the literature review, we found that DT was selected as a methodology framework for similar applications. One example is the design of an organic waste management system solution to evaluate urban recycling in the city of Depok, involving the community and making use of different methods through the various steps of DT such as SCAMPER (Substitute-Combine-Adapt-Modify-Put to another Uses-Eliminate-Reverse), NGT (Nominal Group Technique), and AHP (Analytic Hierarchy Process). DT was validated as an easy-to-use methodology with a fast implementation time and affordable cost [29]. Another example is the application of DT for organic waste production and management, specifically in Bengaluru city. In this technical report, researchers implemented the empathize, define, and ideate phases by identifying problems and innovative solutions to improve city waste management [5].

Moreover, DT principles were applied in Ireland to establish a Connected Health Innovation Framework, aligning the healthcare system needs and the software requirements through four stages: identify the healthcare problems/needs, identify the software capabilities, align the users' requirements, and identify the best solution for healthcare management. This study demonstrates how healthcare innovation supported by DT can complement the engineering process, facilitating the understanding of stakeholders in the context of their day-to-day experiences [19].

The process of innovation in engineering is a necessity in the search for new ideas that generate value for society, adapting and anticipating changes. According to the literature review, using DT in the process of defining requirements allows for identifying the stages in which this technique is applicable and how it improves the identification of problems, user and stakeholder needs, collection of new perspectives of solutions, generation of solution options and prototypes that allow the representation these ideas. Moreover, the DT approach helps to establish the appropriate scope of innovation for sustainability-oriented solutions. Its strong focus on users and stakeholders encourages the development of sustainability innovation that meets the real needs of users. DT focuses on iterative experimentation and ensures positive sustainability effects while reducing the risk of failure and rejection of innovation. In other words, DT is an approach that uses the designer's sensitivity and problem-solving method to meet people's needs in a technologically feasible and commercially viable way.

The previous research provides a foundation that supports the expectations of different investigations about the DT-based approach as an essential but often overlooked asset when addressing sustainability challenges. Lastly, the idea of linking the Sustainability Oriented Innovations (SOI) with DT supports how it can help the organizations with the creation of new technologies, services, and processes intended for environmental and social benefit. SOI challenges like innovation scope, user needs, stakeholder involvement, and assurance of sustainability have specific goals within the DT key principles like problem framing, user focus, diversity, visualization, experimentation, and iteration.

The results of the DT approach can also be seen at the social level for this type of industry; it can help users feel part of the development of a typically rejected technology. DT, in this way, helps to minimize the risk of opposition from the community and phenomena such as Nimby [30].

Therefore, based on the positive results found in the literature and the DT potential for application, we have selected this user-centered methodology as a framework for the conceptual design of the main automatic treatment chamber for infectious waste management in Piedmont.

DT began to be developed theoretically at Stanford University in California (USA) during the 1970s, and its first for-profit application was carried out by the IDEO design consultancy, who then put together a team to create the first d.school in Stanford, being today its principal forerunner and most significant source of experience [31]. Considering its experience and the significant availability of validated case studies, this paper has adopted the DT methodology by d.school from Stanford University structured in five phases [32]: Empathize, Define, Ideate, Prototype, and Test. Since the production costs of a functional prototype are very high and are outside this study's parameters, we intend to generate a design prototype according to the definition proposed by Polydoras et al. [33], representing our concept by a CAD model. Consequently, for the test phase, we will evaluate the conceptual design.

As Chasanidou [34] states, selecting appropriate methods is essential for the correct execution of the DT methodology. The tools can help the team to visualize new perspectives and understand the complexity of the system. Brown [35], defines three key practices for analyzing a need: insight, observations, and empathy. To get to the insights, we must get as much information as possible about our users and stakeholders during the define phase of the DT methodology. Numerous observation techniques come from anthropology and psychology to empathize in the best possible way and obtain relevant data. Some of them are interviews, context exploration, personas technic, and the Customer Journey Map, used typically during the Define phase. During successive phases, other methods such as brainstorming and focus groups will also be a source of insight, which later will become an input for the definition of the conceptual design requirements.

We considered previously validated studies on DT [36], and after analyzing the benefits of each tool, we selected for every phase of the methodology a set of tools reported in Table 1.

This document describes two groups of people: users and stakeholders. Stakeholders refers to a group of people who have a vested interest in the project and are considered key people in the company building a product, often including managers, subject experts, and individuals in charge of verifying compliance with regulations. The user can also be considered a stakeholder; however, not all stakeholders can be considered users. Thus, for this study, the user specifically describes a person who ultimately uses and interacts with the system and who has an actual behavior in the process. Users will be consulted on their views on their needs and desires and will be considered a data resource for technics such as persona and consumer journey map. Opinions from other stakeholders will be also considered during the study.

3. Implementation

The following section will show the implementation of each one of the DT phases described in Table 1 (empathize, define, ideate, prototype, and test) and the application of the selected tools. To complement the DT definitions, it is important to note that in addition to the five phases reported in Table 1, the DT process can also be described through three stages called inspiration, ideation, and implementation [35]. Inspiration is the initial stage. The first part is understanding the problem and the second part is observing how the product is used and how the service is developed. In our scenario, this stage appears during the empathy and define phases. During the ideation stage, it is necessary to generate alternatives and options and to collect ideas to solve the problem by involving users and stakeholders. Notions such as co-creating and brainstorming, typical of the ideate phase, are used in this stage. The implementation stage foresees the prototyping of the possible solution to obtain feedback as soon as possible and, in this way, be able to improve the idea. In our study the implementation stage corresponds to the prototype and test phases.

3.1. Empathize

During the development of new solutions, user interaction can foster the user's future acceptance of new technologies [37]. In the empathize phase, we began by analyzing and defining the context of our case. We then characterized the potential users and interacted with them using the established empathize tools.

Table 1. DT Methodology phases and Selected Tools.

	Objective	Selected Tools for the Study
Empathize	Define the context and users, discover their behaviors, ask open-ended questions. Know the user's needs by putting ourselves in the place of the user.	• Context exploration: explore the current scenario, stakeholders, and challenges. • Personas: this technique models users based on their description and provides information about their characteristics.
Define	Based on the insights gained, summarize the problem to be solved. Use an action-oriented approach to express the stakeholders' requirements effectively.	• Customer journey map: this allows us to shape on a chart each of the phases or stages that a user goes through, from when the user has a requirement until he/she uses a service to solve this need. It involves using the user's thinking and feeling to establish ideas for improvement, rephrasing the user's needs as a well-structured problem statement combining the user's needs and the information we have.
Ideate	Generate ideas to solve the problem. Ideas should be shared and reflected upon openly.	• Benchmarking: investigate technics and systems that are currently being used to solve the problem. • Functional decomposition: decompose the concept into more minor problems for analyzing interactions and relationships. • Brainstorming: think on ideas to solve the problem and share the ideas that arise with the other team members.
Prototype	Concretize the idea and compare solutions. Create one or more concepts with which team members can engage.	• Process and Functional interactions: demonstrates an end-to-end solution for a service scenario through the use of sketches. • Computer Aided Design: create CAD concepts, in order to transform preliminary ideas into visual models
Test	Evaluate the concept and collect feedback. If necessary, go back to previous phases until a feasible solution is found.	• Concept scoring: this uses a weighted selection model to evaluate the concepts.

3.1.1. Piedmont Region Context

In 2018, the Italian production of special waste (produced by industries and companies) amounted to 143.5 million tons. The production of special waste is concentrated in northern Italy, with almost 84.9 million tons/year. Between 2017 and 2018, there was an increase in total production, equal to 3.3%, corresponding to approximately 4.6 million tons/year. In particular, infectious risk wastes produced in Italy amounted to about 144,000 tons/year [13].

From the last data published by the Piano Regionale di Gestione dei Rifiuti Speciali (Regional Management Plan for special waste), we found that the primary producers of infectious waste (Figure 2) are hospitals, with 82% of total production. In second place, subjects who carry out health and assistance activities such as, nursing homes, specialist clinics, research institutes, physiotherapy centers, dental centers, and veterinary services (10.4%). In third place, other health facilities, i.e., outpatient clinics or independent laboratories (5%), and in fourth place, subjects who, while carrying out activities other than "health and assistance" or "veterinary services", have produced infectious waste such as pharmacies and beauty centers (2.6%). [3]

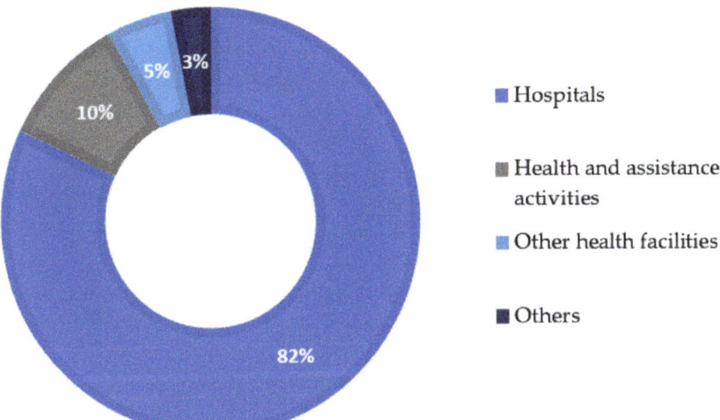

Figure 2. Sources of infectious waste in Piedmont region, North Italy.

In 2018, the Piedmont region produced 9040 tons of infectious waste, but only 3.6% was treated inside Piedmont, in the only incineration facility available and authorized to treat this kind of waste (Figure 3). This incinerator, however, was closed a few years ago. In 2021, all the infectious waste was exported to other regions to be incinerated or sterilized and later destroyed [3,13].

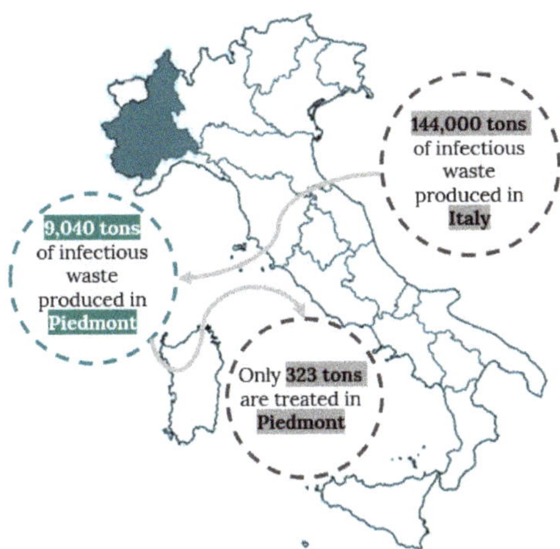

Figure 3. Infectious waste produced and treated in Piedmont in 2018.

Consequently, all the infectious waste produced in Piedmont is sent directly or indirectly (after being in a preliminary warehouse) to other regions, mainly Lombardy, Emilia Romagna, and Liguria. According to the Piedmont regional council [38], each waste trip to other regions has a capacity of 5 tons, so it is necessary to make more than 1800 trips in a year to transport all of the infectious waste produced in the region. Additionally, the average round trip distance is 1240 km, creating a high environmental impact, producing yearly 7,886,400 kg of CO^2, 3,360,000 kg of PM10, 7,100,050 kg of NOx, and 4,318,176 kg of CO, indirectly affecting the health of the population and creating environmental damage. Therefore, it's crucial to develop treatment facilities closer to the infectious waste sources within the Piedmont region to mitigate the environmental and health impacts.

3.1.2. Infectious Risk Waste

In Europe, infectious waste is defined and cataloged by the European Waste Catalogue (EWC) as wastes whose collection and disposal are subject to special requirements to prevent infection, and it is classified under the EWC code 180,103 [39].

Infectious waste in Italy is governed by the regulations on the medical waste management, decree 254/2003; listed in art. 2, it defined "infectious waste" as not only the waste produced by healthcare facilities but also waste produced outside of them with similar infectious risk characteristics [4]. In other words, all materials that have come into contact with infected or presumed infected biological fluids are considered infectious waste. Furthermore, the kind of waste that must be collected and disposed of by applying special precautions to avoid infections is identified as HP9 "infectious danger" according to EU regulation 1357/2014. Waste with HP9 characteristics contains viable microorganisms or toxins which are known to be or are considered causes of diseases in humans or other living organisms, without concentration limits [13,40].

In particular, we aim for a system that can treat primarily hospitals and other health facilities' waste, including, according to the characterization of the World Health Organization, primarily: infectious waste, suspected to contain pathogens (bacteria, viruses, parasites, or fungi) in sufficient concentration or quantity to cause disease in susceptible hosts, e.g., laboratory cultures, waste from isolation wards, tissues, swabs, materials, or equipment that have been in contact with infected patients; pathological waste, human tissues or fluids, e.g., body parts, blood, and other body fluids; and sharp waste, e.g., needles, scalpels, knives, and blades [41,42].

3.1.3. Stakeholders and Personas Technic

The identification of stakeholders allows us to recognize users and groups that may be interested or are involved in the study; identifying and grouping them is part of the empathy phase of DT, since users' needs are the basis of our analysis. In our study, we define primary stakeholders as any user who has direct contact with waste, for example, hospital managers, doctors/nurses, patients, waste handlers, carriers, and waste management company workers. There are also external stakeholders, such as the government authorities, nearby community, and research centers.

Using the personas technique, we proceed to characterize our active stakeholders. This tool helps to cultivate an innovative mindset and encourages the team to produce ideas through the representation of the stakeholders [43]. For the creation of *personas*, we used explorative qualitative interviews. According to Dickinger [44], qualitative interviews have a flexible and continuous design, unlike surveys that have a rigid structure. The interviewer uses open questions to direct the conversation flow to obtain the most significant amount of information. According to Revella [45], to be aware of the user's patterns, a good rule is to complete at least eight to ten interviews for the *personas*. In this order, for our study, we conducted a series of interviews, including three waste management systems managers, two hospitals managers, three workers who are handling the waste, five persons from the nearby community, and ten university students interested in sustainable development, all of them living in Piedmont region. We inquire about their knowledge about infectious waste, waste management systems in Italy, the current Piedmont situation, and techniques. Then we explain our project and ask for opinions.

For this paper, we have depicted in Figure 4 three of the most representative stakeholders, a worker from the company in charge of waste treatment, a hospital director, and a community member. For each one of them, we have reported the gains, pains, jobs to be done, and use cases, according to the user profile suggested by Lewrick [36].

Persona description Franco		Persona description Luisa		Persona description Marisa	
Age: 23 years **Gender:** Male **Job:** Handling staff at Waste management company. He is a student who works to pay his college fees.		**Age:** 40 years **Gender:** Female **Job:** Hospital Director She has a degree in medicine, has been working on health systems management for three years.		**Age:** 78 years **Gender:** Female **Job:** Pensionary She is a pensioner and has lived in the neighborhood for 40 years. She likes to read and participate in community meetings in the neighborhood.	
Gains	**Pains**	**Gains**	**Pains**	**Gains**	**Pains**
Once you have the routine identified, things are easier to do.	The work is exhausting, and all the more so because he must get home to study and fulfill more obligations. Sometimes feels afraid of the contagion of diseases due to the handling of infectious waste.	Keep the hospital in the right conditions. She wants to continue to be an example in the region and contribute to the country. The priority of a hospital is to save lives; that is why all processes and resources must be focused on meeting this objective.	For Luisa, her team is busy focused on processes to save lives. They do not have much time to apply more methodologies. The approach must be innovative; she understands the importance of being sustainable but worries that new methods are costly and not efficient.	It has been a quiet neighborhood. Every project that involves the neighborhood is discussed in the action meeting so that the neighborhood continues to be safe, profitable, and comfortable. We know that waste management is essential.	Marisa cares about the children; she wants them to grow in a healthy and clean environment. She is afraid of new treatment plants near their town.
Jobs to be done	**Use cases**	**Jobs to be done**	**Use cases**	**Jobs to be done**	**Use cases**
It is essential to use a safe container because there are many risks. It is crucial to consider the storage of the waste, keep it for the shortest time while collecting it under optimal conditions. From the beginning, the waste must be packed and sealed correctly. I cannot risk my health.	It should not be just a product but the development of an exemplary process where the correct handling of waste is done having minimal contact with people. It is also essential to coordinate with the hospitals to verify the proper sealing of the containers	Luisa wants a fast and flexible process to be implemented. Does not want to affect process times. The process must be kept for the team and the environment throughout its entire cycle.	Luisa cares about her employees; handling infectious waste must be secure for the operators. It is necessary that the process complies with regulations. It is a process that can transport and treat infectious waste without taking time from nurses and doctors, who should be doing their main activities.	It is important not to generate foul odors, and continuous supervision is needed to ensure no disease. Also, as a community, we want to be considered in any decision that may affect us.	It should be a plant with processes well supervised. It must not make noise when it works. The trucks transporting the waste must be within the corresponding hours, comply with biosecurity measures, and not affect neighborhood peace.

Figure 4. Personas Technic.

From the information gathered through the *personas*, we identified some of our users' main worries and expectations, mostly related to a safer environment, the safety of operations, regulation compliance, and process innovation. Aligned with DT goals, the community manifests the desire to be included in the development of a new technology and its related process.

3.1.4. Local Regulation

Regarding regulations and their influence as external stakeholders for our study, article 8 of Italian Decree 254/2003 [4] defines the characteristics of infectious waste treatment. The elimination of infectious risk can only be achieved through incineration or sterilization. Articles 10 and 11 provide specific indications in this regard:

"Infectious waste disposal must be done in incineration plants. Sterilized waste can be used to produce refuse-derived fuel (RDF) or directly to generate energy. Sterilized waste can be disposed of in incineration plants for solid waste or (if these alternatives are not available) they can be disposed of in landfills for non-hazardous waste, for a limited period".

From this information, we identify the use of sterilization as a potential design strategy and the opportunity to analyze the RDF production for a future scenario.

3.2. *Define*

Based on the insights gained from the empathize phase, we use the consumer journey map (Figure 5) to bring together the context of our research, the activity stages, and the expectations and worries of our stakeholders. It is divided into macro phases, beginning when the wastes are collected from the healthcare unit and ending with the final process in the treatment facility. The visual approach of this tool allows for understanding the overall picture of processes and users' needs. As Lemon and Verhoef [46] indicated, it helps to provide a deep understanding of the user experience and can act as the basis for innovation, improving that experience.

Stages	Transportation outside Piedmont		Handling		Main Process	Handling	Transportation to final disposition
Activities	Recollection of waste containers from hospital	Transport of containers to facilities outside Piedmont	Carrying the containers inside the factory to the main chamber	Load the waste into the chamber for treatment (normally incinerator)	Chamber closing and beginning of the process	Unload the resulting material	Transport of resulting material to final disposition
Thinking and feeling	The operator is concerned about the waste storage in the hospital for a long time. Hospital managers expect a secure procedure that complies with regulations.	Nearby community is worried about long-distance transportation of infectious waste	The operator must perform many journeys inside the facility carrying out the containers, but he is used to it.	The operator is afraid of getting in contact with an infectious risk	The most commonly used method is incineration (outside Piedmont); it produces risks for health and the environment.	The operator works in an "automatic" way, every day, the same steps.	The nearby community is concerned about the final disposition of the resulting material
Ideas for improvement	Develop a standard process capable of collect waste in the shortest possible time	Develop an innovative and closer treatment facility	Automatize the operation; the operator should only verify the proper sealing of containers	Automatize the operation	Consider other methods different from incineration	Automatize the operation	Consider the possibility of using treated waste as a potential RDF

Figure 5. Consumer Journey Map.

Summarizing the problem in Piedmont, there are challenges relating to the absence of incineration plants for infectious waste. The lack of sterilization systems aggravates this issue. Furthermore, the authorized incineration plants are not adjacent to Piedmont, which leads to the need to make long journeys for quite significant distances, even to the order of 400 km.

From the users' point of view, there is concern about the potential risk of new incinerator plants, the safety of operations, compliance with regulation, and the distances traveled to carry the waste. Furthermore, they expressed their satisfaction at being included in the ongoing investigation. They communicated their interest in being informed about what would happen to the final material after the treatment.

Once we identified the main concerns (or insights) of the users, we expressed the users' needs as a requirement, considering the users' thinking and feeling information and the ideas from improvement obtained from the Consumer Journey Map. In Table 2, we establish the requirements for the conceptual design of the main treatment process.

3.3. Ideate

The requirements determine the function representing the intended relationship between inputs and outputs of a system or machine. The goal of the conceptual design ideate phase is to explore the concepts that may address customer needs. It includes a combination of external search employing benchmarking analysis, internal problem solving through brainstorming, and the exploration of solutions. As Pahl et al. [18] indicate, we need a system with a clear and straightforward reproduced relationship between inputs and outputs to solve a technical issue.

Table 2. Conceptual design requirements.

User Need	Requirement
The user expects a safe and fast procedure.	• Develop a standard process capable of treating waste in the shortest possible time.
A procedure compliant with regulations.	• Develop a procedure compliant with regional regulations.
Shorter-distance journeys while transporting waste.	• Develop an innovative and closer treatment facility inside Piedmont.
Inside the facility, the user wants a secure operation with less handling.	• Automatize the operation to guarantee minimum contact and handling.
Users expect an innovative solution, different from the current one (incineration).	• Consider other methods different from incineration.
The user wants to know what happens after the treatment.	• Consider the future possibility of using treated waste as potential RDF.
The user wants to be included.	• Use user-centered design approaches to include users.

To identify the initial relationships from the preliminary requirements established in the define phase, we began by analyzing the requirements related to consider other methods different from incineration and procedures compliant to region regulations.

Italian regulations allow two types of treatment management for infectious waste: incineration or sterilization. Since we want to explore other methods different from incineration, we refer to the active regulation of infectious waste management.

Concerning sterilization, the Italian Environment, Government, and Land Protection Department defines it as a physical (heat, ionizing radiation, microwaves) or chemical treatment that allows for a reduction of the microbial load to guarantee an S.A.L. (Sterility Assurance Level) not less than 10^{-6}, which is a 1 in 1,000,000 chance of a non-sterile unit. Sterilization must be carried out according to the UNI 10384/94 Standards employing a procedure that includes shredding and drying for non-recognition and greater effectiveness of the treatment and waste volume and weight reduction [4].

3.3.1. Standard UNI 10384/94

Within the standards regarding waste management, there is the ISO 14,001 that describes a formal approach to handling waste, the CEN/TS 17,159 that guides the management and safety of high-risk materials and waste, and in Italy, particularly the UNI 10,384.

The UNI 10,384 standard: "Systems and processes for the sterilization of hospital waste," which dates back to 1994, and is managed by the Ente Nazionale Italiano di Unificazione UNI (Italian national unification organization), provides the criteria for the design, operation, and verification of sterilization systems for infectious waste from public and private healthcare facilities. Furthermore, for treatment systems that involve thermal destruction and systems that use ionizing radiation.

Regarding the concept design requirements, the UNI 10,384, 1994 specifies that:

- All the necessary precautions must be taken so that the loading operations in the sterilization system take place in conditions of safety for the operators and the environment.
- The sterilization cycle must be carried out automatically according to a succession of phases.

- The sterilization conditions, within established limits, must be uniformly achieved in the sterilization chamber, kept at the critical point for the pre-established time, and reproducible.
- The sterilization chamber must be equipped with suitable closure or confinement systems. The sterilization cycle must not start until the closing system is closed, blocked, and, where necessary, sealed.
- If a loading device is used, the waste must be cased without interference between the containment systems and the loading door.
- The chamber must be equipped, in addition to the standard operating equipment, with those necessary for carrying out sample tests to detect the appropriate level of sterilization.
- The materials used to build the chamber must resist chemical and physical aggressions deriving from the process and treatable waste.
- The number and position of sensors inside the chamber must be specified.

3.3.2. Functional Decomposition

The next step for establishing the relationships in our conceptual design is decomposing the problem functionally, representing it as a single black box operating on material, energy, and signals, as suggested by Ulrich et al. [47]. The next stage will divide the single black box into sub-functions to obtain a more detailed description of every element and their task oriented to achieve the entire system goal.

After selecting sterilization as a treatment method, considering the requirements and restrictions of regulations, we represented the first black box of our sterilization chamber in Figure 6.

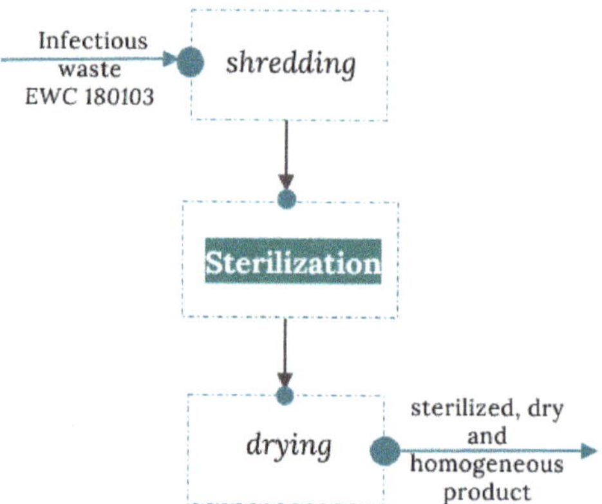

Figure 6. Early black box for the sterilization chamber.

3.3.3. Treatment Methods Benchmark

In concept generation and as part of the DT ideate phase, it is beneficial to investigate current treatment methods used to solve the problem. Ulrich et al. [47] define benchmarking as the exploration of existing processes with similar functionality like that of the system under study. In our case, benchmarking can reveal current concepts that have been implemented and information on the strengths and weaknesses of treatment methods. In this regard, we analyzed currently used sterilization methods and then investigated relevant concepts in Italy for the infectious waste treatment, different from incineration.

Sterilization Methods

Xiao and Hancock, and Lambert [48,49] define some of the most used methods for the disposal of infectious waste

Microwave sterilization: The method refers to the inactivation of infectious bacteria by the heat generated from the microwave vibration of the water molecules, achieving the disinfection of the objects. It is suitable for processing infectious and pathological waste (except human organs and infectious animal carcasses); it is not ideal for the treatment of pharmaceutical waste or chemical waste. In economic terms, it is not feasible due to high construction and operating costs.

Chemical disinfection: This method allows for the disinfection of waste with chemicals that eliminate bacteria and viruses. It uses dry and wet compounds that require the same treatment, and waste must be crushed and disinfected. The disadvantages of using wet chemicals are residual liquid, which can be an environmental pollutant, not suitable for all healthcare waste.

Pyrolysis: Pyrolysis is a thermochemical reaction technology. Suitable for all types of medical waste, but due to operation and maintenance requirements, this method is the most expensive. Furthermore, it is not easy to achieve stable combustion, producing exhaust gases and waste that can generate a secondary source of pollution.

Steam sterilization: This method applies high temperatures to generate steam that kills microorganisms and disinfects waste. The advantages of this technology are low installation and maintenance costs; it does not affect the environment and can be applied to all infectious waste.

Since we aim for a system that can also treat pathological waste (including human anatomical residues), microwave sterilization is not feasible. At the same time, chemical disinfection and pyrolysis can generate residual material (liquid and exhaust gases, respectively), which can be a pollutant by themselves. On the other hand, steam sterilization seems a feasible and secure treatment method suitable for our research.

According to Health Care Without Harm (HCWH) [50], in the European Union some of best practices for infectious waste that use non-incineration technologies range from small units for use at or near the point of generation to high-capacity systems for large medical centers. The majority of non-incineration technologies used in the EU employ low-heat thermal processes and chemical processes. Low-heat thermal processes use thermal energy to decontaminate the waste at temperatures insufficient to cause a chemical breakdown or to support pyrolysis or combustion. In comparison, chemical processes employ disinfectants such as dissolved chlorine dioxide, sodium hypochlorite, peracetic acid, or dry inorganic chemicals. Some of these technologies include autoclaves in the Netherlands, steaming and drying in France, chemical treatment in Spain, steaming and compaction in Germany, as well as vacuum-steam and steam-fragmentation. Microwave treatment in Austria and in other countries such as Italy, France, Germany, Slovenia, Serbia, Slovakia, and Romania. Chemical processes are also used in Italy. Finally, electron beam technologies in Belgium and other countries.

Additionally, in the study "Assessment of Medical Waste Disposal Technologies Based on the AHP" [51], a hierarchical structure model is used to analyze and evaluate five types of disposal technologies quantitatively for infectious waste in China. The classification weights of the optimal and suboptimal alternatives were studied using sensitivity analysis. The results show that the comprehensive benefits of high-temperature steam sterilization in infectious waste treatments are the best compared to the other methods in terms of social, environmental, technological, and economic factors. These overall results allowed us to define steam sterilization as the selected treatment for the main chamber.

3.3.4. Italian Concepts Initiatives

Next, we considered some relevant infectious waste sterilization initiatives in Italy. We looked for systems (including Italian patents) that have a well-established process. In Italy, we found three companies that include within their services the use of sterilization methods. However, we did not find evidence of the use or field implementation of these systems. The processes used are explained in Table 3, according to the public information reported by the companies.

Table 3. Italian Initiatives.

System	Process
A [52]	Waste is pre-treated by grinding and conveying through sealed stainless steel screw conveyors, followed by grinding through shaft shredders and finally refining of the sterilized product utilizing the moist heat sterilization process.
B [53]	The shredding phase starts through a four-shaft shredder. The shredding of the material takes a variable time depending on its consistency. The shredded material is collected in a storage chamber, then transferred in containers to the sterilization chamber; the heating process occurs by injecting hot air into the chamber.
C [54]	The process consists of a combination of dry heat sterilization and microwave. Then the waste is transported to an external facility for the grinding process.

From the different available techniques of sterilization, and considering the initiatives reported in Table 3, we can conclude that most of the systems differ from each other by the sterilization method, the shredding technology, the way of heating up the chamber, and the transportation from the shredding machine to the sterilization chamber.

3.3.5. Brainstorm for Ideation

Convergence is crucial in collaborative problem-solving and decision-making, as the team must focus its resources on the most promising ideas [55].

Brainstorming allows the team to think on ideas to solve the problem and share the thoughts that arise. After analyzing different possibilities for the main chamber, we use the brainstorming technique to converge into preliminary ideas. Figure 7 shows the most discussed methods in the technical brainstorming sessions. For each function, the solution selected is underlined. The selection was made considering the feasibility for construction and operation in Italy, success stories, cost, and originality for the Piedmont region.

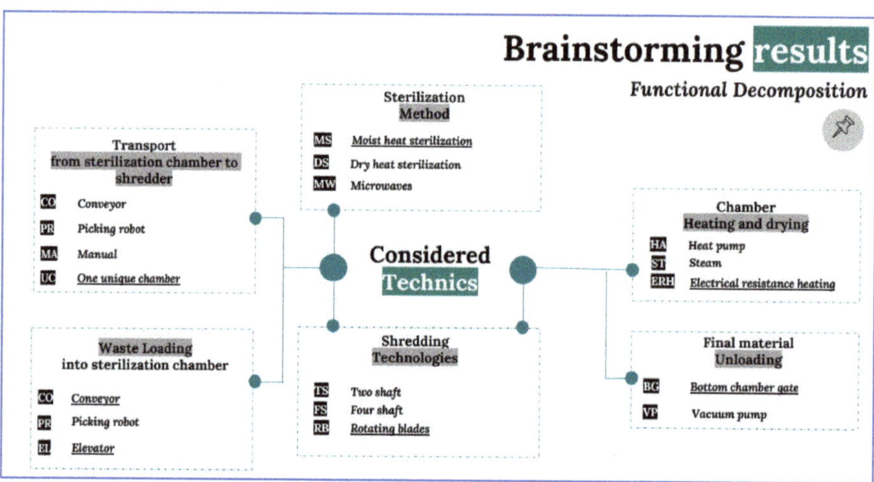

Figure 7. Brainstorming and functional decomposition results.

We decided to use moist heat sterilization for the sterilization method because of its nontoxic characteristics, cycle time, cost, and safe operation [56]. We have also noted that the systems analyzed in Table 3 always transport infectious waste from the sterilization chamber to the shredder or vice versa, incurring more space, logistics, and risks. For that reason, we concluded that a suitable alternative could be the design of a single camera for sterilization and shredding, which will be explained in the next DT phase. Regarding the mechanism for heating and drying, we selected electrical resistance heating because of its efficiency and simplicity by converting nearly 100% of the energy in the electricity to heat. The shredding process will be carried out through rotating blades and counter blades at the bottom of the sterilization chamber. Finally, we decided to explore two options for waste loading, one using a conveyor and a second using an elevator. For the material unloading, we will use a gate at the bottom of the chamber to take advantage of the blades and the force of gravity. All the concepts will be further elaborated on and explained in the next section.

3.4. Prototype: A Conceptual Design

The prototype phase in DT aims to concretize the idea by creating an approximation of the product, reflecting on its features and eventual improvements.

Ulrich et al. [47] define the prototype as an approximation of the product along one or more dimensions of interest. Any entity presenting at least one aspect of the product of interest can be viewed as a prototype. Virtual prototypes can be represented using drawings, mathematical models, simulations, and CAD models.

In our study, we intend to create a design prototype according to the definition given by Polydoras et al. [33]. Prototypes can be classified according to their ability to serve the distinct stages of the design process; a design prototype mostly represents the form and functional relations of the product. A design prototype allows designers to evaluate various aspects of their ideas before committing to the expense and risks of producing a commercial quantity [33,57].

With this aim, according to the previous phases' results, we focused on the conceptual design. To achieve this goal, we defined the process and its functional interactions and the sterilization cycle, and then propose two CAD concepts.

3.4.1. Process and Functional Interactions

The treatment system aims to transform infectious waste into a dry and homogeneous product with no recognizable parts, which is stable over time. The final product could be stored for long periods before its disposal and recovery. The designed process aims to obtain a fine shredding of the waste, the pulverization of glass parts, and the total elimination of liquids, contributing to the reduction of weight and volume.

This system is based on the moist heat sterilization method. It works under the same conditions as steam autoclaves. The combination of time, temperature, and humidity is achieved in a closed environment under vacuum conditions. However, this process peculiarity consists in treating the waste to a temperature up to 151 °C with the presence of humidity and liquid water, operating under vacuum conditions, without diffused pressure as in autoclaves.

The system (Figure 8) consists of a closed cabin (a) with all the devices inside it, the sterilization chamber (b) performs automatically thermal cycles that include shredding and evaporation of liquids, overheating up to 151 °C, sterilization at 151 °C, through continuous dosing of water transformed into steam because of the operative conditions, and for an adequate time to ensure microbial elimination. The process ends with a cooling phase and the discharge/unloading of the dehydrated final material.

The waste is introduced to the system using apposite containers (c). An automatic loader (d) deposits the waste inside the sterilization chamber (b) after the lid (e) has been automatically opened.

The sterilization chamber (b) structure is a vertical cylinder in AISI 306 stainless steel, equipped with a sealed lid (e) made in high-temperature resistant silicone rubber gasket and hydraulic locking systems. All the phases of the sterilization treatment, including loading and unloading, occur automatically inside the cabin (a) without moving the waste outside the chamber until the process end.

In the lower part of the chamber, there is a two-blade Hardox 500 stainless steel rotor (f) driven by an electric motor (g). The blades act like hammers; stationary blades (h) are mounted in correspondence inside the chamber to ensure the necessary contact and, therefore, the material crushing. This friction produces an additional source of heat, valuable for our operation.

In the high, middle, and lower sections and around the perimeter, there are electrical resistances (i) that heat further the chamber (b) in a controlled way. Across the entire chamber height (b), there are ten thermocouples (j) installed for measuring the material temperature, six at the bottom chamber along the perimeter, and four more at the middle and upper section.

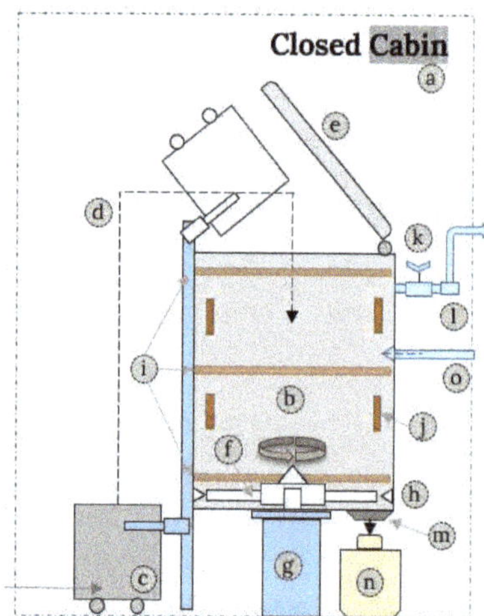

Figure 8. Sterilization chamber—conceptual functional interactions.

The chamber is connected to a throttle valve (k) for vacuum control and then to a vacuum pump (l). At the end of the treatment, the unloading gate (m) automatically opens, and the product is discharged by centrifugal force into apposite bags (n).

The system is kept in slight depression during the entire treatment cycle using a throttle valve (k) and the vacuum pump (l). Considering the variable composition of the incoming waste in terms of typology, and therefore in terms of intrinsic moisture, the system includes a valve (o) for pouring the necessary water into the chamber to reach and guarantee the sterilization conditions.

After the waste is loaded, the chamber lid closes automatically, and the rotor at the bottom starts spinning at a low speed. The waste is shredded while the rotor speed progressively increases up to about 1000 rpm. The mass heats up, thanks to the thermal energy generated by the rotation and friction of the blades. This operation leads to the release of intrinsic moisture from the waste. In this way, the temperature gradually increases until it reaches the value of 151 °C. Extra water can be added in this phase if the intrinsic moisture of the waste would not be sufficient to reach the required sterilization conditions. Compared to the traditional moist heat sterilization process used in autoclaves, this process does not operate with saturated steam but with unsaturated steam on a finely shredded material, operating at higher temperatures but with significantly lower pressures.

3.4.2. Sterilization Cycle

According to the previous description, the sterilization cycle follows a sequence of seven steps:

- Phase 1: Waste loading: The waste is loaded into the sterilization chamber using an automatic loader, then the lid is automatically closed at the end of loading.
- Phase 2: Shredding and Heating: After closing the lid, the rotor begins to rotate until the material is pulverized and the temperature reaches 100 °C. The blade's profile

should be designed to create turbulence on the bottom of the chamber, facilitating the mixing and avoiding depositions on the walls.

- Phase 3: Evaporation and overheating: The heat generated by the friction of the material should cause the vaporization of the humidity contained in the waste, and the temperature should remain stable at 100 °C. Humidity and temperature are constantly measured using a probe designed for test chambers at high temperatures.
- Phase 4: Overheating: Once all moisture has been removed, the generated heat raises the temperature of the material to 151 °C. Extra water can be added if the waste's intrinsic moisture is not enough to reach the sterilization conditions.
- Phase 5: Sterilization: The material temperature is maintained at 151 °C through water dosage and controlled by the thermocouples. When coming into contact with the material, the water evaporates and becomes steam. The water dosage is controlled so that the heat absorbed by the water evaporation balances the heat produced by the material friction. Additionally, there is extra heat coming from the chamber walls using the electrical resistances. This allows keeping the material moist at a high temperature. The whole chamber must be kept at 151 °C for at least 3 min during each operating cycle to ensure the correct functioning of the sterilization system.
- Phase 6: Cooling and drying: The rotor speed decreases and consequently so does the heat generation. If needed, water is sprayed, lowering the temperature to 100 °C. Then, through a vacuum pump, the temperature drops adiabatically until room temperature is reached. During this phase, the heat absorbed by water evaporation exceeds the heat generated by the rotor; therefore, the temperature drops.
- Step 7: Unloading: The treated material is discharged by centrifugal force through the discharge gate located at the bottom chamber. Once the material has been completely discharged, the rotor stops.

3.4.3. Computer-Aided Design (CAD)

During the DT prototype phase, CAD provides a visual manifestation of concepts and supports the generated ideas' transformation into feasible and testable models [58]. CAD concepts can be quickly created and experienced. It can support team communication by facilitating conversations and feedback regarding solutions for a particular product [34].

According to the process and functional interactions, we propose a virtual prototype, represented by two concepts using CAD. Due to the importance of the internal process of the chamber, it remains invariable for both concepts. The main differences between them are related to the uploading transportation system.

- Concept 1: Automatic Elevator Model

The first proposed concept (Figure 9a) consists of a stainless steel trolley to carry all the waste and an automatic uploading system that raises and overturns the loading trolley, allowing the material deposition directly inside the chamber.

- Concept 2: Conveyor Belt Model

The second concept (Figure 9b) consists of a conveyor belt system for loading waste containers. The conveyor system transports the containers automatically up to the open lid. Then, from a closer distance to the lid, the containers automatically open one by one, depositing the waste inside the chamber.

For both concepts, the operations inside the sterilization chamber are the same as reported in the process and functional interaction section.

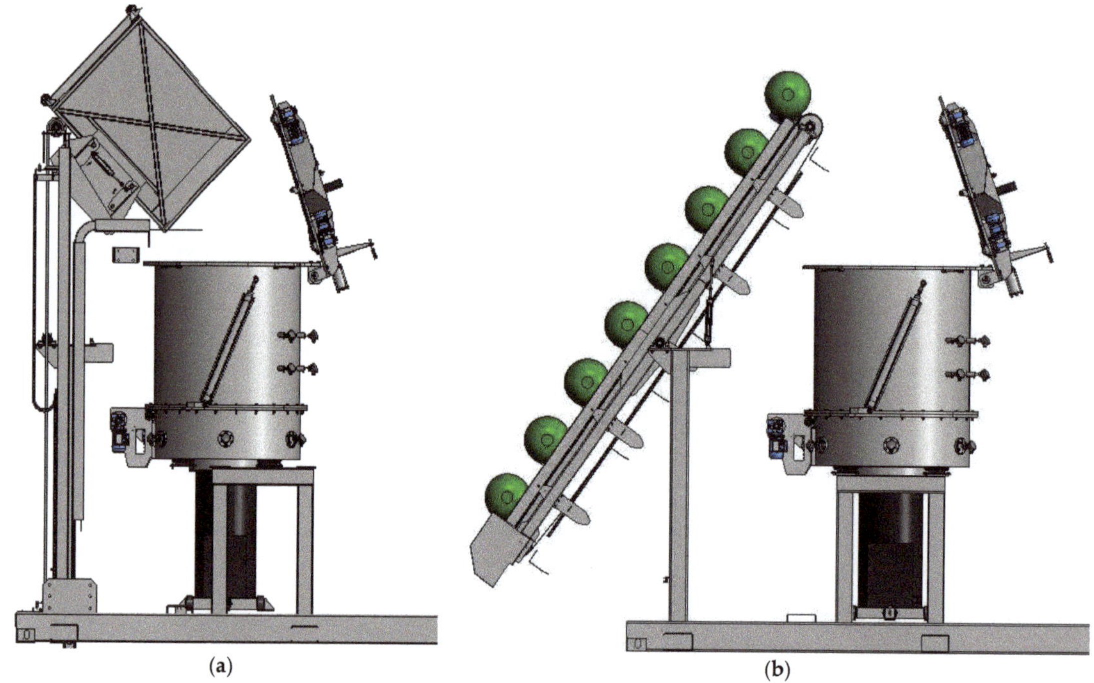

Figure 9. Concept Models. (**a**) Automatic Elevator Model; (**b**) Conveyor Belt Model.

3.5. Test: Concept Evaluation

The fifth and final phase of DT corresponds to the test. According to Bresciani [59], visual representations can guide the designer's work within a design framework. They allow for bringing together academic knowledge and the practice of design, providing a visual reference for users and stakeholders, in this way, they can support the design process.

This section evaluates the virtual prototype (represented by concept models a and b) according to the user's requirements, then collects feedback for further improvements. The test phase results were obtained during focus group meetings that included the development team, experts, users, and other stakeholders. For the test phase we have used the concepts designed in the previous section, represented by their CAD models (Figure 9).

We use the concept scoring technic (Table 4) proposed by Ulrich et al. [38] to evaluate the concepts. It uses a weighted selection model to evaluate the concepts variants against the requirements and other criteria deemed relevant by the team. The selection criteria were determined according to the user's expectations and the Italian regulations described in the previous phases. For every criterion, the team analyzed and agreed on the weight assigned, considering that in this phase we are performing a feasibility analysis for a conceptual design. The same consideration was used for rating the concepts. They were evaluated on a scale of 1 to 5, five being the best and one the worst.

Both concepts used a method different from incineration and were designed to include the users' perspectives. At the beginning of our exploration, stakeholders specified these requirements; both systems are consistent with them, so these parameters are not included in the evaluation.

It is possible to determine the concept ranking using a weighted sum of ratings. After analyzing both concepts, the results show that concept 1 is more promising than concept 2, with a score of 4.0 against 3.3.

Table 4. Concept Scoring.

Selection Criteria	Weight	Concept 1: Elevator		Concept 2: Conveyor Belt	
		Rating	Weighted Score	Rating	Weighted Score
Standard process.	5%	4	0.2	4	0.2
Safe for the environment.	5%	4	0.2	2	0.1
Includes shredding and drying.	10%	5	0.5	5	0.5
Safe operations with minimum handling from workers.	10%	4	0.4	1	0.1
Sterilization cycle carried out automatically according to a succession of phases.	10%	5	0.5	5	0.5
The sterilization conditions are kept at the critical point for the pre-established time and are reproducible.	15%	5	0.75	5	0.75
Loading operations are safe for operators and the environment.	10%	4	0.4	3	0.3
Suitable confinement system. The sterilization cycle must start only after the lid is closed.	10%	4	0.4	4	0.4
The loading device allows the waste to be cased without interference between the containment system and the loading door.	10%	4	0.4	2	0.2
Materials resistant to chemical and physical aggressions.	5%	5	0.25	5	0.25
Total Score			4.0		3.3
Rank			1		2
Continue?			Develop		No

Concept 2 has some downsides due to the loading system that could generate unwanted losses in the environment when opening each of the compartments to deposit the waste inside the chamber. Additionally, concept 2 requires more handling by the operators as they have to locate different compartments on the conveyor. The same happens during cleaning, which should be done one container at a time. On the other hand, concept 1 has a single active trolley, controlled, and transported by the elevator, that later deposits all the waste material inside the chamber, all in a single step, allowing for faster loading times with fewer risks. In this context, we decided to go ahead with concept 1.

After the concept scoring analysis, however, new suggestions for improvement were proposed. The first is the need to add an extractor hood that absorbs any unwanted material escaping from above the chamber. The second is related to vapor and liquid management during the shredding and sterilization process; all the water due to humidity is evaporated. The vapor leaves the chamber through a duct and then is extracted by the vacuum pump. After that, we will include a scrubber for the condensation of water vapor, the absorption of entrained vapors, and dust removal. Finally, regulations request that we have a system for carrying out quick tests.

4. Results

After performing the five phases of DT, we got a final and preliminary concept of the main treatment chamber for an infectious waste management system in Piedmont, Italy. This conceptual design responds to the expectations of the users and is compliant with regulations. When the test phase concluded, some relevant suggestions emerged, leading to the final conceptual design of this study (Figures 10 and 11). It includes an extractor hood to eliminate any possibility of unwanted material dispersion inside the closed chamber and a sampling device for carrying out the test and sampling extraction under contained conditions. For future steps in the general project design, it is planned to add a scrubber for pollutant control and removal. Our concept design is, therefore, delimitated until the phase before the management of liquids and vapors.

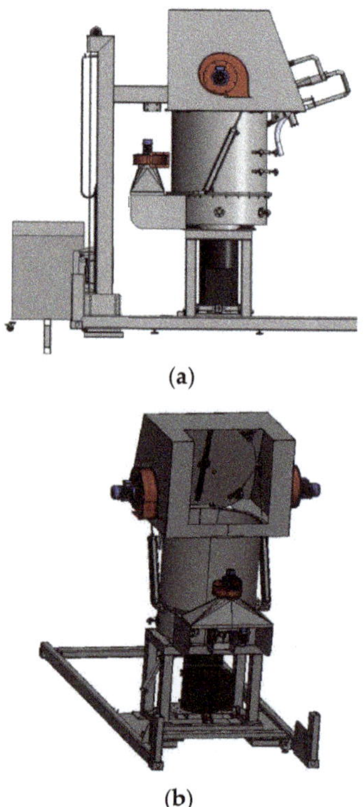

(a)

(b)

Figure 10. Automatic Chamber System. (**a**) Extractor hood. (**b**) Rear view.

Figure 10a shows the entire automatic chamber system with the extractor hood. Figure 10b shows the rearview and allows us to visualize the space where the trolley will enter to discharge the waste material into the chamber.

A section view of the chamber is represented in Figure 11a; the rotor blades are in the lower part, along with the counter blades. It is possible to visualize two of the ten thermocouples and the automatic lid. Figure 11b, shows the safety extraction valve for the sampling withdrawal test, located in the bottom of the sterilization chamber. This valve will allow for the extraction of a sample of final material after the treatment to evaluate its efficacy.

DT seeks to define solutions where three aspects converge: desirability for the end-user, feasibility, and financial viability. DT brings together what is desirable from the point of view of the users and stakeholders with what is technologically feasible and economically viable [35]. To validate the desirability, we spoke with stakeholders and potential users during the empathize phase to understand their concerns and desires; we have summarized these thoughts, commonly referred to as "pains and gains", in Figure 4. In turn, the feasibility involved designing (during the prototype stage) two virtual prototypes of the solution, evaluated later by experts, stakeholders, and users during the test phase. Creating a virtual design prototype enhances the financial viability since the production costs of a functional prototype are very high. In this way, creating a conceptual design minimizes the economic risk by being able to improve the design without incurring the costs that would entail modifying a physical prototype.

This section concludes the conceptual design of the main sterilization chamber. For futures studies, the idea of including handling robots to perform the trolley loading operations could be further analyzed; this would eliminate the workers' handling completely, and this way, they would be in charge only of activation of the automatic cycle and operations control.

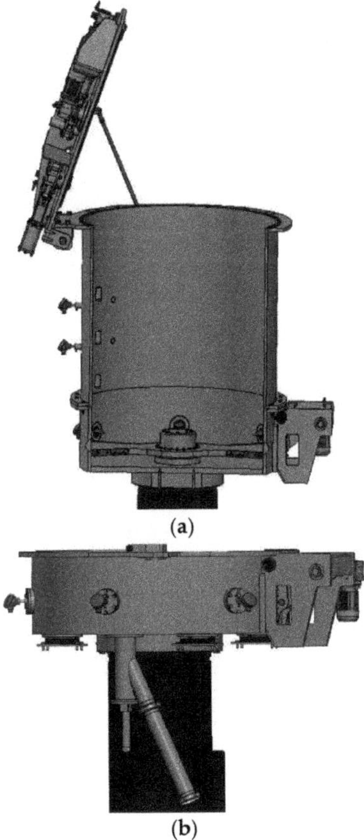

(a)

(b)

Figure 11. Automatic Chamber System. (**a**) Chamber section view. (**b**) Sampling extraction valve.

Finally, another potential future study regards the use of treated waste, since according to Italian regulation [4], the sterilized infectious waste can be used as a means of producing energy adopting the EWC 191210.

5. Discussion and Conclusions

Innovation and new technologies development are essential for most cities that aim to foster an ecofriendly environment. Potential smart cities request more recent systems in shorter time intervals, often customized to their own needs. Companies and research centers have to collaborate closely within their organization and with partners located in various parts of the world to meet these needs. At the same time, companies have to manage increasing technology and manufacturing complexities due to a quickly growing number of environmental and regulatory rules and requirements. Using collaborative tools and strategies, such as DT, for managing product data and integrating and automating business processes generally results in efficiency improvements. In particular, the knowledge in this domain can be applied to the context of eco-sustainable equipment and strategies, working together with users and stakeholders.

This project explored the application of the DT methodology in engineering and sustainability. Its application in technical areas allowed us to explore the methodology and apply it concretely.

The DT approach as methodology allowed the team to work under an established framework, following a series of phases and tools. Design thinking guided the conceptual design, beginning from a general idea to finally be able to produce a CAD prototype. The first phases (empathize and define) allowed us to understand the stakeholder's perspective and context and include their suggestions in the development of the study. The *personas* technique and Customer Journey Map were tools that engaged not only with the users but also with the team members, gathering the stakeholder's points of view effectively.

The ideate phase was essential to transform these ideas into visual models. The functional decomposition acted as a conductor to understand and solve, step-by-step, each subproblem of the entire system, understanding their interactions and relations. The brainstorming particularly helped the team to converge on solutions considering all the opinions.

The prototype phase, including brainstorming and CAD tools, concretized the idea into a conceptual design represented by two virtual concepts. By following the DT directions, these concepts were designed based on the users' needs and the current Piedmont region regulations. Users expected a safe and standard process capable of treating waste quickly, a procedure compliant to regulations different from incineration, minimum handling, and shorter distances journeys while transporting waste.

In this way, the study problem is framed and oriented according to the requirements. Many other concepts could be generated using this DT framework. Still, we decided to focalize our efforts on creating a feasible concept, especially in the design of the sterilization chamber. Due to the cruciality of this operation inside the system, we prefer to be cautious in selecting the technique resulting from the brainstorming session.

In the last DT phase, we analyzed and compared the two concepts using the concept scoring method. We established selection criteria according to the stakeholders' requirements, users and regulations on this matter. We then assigned the weights for each criterion and concept according to the team expertise; it does not mean that the weights are ideal or unique, but that they are certainly a reasonable approximation inside a feasible exploration of concepts. Concluding the test phase, we asked experts for feedback to improve the final concept.

The feedback gave us suggestions on how to improve the concept and insights for the future steps of the project. Our concept "delimits" with the vapor and liquids management system; therefore, for subsequent studies, we propose a scrubber to perform the operations of pollutant control. Also, there is evidence of the potential use of the treated material as RDF. In this regard, for future steps, it could be interesting to consider a pelletizer after sterilization for compressing and molding the final material for later storage. In this way the current paradigm that uses incineration as the primary treatment process within a linear economy approach, will change to a new scenario, using sterilization and a compression process to generate RDF that could be re-integrated into the production cycle within a circular economy approach.

In comparison with other existing facilities, the main improvements of this design are at a local level, the increase in the production and development of our own technologies for the management of infectious waste, looking for new treatment and energy recovery processes. Specifically, in Italy, one of the main improvements in comparison with existing facilities is the use and management of technologies different from the current waste incineration paradigm. Then, an industrial process that guarantees perfect sterilization of the treated waste to obtain a product intended for energy enhancement, through a new completely automatic and locked process for the treatment of infectious waste through shredding and sterilization. The whole process takes place automatically and is isolated inside the main chamber, minimizing the risks for the workers. In addition, it is expected to treat the sterilized waste further to generate derived fuel and enhance the final product.

Moreover, this process operates with unsaturated steam on a finely shredded material, operating at higher temperatures but with significantly lower pressures.

Finally, this investigation is a positive example of how to apply the DT methodology as a framework for new technological designs. With the application of the DT during the development of this study, it was possible to analyze the process holistically, considering the perspectives of the technological and social environment and the opinions of experts from the industrial sector, users, and stakeholders, in order to develop a technologically-based conceptual design, obtaining satisfactory results for the parties interested in this project.

At a social level, the approach of the DT method helped users feel part of the development of a technology which usually is rejected. This study provides a successful case of the application of DT in a sector where few studies have been carried out, leading to an increase in interest and opening possibilities for future research. Additionally, it provides a framework and a demonstration of applied tools for future research that intend to apply DT for similar processes.

Through this research, we found improvement opportunities for infectious waste management in the Piedmont region in Italy. To solve the problem, DT includes the users' needs and involves the community, which is an essential aspect of sustainable developments for eco-friendly cities. The concept result of this study is framed within a feasibility analysis; therefore, it is not definitive. As part of the project, a functional preproduction version of the chamber must be manufactured to verify its performance. This study supports the continuation of the subsequent development phases of the entire system represented in Figure 1.

We expect that this article could be used as a reference for future investigations related to the application of Design Thinking in engineering designs oriented to sustainability, resource efficiency, and waste management.

Author Contributions: Conceptualization, I.A.C.J., S.M. (Stefano Mauro), S.S. and D.N.; data curation, I.A.C.J., S.M. (Stefano Mauro), F.M. and S.S.; formal analysis, I.A.C.J., S.M. (Stefano Mauro), S.S. and E.V.; funding acquisition, D.N. and E.V.; investigation, I.A.C.J., S.M. (Stefano Mauro), S.S., F.M. and M.C.R.T.; methodology, I.A.C.J., S.M. (Stefano Mauro), S.S., F.M. and M.C.R.T.; supervision, D.N., F.M. and E.V.; validation, S.M. (Sandro Moos); visualization, S.M. (Stefano Mauro), S.S., S.M. (Sandro Moos), F.M. and D.N.; writing—original draft, I.A.C.J. and M.C.R.T.; writing—review and editing, I.A.C.J., S.M. (Stefano Mauro), S.S., S.M. (Sandro Moos), F.M. and D.N. All authors have read and agreed to the published version of the manuscript.

Funding: This research received no external funding.

Acknowledgments: Thanks to TWM Company for its support and information provided.

Conflicts of Interest: The authors declare no conflict of interest. The funders had no role in the design of the study; in the collection, analyses, or interpretation of data; in the writing of the manuscript, or in the decision to publish the results.

References

1. Das, S.; Lee, S.H.; Kumar, P.; Kim, K.H.; Lee, S.S.; Bhattacharya, S.S. Solid waste management: Scope and the challenge of sustainability. *J. Clean. Prod.* **2019**, *228*, 658–678. [CrossRef]
2. Keeble, B.R. The Brundtland Report: "Our Common Future". *Med. War* **1988**, *4*, 17–25. [CrossRef]
3. Direzione Ambiente Governo e Tutela del Territorio. *Piano Regionale di Gestione dei Rifiuti Speciali*; Direzione Ambiente Governo e Tutela del Territorio: Turin, Italy, 2018.
4. Italian Presidential Decree DPR n254 07/15/2003; *Recante Disciplina Della Gestione dei Rifiuti Sanitari a Norma Dell'articolo 24 Della Legge 31 Luglio 2002 n. 179. Dpr 254 07/15/2003*; Gazzetta Ufficiale Della Reppublica Italiana: Rome, Italy, 2003; Volume 1, pp. 1–20. Available online: https://www.gazzettaufficiale.it/eli/id/2003/09/11/003G0282/sg (accessed on 13 April 2021).
5. Agrawal, A.; Javaria, G.; Kishor, K.; Mg, B. *Handling Solid Waste Using Design Thinking Principle in Bengaluru*; IJISRT Digital Library: Jaipur, India, 2019.
6. McEvoy, B.; Rowan, N.J. Terminal sterilization of medical devices using vaporized hydrogen peroxide: A review of current methods and emerging opportunities. *J. Appl. Microbiol.* **2019**, *127*, 1403–1420. [CrossRef] [PubMed]
7. Emek, E.; Kara, B.Y. Hazardous waste management problem: The case for incineration. *Comput. Oper. Res.* **2007**, *34*, 1424–1441. [CrossRef]

8. Mohamed, L.F.; Ebrahim, S.A.; Al-thukair, A.A. Hazardous healthcare waste management in the Kingdom of Bahrain. *Waste Manag.* **2009**, *29*, 2404–2409. [CrossRef] [PubMed]
9. Salihoglu, G. Industrial hazardous waste management in Turkey: Current state of the field and primary challenges. *J. Hazard. Mater.* **2010**, *177*, 42–56. [CrossRef]
10. Alam, O.; Mosharraf, A. A preliminary life cycle assessment on healthcare waste management in Chittagong City, Bangladesh. *Int. J. Environ. Sci. Technol.* **2020**, *17*, 1753–1764. [CrossRef]
11. Rada, E.C.; Ragazzi, M.; Marconi, M.; Chistè, A.; Schiavon, M.; Fedrizzi, S.; Tava, M. PCDD/Fs in the soils in the province of Trento: 10 years of monitoring. *Environ. Monit. Assess.* **2015**, *187*, 4114. [CrossRef]
12. Blackman, W. *Basic Hazardous Waste Management*; Routledge: New York, NY, USA, 2001; ISBN 1566705339.
13. ISPRA. *Rapporto Rifiuti Speciali Edizione 2020*; Sistema Nazionale per la Protezione dell'Ambiente: Rome, Italy, 2020. Available online: https://www.isprambiente.gov.it/it/pubblicazioni/rapporti/rapporto-rifiuti-speciali-edizione-2020 (accessed on 26 October 2021).
14. ISPRA. Rapporto Rifiuti Speciali Edizione 2018. In *Rapporto Rifiuti Speciali Edizione 2018*; Sistema Nazionale per la Protezione dell'Ambiente: Rome, Italy, 2018. Available online: https://www.isprambiente.gov.it/it/pubblicazioni/rapporti/rapporto-rifiuti-urbani-edizione-2018 (accessed on 17 September 2020).
15. Italian Presidential Decree n152 04/2006; *Codice del Ambiente: Decreto Legislativo 152/2006*; Gazzetta Ufficiale Della Reppublica Italiana: Rome, Italy, 2006; Volume 1, pp. 1–861.
16. Liedtka, J. Perspective: Linking Design Thinking with Innovation Outcomes through Cognitive Bias Reduction. *J. Prod. Innov. Manag.* **2015**, *32*, 925–938. [CrossRef]
17. Buhl, A.; Schmidt-keilich, M.; Muster, V.; Blazejewski, S.; Schrader, U.; Harrach, C.; Schäfer, M.; Süßbauer, E. Design thinking for sustainability: Why and how design thinking can foster sustainability-oriented innovation development. *J. Clean. Prod.* **2019**, *231*, 1248–1257. [CrossRef]
18. Pahl, G.; Beitz, W.; Feldhusen, J.; Grote, K.-H. *Engineering Design A systematic Approach*, 3rd ed.; Wallace, K., Blessing, L., Eds.; Springer: London, UK, 2007; Volume 53, ISBN 1846283183.
19. Carroll, N.; Richardson, I. Aligning healthcare innovation and software requirements through design thinking. In *Proceedings of the International Workshop on Software Engineering in Healthcare Systems (SEHS), Austin, TX, USA, 14–16 May 2016*; Association for Computing Machinery, Inc.: New York, NY, USA, 2016; pp. 1–7.
20. Valentine, L.; Kroll, T.; Bruce, F.; Lim, C.; Mountain, R. Design Thinking for Social Innovation in Health Care. *Des. J.* **2017**, *20*, 755–774. [CrossRef]
21. Lake, D.; McFarland, A.; Vogelzang, J. Creating resilient interventions to food waste: Aligning and leveraging systems and design thinking. In *Food Waste Management: Solving the Wicked Problem*; Springer International Publishing: Berlin/Heidelberg, Germany, 2019; pp. 193–221. ISBN 9783030205614.
22. Kwon, J.; Choi, Y.; Hwang, Y. Enterprise design thinking: An investigation on user-centered design processes in large corporations. *Designs* **2021**, *5*, 43. [CrossRef]
23. Chou, D.C. Applying design thinking method to social entrepreneurship project. *Comput. Stand. Interfaces* **2018**, *55*, 1339–1351. [CrossRef]
24. Acar, A.E.; Rother, D. Design Thinking in Engineering Education and its Adoption in Technology-driven Startups. *Adv. Sustain. Manuf.* **2011**, *1*, 57–62. [CrossRef]
25. Skogstad, P.; Leifer, L. A Unified Innovation Process Model for Engineering Designers and Managers. In *Design Thinking: Understand—Improve—Apply*; Springer: Berlin/Heidelberg, Germany, 2011; pp. 19–43.
26. Buhl, A.; Blazejewski, S.; Dittmer, F. The More, the Merrier: Why and How Employee-Driven Eco-Innovation Enhances Environmental and Competitive Advantage. *Sustainability.* **2016**, *8*, 946. [CrossRef]
27. Alves, R.; Jardim Nunes, N. Towards a taxonomy of service design methods and tools. In *Lecture Notes in Business Information Processing*; Springer: Berlin/Heidelberg, Germany, 2013; Volume 143, pp. 215–229.
28. Kumar, K.; Zindani, D.; Davim, J.P. *Design Thinking to Digital Thinking*; SpringerBriefs in Applied Sciences and Technology; Springer International Publishing: Cham, Switzerland, 2020; ISBN 978-3-030-31358-6.
29. Rois, M.; Mubarak, A.; Suzianti, A. Designing Solution for Organic Waste Management System with Design Thinking Approach (Case Study in Depok). In *IOP Conference Series: Earth and Environmental Science*; Institute of Physics Publishing: Bristol, UK, 2020; Volume 464.
30. Bishop, P. Urban Design in the Fragmented City. In *Contemporary Urban Design Thinking*; Roggema, R., Ed.; Springer: Cham, Switzerland, 2018; pp. 71–93. ISBN 978-3-319-91949-2.
31. Kelley, T. *The Art of Innovation: Lessons in Creativity from IDEO, America's Leading Design Firm Tom Kelley with Jonathan Littman; New York: Doubleday, 2001, 308 + xii Pages, $26.00*, 1st ed.; Doubleday: New York, NY, USA, 2001; Volume 1, ISBN 0-385-49984-1.
32. Stanford University School. *Design Thinking Bootleg*; Stanford University: Stanford, CA, USA, 2013; Volume 90.
33. Polydoras, S.; Sfantsikopoulos, M.; Provatidis, C. Rational Embracing of Modern Prototyping Capable Design Technologies into the Tools Pool of Product Design Teams. *ISRN Mech. Eng.* **2011**, *2011*, 739892. [CrossRef]
34. Chasanidou, D.; Gasparini, A.A.; Lee, E. Design Thinking Methods and Tools for Innovation in Multidisciplinary Teams. In *Innovation in HCI: What Can We Learn from Design Thinking?* NordiCHI: Helsinki, Finland, 2014; Volume 1, pp. 27–30. [CrossRef]

35. Brown, T. *Change by Design: How Design Thinking Transforms Organizations and Inspires Innovation*; Harperbusiness: New York, NY, USA, 2019; ISBN 0062856626.
36. Lewrick, M. *The Design Thinking Playbook: Mindful Digital Transformation of Teams, Products, Services*; Wiley: Hoboken, NJ, USA, 2018; ISBN 9781119467472.
37. Castiblanco Jimenez, I.A.; Cepeda Garcia, L.C.; Marcolin, F.; Violante, M.G.; Vezzetti, E. Validation of a tam extension in agriculture: Exploring the determinants of acceptance of an e-learning platform. *Appl. Sci.* **2021**, *11*, 4672. [CrossRef]
38. Consiglio Regionale del Piemonte Ordine del giorno n.130. In *Atti del Cons*; Consiglio Regionale del Piemonte Ordine: Turin, Italy, 2019.
39. European Commission. *European Commission European Waste Catalogue*; European Commission: Brussels, Belgium, 2015; pp. 1–46.
40. European Commission. *Regolamento (UE) N. 1357/2014 Della Commissione del 18 Dicembre 2014 che Sostituisce L'allegato III Della Direttiva 2008/98/CE del Parlamento Europeo e del Consiglio Relativa ai Rifiuti e che Abroga Alcune Direttive*; European Commission: Brussels, Belgium, 2014; Volume 2014, p. 8.
41. Nuzzo, G.; Engineering, F.; David, R.; Am, C.O.P.R.; Blasio, D. Optimization of Infectious Hospital Waste Management in Italy: Part 1—Wastes Production and Characterization Study. *Waste Manag. Res.* **1994**, *12*, 373–385.
42. World Health Organization. *Safe Management of Wastes from Healthcare Activities*; Pruss, A., Giroult, E., Rushbrook, P., Eds.; WHO Library Cataloguing-in-Publication Data: Geneve, Switzerland, 1999; ISBN 92.4.154525.9.
43. Bonakdar, A.; Gassmann, O. Design thinking for revolutionizing your business models. In *Design Thinking for Innovation: Research and Practice*; Springer International Publishing: Berlin/Heidelberg, Germany, 2016; pp. 57–66. ISBN 9783319261003.
44. Dickinger, A. *Perceived Quality of Mobile Services: The Explorative Research*, 18th ed.; Peter Lang AG: Bern, Switzerland, 2007.
45. Revella, A. *Buyer Personas*, 1st ed.; Wiley: Hoboken, NJ, USA, 2015; ISBN 978-1-118-96150-6.
46. Lemon, K.N.; Verhoef, P.C. Understanding customer experience throughout the customer journey. *J. Mark.* **2016**, *80*, 69–96. [CrossRef]
47. Ulrich, K.T.; Eppinger, S.D.; Yang, M.C. *Product Design and Development*, 7th ed.; McGraw-Hill Education: New York, NY, USA, 2020; ISBN 9781260043655.
48. Xiao, F. Engineering Applications of Artificial Intelligence A novel multi-criteria decision making method for assessing healthcare waste treatment technologies based on D numbers. *Eng. Appl. Artif. Intell.* **2018**, *71*, 216–225. [CrossRef]
49. Hancock, C.; Lambert, P. *Sterilization Processes*, 5th ed.; Fraise, A., Maillard, J., Sattar, S., Eds.; Wiley-Blackwell: Oxford, UK, 2013; ISBN 9781444333251.
50. Health Care Without Harm Europe. *Non-Incineration Medical Waste Treatment Technologies in Europe: A Recource for Hospital Administrators, Facility Managers, Health Care Professionals, Environmental Advocates, and Community Members*; Health Care Without Harm Europe: Prague, Czech Republic, 2004; p. 44.
51. Xu, X.-F.; Tan, Q.-Y.; Liu, L.-L.; Li, J.-H. Assessment of Medical Waste Disposal Technologies Based on the AHP. *Huanjing Kexue/Environ. Sci.* **2018**, *39*, 5717–5722. [CrossRef]
52. Forrec Forrec Srl Trituratori. Available online: https://www.forrec.it/ (accessed on 16 August 2020).
53. Ecosyst Eco.System S.u.r.l. Available online: https://www.ecosyst.it/ (accessed on 18 August 2020).
54. Ciclia ZEF, s.r.l. SB Unipersonale. Available online: https://www.ciclia.it/ (accessed on 14 September 2020).
55. Seeber, I.; Maier, R.; Weber, B.; De Vreede, G.J.; De Vreede, T.; Alothaim, A. Brainstorming is just the beginning: Effects of convergence techniques on satisfaction, perceived usefulness of moderation, and shared understanding in teams. In Proceedings of the 2015 48th Hawaii International Conference on System Sciences, Kauai, HI, USA, 5–8 January 2015; pp. 581–590. [CrossRef]
56. Vella, L.; Napoli, D.; Russo, S.; Micozzi, I.; Specchia, V.; Demartini, C.; Specchia, S. "Zero Km Med-Wastes": Smart Collection, Sterilization and Energetic Valorisation. *Récents Prog. Génie Procédés* **2014**, *104*, 1–9.
57. De Beer, D.J.; Barnard, L.J.; Booysen, G.J. Three-dimensional plotting as a visualization aid for architectural use. *Rapid Prototyp. J.* **2004**, *10*, 146–151. [CrossRef]
58. Liedtka, J.; Ogilvie, T. *Designing for Growth: A Design Thinking Tool Kit for Managers*; Columbia University Press: New York, NY, USA; 2011
59. Bresciani, S. Visual Design Thinking: A Collaborative Dimensions framework to profile visualizations. *Des. Stud.* **2019**, *63*, 92–124. [CrossRef]

MDPI

St. Alban-Anlage 66

4052 Basel

Switzerland

Tel. +41 61 683 77 34

Fax +41 61 302 89 18

www.mdpi.com

Electronics Editorial Office

E-mail: electronics@mdpi.com

www.mdpi.com/journal/electronics